Advances in Deterministic and Stochastic Analysis

Advances in Deterministic and Stochastic Analysis

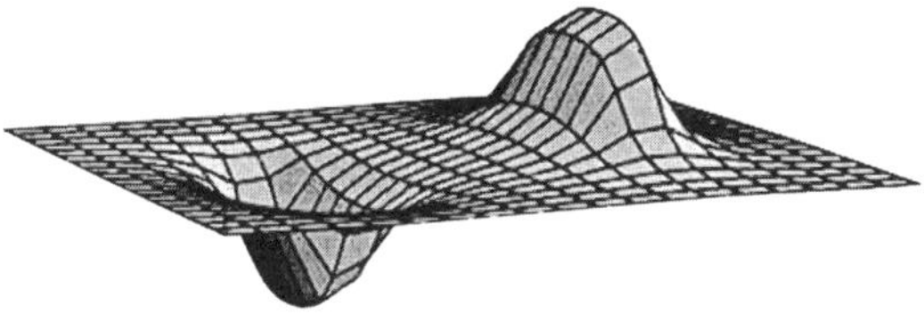

Editors

N M Chuong
Institute of Mathematics, Vietnamese Acad. of Sci. & Tech., Vietnam

P G Ciarlet
City University of Hong Kong, Hong Kong

P Lax
Courant Institute, USA

D Mumford
Brown University, USA

D H Phong
Columbia University, USA

World Scientific

NEW JERSEY • LONDON • SINGAPORE • BEIJING • SHANGHAI • HONG KONG • TAIPEI • CHENNAI

Published by

World Scientific Publishing Co. Pte. Ltd.

5 Toh Tuck Link, Singapore 596224

USA office: 27 Warren Street, Suite 401-402, Hackensack, NJ 07601

UK office: 57 Shelton Street, Covent Garden, London WC2H 9HE

British Library Cataloguing-in-Publication Data
A catalogue record for this book is available from the British Library.

ADVANCES IN DETERMINISTIC AND STOCHASTIC ANALYSIS

ISBN-13 978-981-270-550-1
ISBN-10 981-270-550-3

Printed in Singapore by World Scientific Printers (S) Pte Ltd

PREFACE

In June 4–9, 2005, the Second International Conference on Abstract and Applied Analysis was held at Quy Nhon, Vietnam. The conference brought together Vietnamese mathematicians working on analysis, as well as many distinguished foreign visitors. It covered all areas of analysis where Vietnamese mathematicians have been playing an active role, including pure and applied analysis, and deterministic and stochastic approaches. The proceedings of the conference are gathered in the present volume. The specific topics discussed at the conference reflect the most active research directions in Vietnam today, and as such, are quite varied. Seventeen papers appear here. For convenience, they have been loosely organized into four different chapters.

Chapter I, "Integral and Pseudodifferential Operators", includes the papers where these operators play a major role, particularly in the theory of partial differential equations. The paper by Nguyen Minh Tri provides a criterion for the hypoellipticity for a class of operators with sign-changed characteristic form, generalizing earlier works of Beals and Fefferman. The paper by Nguyen Minh Chuong and Dang Anh Tuan deals with generalizations of the oblique derivative problem in several directions, including non-linear settings. The last two papers in the chapter deal more with the properties of the operators themselves, with the boundedness of commutator integrals of mixed homogeneity studied by Lubomina Softova, and a classification of integral transforms of both convolution and non-convolution types provided by Vu Kim Tuan.

Chapter II, "Partial Differential Equations", deals of course also with partial differential equations, but with essentially different methods than the pseudodifferential and integral operator methods of Chapter I. The paper by Martin Schechter provides a unified min-max approach to variational problems, through the idea of linking sets separating a given functional. The paper by Mikio Tsuji and Peter Wagner studies generalized solutions and the possible appearance of shock waves to a class of equations generalizing Burgers' equation. Boundary value problems for the biharmonic equation are studied in the paper by Dang Quang A and Le Tung Son, by a reduction to a boundary value problem for a second order equation, combined with an iterative scheme. A parabolic boundary value problem, arising from an inverse problem in engineering for finding the heating regime for a given desired final temperature, is studied in the paper by Dinh Nho Hao, Nguyen Trung Thanh, and H. Sahli, by the DC ("difference of two convex functions") method. The paper by Peter Massopust describes explicit solutions, in various coordinate systems, for the Maxwell equations describing magnetic fields generated by a hard ferromagnet. The paper by Ha Tien Ngoan and Nguyen Thi Nga deals with the local solvability of the non-characteristic Cauchy problem for weakly hyperbolic classical Monge-Ampere equations. The asymptotic analysis of the solutions of the Navier-Stokes problem is considered by Makram Hamouda and Roger Temam when the viscosity goes to zero and the flow in a channel of R^3 is also considered by them in the case when the boundary is non-characteristic.

Chapter III, "Geometric Analysis", groups together the papers with a more pronounced geometric flavor. The paper by Le Hong Van provides a construction of obstructions to imbedding a given statistical manifold into another one. A statistical manifold is a Riemannian manifold with an equivalence class of symmetric 3-tensors, a geometric concept which arises from information geometry. The paper by Do Ngoc Diep constructs a Z2-graded Cech cohomology, building in ideas of Rosenberg and Kontsevich from noncommutative geometry. The paper by Nguyen Minh Chuong and Le Duc Thinh studies Sobolev spaces with weights on Riemannian manifolds. Finally, as its title indicates, Chapter IV, "Stochastic and Infinite-Dimensional Analysis", groups together papers more related to either probability or analysis in infinite-dimensional spaces. They include the paper by Situ Rong, on the existence of solutions to some reflexive stochastic differential equations with jumps and its applications to stochastic population control; the paper by Shigeyoshi Ogawa, which provides a survey of developments related to a notion introduced by the author; the paper by Andrei Khrennikov, which discusses the foundations of quantum mechanics from the point of view of infinite-dimensional spaces, and the paper by Karl Gustafson, which discusses the ideas of the author on Bell's inequality and contraction semigroups.

The Editors

CONTENTS

Preface v

Chapter I Integral and Pseudodifferential Operators

§1. Pseudodifferential Operators of Second Order with Sign-Changed Characteristic Form 3
Nguyen Minh Tri

§2. A Semilinear Nonclassical Pseudodifferential Boundary Value Problem in Sobolev Spaces $H_{\ell,p}$, $1 < p < \infty$ 15
Nguyen Minh Chuong and Dang Anh Tuan

§3. Singular Integral Operators in Functional Spaces of Morrey Type 33
Lubomira Softova

§4. Classification of Integral Transforms 43
Vu Kim Tuan

Chapter II Partial Differential Equations

§5. Unified Minimax Methods 75
Martin Schechter

§6. Some Remarks on Single Conservation Laws 91
Mikio Tsuji and Peter Wagner

§7. Iterative Method for Solving a Mixed Boundary Value Problem for Biharmonic Type Equation 103
Dang Quang A and Le Tung Son

§8. Numerical Solution to a Non-Linear Parabolic Boundary Control Problem 115
Dinh Nho Hao, Nguyen Trung Thanh and H. Sahli

§9. A Class of Solutions to Maxwell's Equations in Matter and Associated Special Functions 131
Peter Massopust

§10. On the Cauchy Problem for a Quasilinear Weakly Hyperbolic System in Two Variables and Applications to that for Weakly Hyperbolic Classical Monge-Ampère Equations 177
Ha Tien Ngoan and Nguyen Thi Nga

§11. Some Singular Perturbation Problems Related to the Navier-Stokes Equations 197
Makram Hamouda and Roger Temam

Chapter III Geometric Analysis

§12. Monotone Invariants and Embeddings of Statistical Manifolds 231
Le Hong Van

§13. Graded Čech Cohomology in Noncommutative Geometry 255
Do Ngoc Diep

§14. Sobolev Spaces with Weight on Riemannian Manifolds 269
Nguyen Minh Chuong and Le Duc Thinh

Chapter IV Stochastic and Infinite-Dimentional Analysis

§15. Stochastic Population Control and RSDE with Jumps 281
Situ Rong

§16. Noncausal Stochastic Calculus Revisited — Around the So-called Ogawa Integral 297
Shigeyoshi Ogawa

§17. Infinite-Dimensional Stochastic Analysis and Foundations of Quantum Mechanics 321
Andrei Khrennikov

§18. Noncommutative Trigonometry and Quantum Mechanics 341
Karl Gustafson

Chapter I

INTEGRAL AND PSEUDODIFFERENTIAL OPERATORS

Advances in Deterministic and Stochastic Analysis
Eds. N. M. Chuong *et al.* (pp. 3–13)

§1. PSEUDODIFFERENTIAL OPERATORS OF SECOND ORDER WITH SIGN-CHANGED CHARACTERISTIC FORM

NGUYEN MINH TRI

Institute of Mathematics
18 Hoang Quoc Viet, Cau Giay, 10307 Hanoi, Vietnam
E-mail: triminh@math.ac.vn

In this paper we prove hypoellipticity for pseudodifferential operators of second order with sign-changed characteristic form. The obtained results extend the previous ones of Beals-Fefferman and Ganja for differential operators.

In this paper we consider operator of the form

$$P = -\frac{\partial}{\partial t} + t\sum_{j=1}^{k} X_j^2 + c(x,t) \tag{1}$$

in a bounded domain $\Omega \subset \mathbb{R}^{n+1}_{x,t} = \mathbb{R}^n_x \times \mathbb{R}^1_t$, where $X_1, \ldots, X_k$ are proper pseudo differential operators (p. d. o.) of first order with real principle symbols $X_j(x,t,\xi,\tau), j = \overline{1,k}$ and $c(x,t)$ is a smooth function in Ω. In Ref. 1 Hörmander proved that for a differential operator of second order to be hypoelliptic it is necessary that its characteristic form does not change sign at a fixed point in Ω. However, Kannai's example in Ref. 2 showed that characteristic form of a hypoelliptic operators can change sign from point to point in Ω. Necessary conditions for hypoellipticity of operators generalizing the Kannai example were studied in papers of Helffer and Zuily,[3] and Beals and Fefferman.[4] In Ref. 4 it is proved that for the differential operator P in the form (1) to be hypoelliptic it is necessary that $X_1, \ldots, X_k$ satisfy the following conditions

$$X_j(x,0,0,1) = 0, \quad j = 1, \ldots, k.$$

2000 *Mathematics Subject Classification* Primary 35H10; Secondary 35A08, 35B65, 35C15, 35D10.

In Ref. 4 Beals and Fefferman also established some sufficient conditions for the differential operator P in the form (1) to be hypoelliptic. In Refs. 5,6 Ganja obtained further sufficient conditions for the differential operator P in the form (1) to be hypoelliptic. In this note we extend the results of Beals and Fefferman, and Ganja for the pseudodifferential operator P in the form (1). Now we introduce the transformation:

$$\Omega_{(x,t)} \to \Omega'_{(x',t')}$$
$$(x,t) \to (x',t');\ x' = x, t' = \delta t;\ \delta \in (0,1].$$

For a function $u \in C_0^\infty(\Omega)$ we can define the function $u_\delta \in C_0^\infty(\Omega)$ as follows:

$$u_\delta(x',t') = u(x, t'\delta).$$

Note that when $\delta = 1$ we can identify the coordinates (x,t) and (x',t').

Lemma 1. *The following estmates hold*

$$\delta^{2s-1}\|u\|^2_{H^s(\Omega)} \leqslant \|u_\delta\|^2_{H^s(\Omega')} \leqslant \frac{1}{\delta}\|u\|^2_{H^s(\Omega)},\ \text{for } s \geq 0;$$
$$\frac{1}{\delta}\|u\|^2_{H^s(\Omega)} \leqslant \|u_\delta\|^2_{H^s(\Omega')} \leqslant \delta^{2s-1}\|u\|^2_{H^s(\Omega)},\ \text{for } s \leqslant 0.$$

Proof. Since $\tilde{u}_\delta(\xi,\tau) = \frac{1}{\delta}\tilde{u}(\xi,\frac{\tau}{\delta})$, using the inequalities

$$\delta^{2s}(1+|\xi|^2+\tau^2)^s \leqslant (1+|\xi|^2+\tau^2\delta^2)^s \leqslant (1+|\xi|^2+\tau^2)^s \quad \text{for } s \geq 0;$$
$$\delta^{2s}(1+|\xi|^2+\tau^2)^s \geq (1+|\xi|^2+\tau^2\delta^2)^s \geq (1+|\xi|^2+\tau^2)^s \quad \text{for } s \leqslant 0$$

we can get easily the desired estimates. □

Let P_δ be the operator in the coordinates (x',t') defined by the formula:

$$P_\delta = -\frac{1}{\delta}\frac{\partial}{\partial t'} + t'\delta \sum_{j=1}^{k} X^2_{j,\delta} + c_\delta(x',t'), \tag{2}$$

where $X_{j,\delta}(x',t',\xi',\tau') = X_j(x',t'\delta,\xi',\frac{\tau'}{\delta}), c_\delta(x',t') = c(x',t'\delta)$. It is easy to see that $P_\delta u_\delta = (Pu)_\delta$. Therefore, by Lemma 1, we deduce that

$$\delta^{2s-1}\|Pu\|^2_{H^s(\Omega)} \leqslant \|P_\delta u_\delta\|^2_{H^s(\Omega')} \leqslant \frac{1}{\delta}\|Pu\|^2_{H^s(\Omega)},\ \text{for } s \geq 0;$$
$$\frac{1}{\delta}\|Pu\|^2_{H^s(\Omega)} \leqslant \|P_\delta u_\delta\|^2_{H^s(\Omega')} \leqslant \delta^{2s-1}\|Pu\|^2_{H^s(\Omega)},\ \text{for } s \leqslant 0.$$

Lemma 2. *Let the operator P_δ have the form* (2). *Then for arbitrary $s \in \mathbb{R}$ and for arbitrary $K \Subset \Omega$ there exists a constant $C(\delta) = C(s, K, \delta)$ such that*

$$\sum_{j=1}^{k} \|\delta t' X_{j,\delta} g\|_s^2 \leqslant 2|(\delta t' P_\delta g, g)_s| + C(\delta)\|g\|_s^2, \quad \forall g \in C_0^\infty(K), \tag{3}$$

where $\|\cdot\|_s, (,)_s$ are taken in the (x', t') coordinates.

Proof. Put $B_{j,\delta} = \delta t' X_{j,\delta} - (\delta t' X_{j,\delta})^*$. Obviously, ord $B_j = 0$ (since $X_{j,\delta}$ has real principle symbol). Note that

$$\delta t' P_\delta = \sum_{j=1}^{k} (\delta t' X_{j,\delta})^2 + Q_\delta,$$

where

$$Q_\delta = -t' \frac{\partial}{\partial t'} + \delta t' c_\delta(x', t') - \sum_{j=1}^{k} \delta^2 t' X_{j,\delta}(t') X_{j,\delta}; \quad \text{ord } Q_\delta = 1.$$

We obtain

$$(\delta t' P_\delta g, g)_s = (\delta t' P_\delta g, \Lambda^{2s} g)_0 = \sum_{j=1}^{k} ((\delta t' X_{j,\delta})^2 g, \Lambda^{2s} g)_0 + (Q_\delta g, \Lambda^{2s} g)_0$$

$$= \sum_{j=1}^{k} (\delta t' X_{j,\delta} g, \delta t' X_{j,\delta} \Lambda^{2s} g)_0 + (Q_{1,\delta} g, \Lambda^{2s} g)_0,$$

where $Q_{1,\delta} = Q_\delta - \sum_{j=1}^{k} \delta B_{j,\delta}^* t' X_{j,\delta}$; Λ^s is the standard p. d. o. with symbol $(1 + |\xi|^2 + \tau^2)^{s/2}$. It follows immediately that

$$\sum_{j=1}^{k} \|\delta t' X_{j,\delta} g\|_s^2 \leqslant 2|(\delta t' P_\delta g, g)_s| + C(\delta)\|g\|_s^2.$$

Thus, Lemma 2 is proved. □

Corollary 1. *From the inequality* (3) *it implies that*

$$\sum_{j=1}^{k} \|t' X_{j,\delta} g\|_s^2 \leqslant 2\|t' P_\delta g\|_s^2 + C'(\delta)\|g\|_s^2.$$

Lemma 3. *Let X_j be the proper homogeneous p. d. o. with real principal symbol $X_j(x, t, \xi, \tau)$. Moreover assume that*

$$\frac{\partial X_j}{\partial \tau}(x, t, \xi, \tau) = t\varphi_j(x, t, \xi, \tau) \tag{4}$$

for $j=1,\dots,k$; ord $\varphi_j=0$. *Then operator* P_δ *can be written in the form*

$$P_\delta = -\frac{1}{\delta}\frac{\partial}{\partial t'} + \delta\sum_{j=1}^{k} X_{j,\delta}t'X_{j,\delta} + \delta t'\sum_{j=1}^{k}\varphi_{j,\delta}X_{j,\delta} + c_\delta(x',t'),$$

where $\varphi_{j,\delta}(x',t',\xi',\tau') = \varphi_j(x',t'\delta,\xi',\frac{\tau'}{\delta})$.

Proof. We have

$$t'X_{j,\delta}^2 = X_{j,\delta}t'X_{j,\delta} + [t',X_{j,\delta}]X_{j,\delta}.$$

By a theorem on composition of p. d. o. it is easy to see that

$$\sigma([t',X_{j,\delta}]) = \frac{\partial X_{j,\delta}}{\partial \tau'} = t'\varphi_{j,\delta},$$

where $\sigma([t',X_{j,\delta}])$ denotes the symbol of the operator $[t',X_{j,\delta}]$. From above we get the desired equality. □

Lemma 4. *Assume that* X_j *satisfies the condition of Lemma 3. For every* $s\in\mathbb{R}$, $K'\Subset\Omega'$, *there exists a constant* $C=C(s,K',X_j)$ *such that*

$$\|X_{j,\delta}g\|_s^2 \leqslant C\|g\|_{s+1}^2 + C(\delta)\|g\|_{s-1}^2, \quad \forall g\in C_0^\infty(K').$$

The constant C *is independent of* δ.

Proof. We have

$$X_{j,\delta}(x',t',\xi',\tau') = X_j\left(x',t'\delta,\xi',\frac{\tau'}{\delta}\right).$$

By the assumption of Lemma 3, we have $\frac{\partial X_j}{\partial\tau}(x,0,0,1)=0$. Since

$$X_j(x_1,\dots,x_n,t,\xi_1,\dots,\xi_n,\tau)$$

is a homogeneous function of degree one in $(\xi_1,\dots,\xi_n,\tau)$ we have the following Euler formula

$$\sum_{i=1}^{n}\xi_i\frac{\partial X_j}{\partial\xi_i}(x,t,\xi,\tau) + \tau\frac{\partial X_j}{\partial\tau}(x,t,\xi,\tau) = X_j(x,t,\xi,\tau). \tag{5}$$

Putting in Eq. (5) $\xi_1=\dots=\xi_n=0, \tau=1$ we get

$$0 = \frac{\partial X_j}{\partial\tau}(x,0,0,1) = X_j(x,0,0,1).$$

Consequently

$$\begin{aligned} X_{j,\delta}(x',t',\xi',\tau') &= \frac{1}{\delta}X_j(x',t'\delta,\xi'\delta,\tau') \\ &= \frac{1}{\delta}\Big[X_j(x',0,\delta\xi',\tau') + \delta t'\frac{\partial X_j}{\partial t}(x',0,\delta\xi',\tau') + O(\delta^2)\Big]. \end{aligned}$$

Therefore

$$|X_{j,\delta}(x',t',\xi',\tau')| \leqslant C(1+|\xi'|^2+\tau'^2)^{\frac{1}{2}}.$$

It follows that

$$\|X_{j,\delta}g\|_s^2 \leqslant C_{s,K',X_j}\|g\|_{s+1}^2 + C(\delta)\|g\|_{s-1}^2. \qquad \square$$

Note that if X_j satisfy the condition (4) in Lemma 3, then, equivalently, X_j can be written in the following form

$$X_j(x,t,\xi,\tau) = tY_j(x,t,\xi,\tau) + Z_j(x,t,\xi).$$

Lemma 5. *Let the operator* X_j *have the form*

$$X_j(x,t,\xi,\tau) = tY_j(x,\xi,\tau) + Z_j(x,\xi),$$

where $Y_j(x,\xi,0) = 0, j = \overline{1,k}$. *Then for every* $s \in \mathbb{R}$ *and an arbitrary compact* K' *in* Ω' *there exist constants* $C = C(s,K), \delta_0$ *such that for* $\delta \in (0,\delta_0)$

$$\begin{aligned} &\left\|\frac{\partial g}{\partial t'}\right\|_{s-\frac{1}{2}}^2 + \sum_{j=1}^k \|X_{j,\delta}g\|_{s-\frac{1}{2}}^2 \\ &\leqslant C(\delta)\left\{\left|\left(P_\delta g, -\Lambda'^{-1}\frac{\partial g}{\partial t'}\right)_s\right| + |(t'P_\delta g, g)_{s-\frac{1}{2}}| + \|g\|_{s-\frac{1}{2}}^2\right\}, \end{aligned} \tag{6}$$

where $\|\cdot\|_s, (\cdot,\cdot)_s$ *are taken in the coordinates* (x',t').

Proof. Consider the expression

$$\left(P_\delta g, -\Lambda'^{-1}\frac{\partial g}{\partial t'}\right)_s = \left(P_\delta g, -\Lambda'^{2s-1}\frac{\partial g}{\partial t'}\right)_0.$$

By Lemma 3, the operator P_δ may be rewritten in the form

$$P_\delta = -\frac{1}{\delta}\frac{\partial}{\partial t'} + \delta\sum_{j=1}^k X_{j,\delta}t'X_{j,\delta} + \delta t'\sum_{j=1}^k \varphi_{j,\delta}X_{j,\delta} + c_\delta(x',t'),$$

where ord $\varphi_{j,\delta} = 0$. It follows that

$$\begin{aligned} \left(P_\delta g, -\Lambda'^{2s-1}\frac{\partial g}{\partial t'}\right)_0 &= \left(-\frac{1}{\delta}\frac{\partial g}{\partial t'}, -\Lambda'^{2s-1}\frac{\partial g}{\partial t'}\right)_0 \\ &\quad - \delta\sum_{j=1}^k \left(t'X_{j,\delta}g, X_{j,\delta}\Lambda'^{2s-1}\frac{\partial g}{\partial t'}\right)_0 + \Phi_\delta, \end{aligned} \tag{7}$$

where Φ_δ may be estimated by

$$C(\delta)\Big(\sum_{j=1}^{k}\|t'X_{j,\delta}g\|^2_{s-\frac{1}{2}}+\|g\|^2_{s-\frac{1}{2}}\Big)+\frac{\delta}{8}\Big\|\frac{\partial g}{\partial t'}\Big\|^2_{s-\frac{1}{2}}. \tag{8}$$

Note that

$$\left(-\frac{1}{\delta}\frac{\partial g}{\partial t'},-\Lambda'^{2s-1}\frac{\partial g}{\partial t'}\right)_0=\frac{1}{\delta}\Big\|\frac{\partial g}{\partial t'}\Big\|^2_{s-\frac{1}{2}},\quad \text{for } s\in\mathbb{R}. \tag{9}$$

Let us estimate the term $-\delta\sum_{j=1}^{k}\left(t'X_{j,\delta}g,X_{j,\delta}\Lambda'^{2s-1}\frac{\partial g}{\partial t'}\right)_0$. We have

$$-\delta\sum_{j=1}^{k}\left(t'X_{j,\delta}g,X_{j,\delta}\frac{\partial}{\partial t'}\Lambda'^{2s-1}g\right)_0=-\delta\sum_{j=1}^{k}\left(t'X_{j,\delta}g,\Big[X_{j,\delta},\frac{\partial}{\partial t'}\Big]\Lambda'^{2s-1}g\right)_0$$
$$-\delta\sum_{j=1}^{k}\left(t'X_{j,\delta}g,\frac{\partial}{\partial t'}X_{j,\delta}\Lambda'^{2s-1}g\right)_0. \tag{10}$$

Since $\sigma\Big[X_{j,\delta},\frac{\partial}{\partial t'}\Big](x',t',\zeta',\tau')=Y_j(x',\delta\xi',\tau')$, the first term in Eq. (10) will be bounded by

$$C(\delta)\left(\sum_{j=1}^{k}\|t'X_{j,\delta}g\|^2_{s-\frac{1}{2}}+\|g\|^2_{s-\frac{1}{2}}\right)+\frac{\delta}{8}\Big\|\frac{\partial g}{\partial t'}\Big\|^2_{s-\frac{1}{2}}. \tag{11}$$

It remains to estimate the second term in Eq. (10)

$$-\delta\sum_{j=1}^{k}\left(t'X_{j,\delta}g,\frac{\partial}{\partial t'}X_{j,\delta}\Lambda'^{2s-1}g\right)_0=-\delta\sum_{j=1}^{k}\left(t'X_{j,\delta}g,\frac{\partial}{\partial t'}\Lambda'^{2s-1}X_{j,\delta}g\right)_0$$
$$-\delta\sum_{j=1}^{k}\left(t'X_{j,\delta}g,\frac{\partial}{\partial t'}[X_{j,\delta},\Lambda'^{2s-1}]g\right)_0. \tag{12}$$

Put $D_{j,s,\delta}=[X_{j,\delta},\Lambda'^{2s-1}]$. Since ord $D_{j,s,\delta}=2s-1$, the second term in Eq. (12) will be estimated by

$$C(\delta)\Big(\sum_{j=1}^{k}\|t'X_{j,\delta}g\|^2_{s-\frac{1}{2}}+\|g\|^2_{s-\frac{1}{2}}\Big)+\frac{\delta}{8}\Big\|\frac{\partial g}{\partial t'}\Big\|^2_{s-\frac{1}{2}}.$$

It remains to estimate the first term in Eq. (12)

$$-\delta\sum_{j=1}^{k}\left(t'X_{j,\delta}g,\frac{\partial}{\partial t'}\Lambda'^{2s-1}X_{j,\delta}g\right)_0=-\delta\sum_{j=1}^{k}\left(t'\Lambda'^{2s-1}X_{j,\delta}g,\frac{\partial}{\partial t'}X_{j,\delta}g\right)_0$$
$$-\delta\sum_{j=1}^{k}\left([\Lambda'^{2s-1},t']X_{j,\delta}g,\frac{\partial}{\partial t'}X_{j,\delta}g\right)_0.$$

We have

$$-\delta\sum_{j=1}^{k}\left(t'\Lambda'^{2s-1}X_{j,\delta}g,\frac{\partial}{\partial t'}X_{j,\delta}g\right)_0=\delta\sum_{j=1}^{k}\left(\Lambda'^{2s-1}X_{j,\delta}g,X_{j,\delta}g\right)_0$$
$$+\delta\sum_{j=1}^{k}\left(t'\Lambda'^{2s-1}\frac{\partial}{\partial t'}X_{j,\delta}g,X_{j,\delta}g\right)_0.$$

Consequently

$$-2\delta\mathrm{Re}\sum_{j=1}^{k}\left(t'\Lambda'^{2s-1}X_{j,\delta}g,\frac{\partial}{\partial t'}X_{j,\delta}g\right)_0=\delta\sum_{j=1}^{k}\|X_{j,\delta}g\|^2_{s-\frac{1}{2}}$$
$$+\delta\sum_{j=1}^{k}\left(\frac{\partial}{\partial t'}X_{j,\delta}g,[\Lambda'^{2s-1},t']X_{j,\delta}g\right)_0. \tag{13}$$

By virtue of the formula

$$\sigma([\Lambda'^{2s-1},t'])(x',t',\xi',\tau')=\frac{1}{\mathrm{i}}\frac{\partial}{\partial\tau}(1+|\xi'|^2+\tau'^2)^{\frac{2s-1}{2}}$$
$$=\frac{\tau'(2s-1)}{\mathrm{i}(1+|\xi'|^2+\tau'^2)^{-\frac{2s-3}{2}}},$$

we get

$$\delta\left|\sum_{j=1}^{k}\left(\frac{\partial}{\partial t'}X_{j,\delta}g,[\Lambda'^{2s-1},t']X_{j,\delta}g\right)_0\right|\leqslant\delta(2s-1)\sum_{j=1}^{k}\left\|\frac{\partial}{\partial t'}X_{j,\delta}g\right\|^2_{s-\frac{3}{2}}.$$

By Lemma 4

$$\delta(2s-1)\sum_{j=1}^{k}\left\|\frac{\partial}{\partial t'}X_{j,\delta}g\right\|^2_{s-\frac{3}{2}}\leqslant C(2s-1)\delta\left\|\frac{\partial g}{\partial t'}\right\|^2_{s-\frac{1}{2}}+C(\delta)\|\|^2_{s-\frac{1}{2}}, \tag{14}$$

where C is the constant appearing in Lemma 4. By choosing $\delta_0=\min\{1,\frac{1}{8C(2s-1)}\}$ and combining (7) - (14) we get (5). □

Corollary 2. *Under the hypothesis of Lemma 5 the following inequality holds*

$$\left\|\frac{\partial g}{\partial t'}\right\|^2_{s-\frac{1}{2}}+\sum_{j=1}^{k}\|X_{j,\delta}g\|^2_{s-\frac{1}{2}}\leqslant C(\delta)\left(\|P_\delta g\|^2_{s-\frac{1}{2}}+\|g\|^2_{s-\frac{1}{2}}\right), \tag{15}$$

for all $g\in C_0^\infty(K')$.

Corollary 3. *Under the hypothesis of Lemma 5 the following estimate holds*

$$\left\|\frac{\partial f}{\partial t}\right\|^2_{s-\frac{1}{2}} + \sum_{j=1}^{k} \|X_j f\|^2_{s-\frac{1}{2}} \leqslant C\left(\|Pf\|^2_{s-\frac{1}{2}} + \|f\|^2_{s-\frac{1}{2}}\right), \ \forall f \in C_0^\infty(K),$$

for some constant $C = C(s, K)$.

Proof. First choose $\delta = \frac{\delta_0}{2}, g = f_\delta$ in Eq. (15), then apply Lemma 1 we immediately get the result. □

Corollary 4. *Let the hypothesis of Lemma 5 be fulfilled and the algebra Lie generated by* $\{\frac{\partial}{\partial t}, X_1, \ldots, X_k\}$ *has rank* $n+1$ *in each point of* Ω. *There exists a number* $\sigma > 0$ *such that following estimate holds*

$$\|f\|_{s+\sigma} \leqslant C\left(\|Pf\|_s + \|f\|_s\right), \ \forall f \in C_0^\infty(K), \tag{16}$$

for some constant $C = C(s, K)$.

Proof. We can easily deduce the desired estimate from Corollary 3 and the apriori estimate achieved in Ref. 1 (see also Refs. 7–9). □

Theorem 1. *Let* P *have the form* (1). *We impose on the operators* X_j *the following conditions*

- $X_j(x, t, \xi, \tau) = tY_j(x, \xi, \tau) + Z_j(x, \xi)$, *where* $Y_j(x, \xi, 0) = 0, j = \overline{1, k}$.
- *The algebra Lie generated by* $\{\frac{\partial}{\partial t}, X_1, \ldots, X_k\}$ *has rank* $n+1$ *in each point of* Ω.

Then P *is a hypoelliptic operator.*

Proof. We shall show that if $f \in \mathcal{D}'(\Omega), Pf \in H^s_{\text{loc}}(\Omega)$, then $f \in H^{s+\sigma}_{\text{loc}}(\Omega)$, $X_j f \in H^{s+\frac{\sigma}{2}}_{\text{loc}}(\Omega), j = \overline{1, k}$ for some $\sigma > 0$. First we shall prove this assertion in the case $f \in \mathcal{E}'(\Omega)$. Since $f \in \mathcal{E}'(\Omega)$, there exists $\gamma \in \mathbb{R}$ such that $f \in H^\gamma(\mathbb{R}^{n+1})$. We can assume that $\gamma \leqslant s + \sigma$. Set $\eta = \inf(s, \gamma)$. Then $f, Pf \in H^\eta$. For the reason of convenience we will work in the coordinates (x', t') for $\delta < \delta_0$. Instead of f we write g in the (x', t') coordinates. Let

$\chi_\varepsilon *$ be an averaging kernel. We have

$$
\begin{aligned}
P_\delta(\chi_\varepsilon * g) &= -\frac{1}{\delta}\frac{\partial(\chi_\varepsilon * g)}{\partial t'} + \delta t'\Big(\sum_{j=1}^{k} X_{j,\delta}^2\Big)(\chi_\varepsilon * g) + c_\delta(x',t')(\chi_\varepsilon * g) \\
&= \chi_\varepsilon * P_\delta g - \Big[\frac{1}{\delta}\frac{\partial}{\partial t'}, \chi_\varepsilon *\Big] g + \delta \sum_{j=1}^{k}[t', \chi_\varepsilon *]X_{j,\delta}^2 g \\
&+ 2\delta t' \sum_{j=1}^{k} X_{j,\delta}[X_{j,\delta}, \chi_\varepsilon *]g + \delta t' \sum_{j=1}^{k}[[X_{j,\delta}, \chi_\varepsilon *], X_{j,\delta}]g + [c_\delta, \chi_\varepsilon *]g.
\end{aligned}
$$

Let us compute $(t'P_\delta(\chi_\varepsilon * g), \chi_\varepsilon * g)_\eta$ and $(P_\delta(\chi_\varepsilon * g), \Lambda^{-1}\frac{\partial(\chi_\varepsilon * g)}{\partial t'})_{\eta+\frac{1}{2}}$. We have

$$
\begin{aligned}
&(t'P_\delta(\chi_\varepsilon * g), \chi_\varepsilon * g)_\eta \\
&= (\chi_\varepsilon * P_\delta g, [t', \chi_\varepsilon *]g)_\eta + (\chi_\varepsilon * P_\delta g, \chi_\varepsilon * t'g)_\eta \\
&+ (T_{\varepsilon,\delta} g, \chi_\varepsilon * t'g)_\eta + (T_{\varepsilon,\delta} g, [t', \chi_\varepsilon] * g)_\eta \\
&+ \delta \sum_{j=1}^{k}([X_{j,\delta}, \chi_\varepsilon *]g, [X_{j,\delta}, t'^2](\chi_\varepsilon * g))_\eta \\
&+ \delta \sum_{j=1}^{k}([X_{j,\delta}, \chi_\varepsilon *]g, t'^2 X_{j,\delta}(\chi_\varepsilon * g))_\eta + \delta \sum_{j=1}^{k}(t'[t', \chi_\varepsilon *]X_{j,\delta}^2 g, \chi_\varepsilon * g)_\eta; \\
&\Big(P_\delta(\chi_\varepsilon * g), \Lambda^{-1}\frac{\partial(\chi_\varepsilon * g)}{\partial t'}\Big)_{\eta+\frac{1}{2}} \\
&= \Big(\chi_\varepsilon * P_\delta(g), \Lambda^{-1}\frac{\partial(\chi_\varepsilon * g)}{\partial t'}\Big)_{\eta+\frac{1}{2}} + \Big(T'_{\varepsilon,\delta} g, \Lambda^{-1}\frac{\partial(\chi_\varepsilon * g)}{\partial t'}\Big)_{\eta+\frac{1}{2}} \\
&+ \delta \sum_{j=1}^{k}\Big([t', \chi_\varepsilon *]X_{j,\delta}^2 g, \Lambda^{-1}\frac{\partial(\chi_\varepsilon * g)}{\partial t'}\Big)_{\eta+\frac{1}{2}} \\
&- 2\delta \sum_{j=1}^{k}\Big(t'[X_{j,\delta}, \chi_\varepsilon *]X_{j,\delta} g, \Lambda^{-1}\frac{\partial(\chi_\varepsilon * g)}{\partial t'}\Big)_{\eta+\frac{1}{2}}
\end{aligned}
$$

where all $T_{\varepsilon,\delta}, T'_{\varepsilon,\delta}, 0 < \varepsilon \leqslant \varepsilon_0$ belong to a bounded set in the spase of p. d. o. of order zero, if ε_0 is sufficiently small. By choosing an appropriate χ_ε from Lemma 5 we obtain

$$
\begin{aligned}
&\Big\|\frac{\partial(\chi_\varepsilon * g)}{\partial t'}\Big\|_\eta^2 + \sum_{j=1}^{k}\|X_{j,\delta}(\chi_\varepsilon * g)\|_\eta^2 \\
&\leqslant C(\Big(P_\delta(\chi_\varepsilon * g), \Lambda^{-1}\frac{\partial(\chi_\varepsilon * g)}{\partial t'}\Big)_{\eta+\frac{1}{2}} + |(t'P_\delta(\chi_\varepsilon * g), \chi_\varepsilon * g)_\eta| + \|\chi_\varepsilon * g\|_\eta^2)
\end{aligned}
$$

$$\leqslant \frac{1}{2}\Big\|\frac{\partial(\chi_\varepsilon * g)}{\partial t'}\Big\|_\eta^2 + \frac{1}{2}\sum_{j=1}^k \|X_{j,\delta}(\chi_\varepsilon * g)\|_\eta^2 + C(\|\chi_\varepsilon * P_\delta g\|_\eta^2 + \|\chi_\varepsilon * g\|_\eta^2).$$

Therefore

$$\Big\|\frac{\partial(\chi_\varepsilon * g)}{\partial t'}\Big\|_\eta^2 + \sum_{j=1}^k \|X_{j,\delta}(\chi_\varepsilon * g)\|_\eta^2 \leqslant M$$

uniformly in ε. Passing to the limit as $\varepsilon \to 0$ we obtain $\frac{\partial g}{\partial t'}, X_{j,\delta} g \in H^\eta$. That means that $\frac{\partial f}{\partial t}, X_j f \in H^\eta$. From (16) it follows that

$$\|\chi_\varepsilon * f\|_{\eta+\sigma} \leqslant C\Big(\sum_{j=1}^k \|X_j(\chi_\varepsilon * f)\|_\eta + \Big\|\frac{\partial(\chi_\varepsilon * f)}{\partial t}\Big\|_\eta + \|\chi_\varepsilon * f\|_\eta\Big)$$

$$\leqslant \sum_{j=1}^k \|\chi_\varepsilon * X_j f\|_\eta + \sum_{j=1}^k \|[\chi_\varepsilon *, X_j] f\|_\eta + \Big\|\chi_\varepsilon * \frac{\partial f}{\partial t}\Big\|_\eta + \Big\|\Big[\chi_\varepsilon *, \frac{\partial}{\partial t}\Big] f\Big\|_\eta$$

$$+ \|\chi_\varepsilon * f\|_\eta \leqslant M_1$$

uniformly in ε. Passing to the limit as $\varepsilon \to 0$ we obtain $f \in H^{\eta+\sigma}$. By the Schwarz inequality

$$\big|(\chi_\varepsilon * tPf, \chi_\varepsilon * f)_{\eta+\frac{\sigma}{2}}\big| \leqslant C(\|\chi_\varepsilon * tPf\|_\eta^2 + \|\chi_\varepsilon * f\|_{\sigma+\eta}^2).$$

Hence

$$\Big\|\frac{\partial(\chi_\varepsilon * f)}{\partial t'}\Big\|_{\eta+\frac{\sigma}{2}}^2 + \sum_{j=1}^k \|X_j(\chi_\varepsilon * f)\|_{\eta+\frac{\sigma}{2}}^2 \leqslant C(\|\chi_\varepsilon * tPf\|_\eta^2 + \|\chi_\varepsilon * f\|_{\sigma+\eta}^2) \leqslant M_2$$

uniformly in ε. Passing to the limit as $\varepsilon \to 0$ we obtain $\frac{\partial f}{\partial t}, X_j f \in H^{\eta+\frac{\sigma}{2}}$. We can repeat our argument with $\inf(s, \eta+\sigma)$ instead of η and get a similar assertion, the replace $\inf(s, \eta+\sigma)$ by $\inf[s, \inf(s, \eta+\sigma)+\sigma], \ldots$. At last we obtain the needed assertion for all $f \in \mathcal{E}'(\Omega)$. Now let $\tilde{\Omega}$ be a relative compact set in Ω, and f be the distribution on Ω, not necessarily with compact support. By the general theory of distributions it follows that the restriction of f to $\tilde{\Omega}$ belongs to $H^\gamma_{\mathrm{loc}}(\tilde{\Omega})$ for some $\gamma \in \mathbb{R}$. Consider a sequence of C^∞-function $\{g_i\}$ with compact support in $\tilde{\Omega}$ such that $g_{i+1} \equiv 1$ in a neighbourhood of supp g_i, $(i = 0, 1, \ldots,)$. We have

$$P(g_i f) = g_i Pf + 2\sum_{j=1}^k [X_j, g_i] t X_j(g_{i+1} f) - \Big[\frac{\partial}{\partial t} + c, g_i\Big] f$$

$$+ t\sum_{j=1}^k [X_j, [X_j, g_i]](g_{i+1} f) + 2\sum_{j=1}^k [t, [X_j, g_i]] X_j(g_{i+1} f). \qquad \text{(i)}$$

By a theorem on order of commutators we get

$$\text{ord}\,[X_j, g_i] = \text{ord}\,\left[\frac{\partial}{\partial t} + c, g_i\right] = \text{ord}\,[X_j, [X_j, g_i]] = \text{ord}\,[t, [X_j, g_i]] = 0.$$

We assume as above that $Pf \in H^s_{\text{loc}}(\Omega)$. Set $\eta = \inf(s, \gamma)$. Then the right hand side in the $(i+1)$th inequality belongs to $H^{\eta-1}$, i. e. $P(g_{i+1}f) \in H^{\eta-1}$. Hence by the previous step we have

$$X_j(g_{i+1}f) \in H^{\eta-1+\frac{\sigma}{2}}.$$

Therefore the right hand in Eq. (i) belongs to $H^{\eta-1+\frac{\sigma}{2}}$ whence $P(g_i f) \in H^{\eta-1+\frac{\sigma}{2}}$. But by result proved above $X_j(g_i f) \in H^{\eta-1+\sigma}, g_i f \in H^{\eta-1+\frac{3\sigma}{2}}$. Repeating this argument again and again we get $P(g_0 f) \in H^{\eta}$. It follows that $X_j(g_0 f) \in H^{\eta+\frac{\sigma}{2}}, g_0 f \in H^{\eta+\sigma}$. But we can choose function $g_0 \in C_0^\infty(\tilde{\Omega})$ arbitrarily. Therefore

$$f \in H^{\eta+\sigma}_{\text{loc}}(\Omega'), X_j f \in H^{\eta+\frac{\sigma}{2}}_{\text{loc}}(\Omega'), j = \overline{1,k}.$$

Using this assertion again and again we shall come to $t = s$ and thus we obtain the desired assertion. Our theorem is proved completely. □

References

1. L. Hörmander, *Acta Math.*, **119**, 147 (1967).
2. Y. Kannai, *Isr. J. Math.*, **9**, 306 (1971).
3. B. Helffer, C. Zuily, *C. R. Acad. Sci. Paris Sér. A-B*, **277**, 1061 (1973).
4. R. Beals, C. Fefferman, *Comm. Part. Diff. Equat.*, **1**, 73 (1976).
5. И. Е. Ганжа, *Вестник МГУ*, (1985), pp. 96-99.
6. И. Е. Ганжа, УМН, **41**, 217 (1986).
7. J. J. Kohn, *Proc. Symp. Pure Math.*, **23**, 61-69 (1973).
8. О. А. Олейник, Е. Б. Радкевич, *Уравнения Второго Порядка с Неотрицательной Характеристической Формой*, Итоги Науки. Мат. анализ. -М.: ВИНИТИ, 1971.
9. F. Treves, *Pseudo-Differential Operators and Fourier Integral Oprators*, (Plenum Press, 1982).

Advances in Deterministic and Stochastic Analysis
Eds. N. M. Chuong *et al.* (pp. 15–32)

§2. A SEMILINEAR NONCLASSICAL PSEUDODIFFERENTIAL BOUNDARY VALUE PROBLEM IN SOBOLEV SPACES $H_{\ell,p}$, $1 < p < \infty$

NGUYEN MINH CHUONG[‡] and DANG ANH TUAN

Institute of Mathematics
18 Hoang Quoc Viet, Cau Giay, 10307 Hanoi, Vietnam
[‡]*E-mail: nmchuong@math.ac.vn*

We study a semilinear nonclassical pseudodifferential value problem in Sobolev spaces $H_{\ell,p}$, $1 < p < \infty$. This paper is the continuation of Refs. 7,8,10,12, where the problem was investigated in $H_{\ell,2}$.

1. Introduction

In Ref. 5 A. V. Bicatze firstly on the world dropped the classical boundary condition imposed on the oblique derivative $D_\nu u = \frac{\partial u}{\partial v}$, namely, the condition that the vector field D_ν must be not tangent to the boundary. He investigated such problem, event for Laplace equations in three dimensional region with the assumption that the vector field D_ν may be tangent to the boundary. Then many mathematicians such as R. Borelli,[6] L. Hörmander,[14] Yu. V. Egorov and V. A. Kondratiev,[10] Yu. V. Egorov and Nguyen Minh Chuong,[9] Le Quang Trung,[19] Nguyen Minh Chuong and Tran Tri Kiet,[8] Nguyen Minh Chuong and Dang Anh Tuan,[7] etc., continue to extend this problem in several directions. For instance in Ref. 10 the problem was extended to an elliptic differential operator of second order for a bounded domain Ω in $\mathbb{R}^n$ with a smooth boundary $\partial\Omega$, in Ref. 9 the problem was studied for a parabolic differential operator of second order, in Ref. 11 a non-classical boundary value problem for a second order elliptic equation in Sobolev space of variable order, in Ref. 12 a semilinear boundary value problem for an elliptic singular integro - differential equation of order 2m in $H_{\ell,2}$ was investigated, in Ref. 8 a priori estimates were established for non-classical derivative problems for elliptic and parabolic differential operators of second order in Sobolev spaces $H_{\ell,p}$, $1 < p < \infty$.

The purpose of this paper is to investigate a semilinear non classical pseudodifferential boundary value problem in Sobolev spaces $H_{\ell,p,q}$, $1 < p < \infty, q$ is a complex parameter, $\alpha_0 \leq arg\ q \leq \beta_0$

This problem was investigated in Ref. 12 by Yu. V. Egorov and by the second author of this paper, with the singular integro - differential opetators of M. S. Agranovich type[3] and the Sobolev spaces $H_{\ell,2}$. In Ref. 3 M. S. Agranovich used the singular integral operators corresponding to spherical functions on a unit sphere $\mathbb{S}^{n-1}$ in $\mathbb{R}^n$. He used the symbols expanded into sequences of spherical functions converging in $L_2(\mathbb{S}^{n-1})$. Here by means of a new class of singular integral operators it is able to solve the non classical pseudodifferential boundary value problem in Ref. 12 in Sobolev spaces $H_{\ell,p.q}, 1 < p < \infty$. The approach used in this paper is quite different from Ref. 3, even from Ref. 12.

It is noticed that, by writing a non classical boundary value problem for a Laplace equation in pseudodifferential operator form, in Ref. 13 by the first author Yu. V. Egorov, an interesting estimate of subelliptic type was obtained.

2. Sobolev Spaces

- Let $H_{\ell,p}(\mathbb{R}^n), \ell \in \mathbb{R}, 1 \leq p < \infty$, be the completion of the space $C_0^\infty(\mathbb{R}^n)$ of functions with compact support with respect to the norm

$$||u||_{\ell,p} = ||u||_{\ell,p,\mathbb{R}^n} = \left(\int_{\mathbb{R}^n_\xi} (1 + |\xi|)^{pl} |Fu(\xi)|^p d\xi \right)^{\frac{1}{p}},$$

  where $Fu = \hat{u}(\xi)$ is the Fourier transform of function $u(x), |\xi| = \sqrt{\xi_1^2 + \ldots + \xi_n^2}$.
- Let p, ℓ be in $\mathbb{R}$ such that $0 \leq \ell$, $1 \leq p < \infty$. We denote by $H_{\ell,p}(\mathbb{R}^n_+)$ the completion of the space $C_0^\infty(\mathbb{R}^n_+)$ with respect to the norm:

$$\|u\|_{\ell,p} = \|u\|_{\ell,p,\mathbb{R}^n_+} = \inf \|lu\|_{\ell,p,\mathbb{R}^n},$$

  when the infimum is taken over all extension l from $\mathbb{R}^n_+$ to $\mathbb{R}^n$.
- Let Ω be a compact domain in $\mathbb{R}^n$ with $(n-1)$-dimensional smooth boundary $\partial\Omega$, $\{\varphi_j\}_{j=1}^N$ a partition of identity on Ω and $\{\psi_j\}_{j=1}^M$ a partition of identity on $\partial\Omega$ such that for each support of ψ_j, $j = 1, \ldots, M, \partial\Omega$ can be locally transformed into a superplane.
  Let p, ℓ be in $\mathbb{R}$ such that $0 \leq \ell$, $1 \leq p < \infty$. We denote by $H_{\ell,p}(\Omega)$ the completion of the space $C^\infty(G)$ with respect to the norm

$$\|u\|_{\ell,p} = \|u\|_{\ell,p,\Omega} = \left(\sum{}' \|\varphi_j u\|^p_{\ell,p,\mathbb{R}^n} + \sum{}'' \|\varphi_j u\|^p_{\ell,p,\mathbb{R}^n_+} \right)^{1/p},$$

where $\sum'$ denotes the sum over all j such that the neighbourhoods U_j do not intersect the boundary $\partial\Omega$, $\sum''$ denotes the sum over all j such that the neighbourhoods U_j intersect the boundary $\partial\Omega$.
Let p, ℓ be in $\mathbb{R}$ such that $0 \le \ell$, $1 \le p < \infty$. Let $H_{\ell,p}(\partial\Omega)$ the completion of the space $C^\infty(\partial\Omega)$ with respect to the norm

$$\|u\|_{\ell,p} = \|u\|_{\ell,p,\partial\Omega} = \Big(\sum_{j=1}^{M} \|\psi_j u\|^p_{\ell,p,\mathbb{R}^{n-1}}\Big)^{1/p},$$

where $\|\psi_j u\|_{\ell,p,\mathbb{R}^{n-1}}$ is the norm of the space $H_{\ell,p}(\mathbb{R}^{n-1})$ of $\psi_j u$ in the locally coordinate system.

- Let q be in $\mathbb{C}$, p, ℓ be in $\mathbb{R}$ such that $1 \le p < \infty$. In the spaces $H_{\ell,p,q}(\mathbb{R}^n)$, $H_{\ell,p,q}(\Omega)$ or $H_{\ell,p,q}(\mathbb{R}^n_+)$, $H_{\ell,p,q}(\partial\Omega)$ $(\ell \ge 0)$ we define a respective norm with parameter q:

$$|||u|||_{\ell,p} = \Big(\|u\|^p_{\ell,p} + |q|^{p\ell}\|u\|^p_{0,p}\Big)^{1/p}.$$

We give now some properties of the spaces $H_{\ell,p}$.

Proposition 2.1. *Let ℓ, k, p be in $\mathbb{R}$ such that $1 \le p < \infty$, $k < \ell$. The following bounded imbeddings hold true*

$$H_{\ell,p}(\mathbb{R}^n) \hookrightarrow H_{k,p}(\mathbb{R}^n),$$
$$H_{\ell,p}(\mathbb{R}^n_+) \hookrightarrow H_{k,p}(\mathbb{R}^n_+).$$

In particular, the imbedding $H_{\ell,p}(\Omega) \hookrightarrow H_{k,p}(\Omega)$ is compact.

Proposition 2.2. *Let p, ℓ be in $\mathbb{R}$ such that $1 \le p < \infty$, $0 \le \ell$. The restricted operator M from $\mathbb{R}^n$ to $\mathbb{R}^m_+$ is a bounded linear operator from $H_{\ell,p}(\mathbb{R}^n)$ to $H_{\ell,p}(\mathbb{R}^n_+)$. There exists extended bounded linear operator L from $H_{\ell,p}(\mathbb{R}^n_+)$ to $H_{\ell,p}(\mathbb{R}^n)$.*

Proposition 2.3. *Let p, ℓ be in $\mathbb{R}$ such that $1 \le p < \infty$, $(1-\frac{1}{p}) < \ell$. The operator defined by $u \mapsto u\big|_{x_n=0}$ is a bounded linear operator from $H_{\ell,p}(\mathbb{R}^n)$ (or $H_{\ell,p}(\mathbb{R}^n_+)$) to $H_{\ell-(1-\frac{1}{p}),p}(\mathbb{R}^{n-1})$. The operator defined by $u \mapsto u\big|_\Gamma$ is a bounded linear operator from $H_{\ell,p}(\Omega)$ to $H_{\ell-(1-\frac{1}{p}),p}(\partial\Omega)$.*

Proof. In fact, we have only to prove the following inequality

$$\|u|_{x_n=0}\|_{H_{\ell-(1-\frac{1}{p}),p}(\mathbb{R}^{n-1})} \le C\|u\|_{H_{\ell,p}(\mathbb{R}^n)}, \forall u \in C_0^\infty(\mathbb{R}^n), \tag{1}$$

where C is a positive constant not depending on u.

Note that $F'(u|_{x_n=0})(\xi',0) = (2\pi)^{-\frac{1}{2}} \int_{\mathbb{R}} Fu(\xi)d\xi_n$ for each $u \in C_0^\infty(\mathbb{R}^n)$ and the integral

$$\int_{\mathbb{R}} (1+|\xi'|+|\xi_n|)^{-\frac{\ell p}{p-1}} d\xi_n = (1+|\xi'|)^{-\frac{\ell p}{p-1}+1} \int_{\mathbb{R}} (1+|t|)^{-\frac{\ell p}{p-1}} dt,$$
$$(t = (1+|\xi'|^{-1}\xi_n),$$

converges for $(1-\frac{1}{p}) < \ell$, so by Hölder inequality we have

$$\begin{aligned} |F'(u|_{x_n=0})(\xi',0)|^p &\le (2\pi)^{-\frac{p}{2}} \Big(\int_{\mathbb{R}} (1+|\xi'|+|\xi_n|)^{-\frac{\ell p}{p-1}} d\xi_n\Big)^{p-1} \\ &\quad\times \int_{\mathbb{R}} (1+|\xi'|+|\xi_n|)^{\ell p} |Fu(\xi)|^p d\xi_n \\ &\le C(1+|\xi'|)^{-(lp-p+1)} \int_{\mathbb{R}} (1+|\xi'|+|\xi_n|)^{\ell p} |Fu(\xi)|^p d\xi_n \end{aligned}$$

hence,

$$(1+|\xi'|)^{(\ell-(1-\frac{1}{p}))p} |F'(u|_{x_n=0})(\xi',0)|^p \le C \int_{\mathbb{R}} (1+|\xi|)^{\ell p} |Fu(\xi)|^p d\xi_n,$$

by integrating both side of the above inequality we get the inequality (1). □

3. Pseudodifferential Operators in $\mathbb{R}^n$

Definition 3.1. Let m be in $\mathbb{Z}_+$, γ_1, γ_2 be in $\mathbb{R}$ and

$$Q = \{z \in \mathbb{C} | \gamma_1 \le \arg z \le \gamma_2\}.$$

The following operator is called a *pseudodifferential operator* in $\mathbb{R}^n$ of order m, with parameter q if it is defined by

$$\begin{aligned} Au(x,q) &= A(x,D,q)u(x,q) \\ &= (2\pi)^{-\frac{n}{2}} \Big[\sum_{|\alpha|+\beta \le m} q^\beta \int_{\mathbb{R}^n_y} K_{\alpha\beta}(x, x-y) D_y^\alpha u(y) dy \Big], \end{aligned} \tag{2}$$

with

$$q \in Q, \quad K_{\alpha\beta}(x,z) = H_{\alpha\beta}(z) + k_{\alpha\beta}(x,z),$$

where

(i) $H_{\alpha\beta}(z)$, $k_{\alpha\beta}(x,z)$ are homogeneous of order $(-n)$ with respect to z,
(ii) $k_{\alpha\beta}(x,z)$ belongs to $C^\infty(\mathbb{R}^n)$ with respect to x,
$D_x^\gamma k_{\alpha\beta}(x,z) \to 0$ when $|x| \to \infty$, for all γ (multi-index)

(iii)

$$\int\limits_{|z|=1} H_{\alpha\beta}(z)d\sigma(z) = 0, \qquad \int\limits_{|z|=1} k_{\alpha\beta}(x,z)d\sigma(z) = 0,$$

(iv)

$$\int\limits_{|z|=1} |H_{\alpha\beta}(z)|^2 d\sigma(z) < +\infty, \quad \int\limits_{\mathbb{R}^n_x}\int\limits_{|z|=1} |D^\gamma k_{\alpha\beta}(x,z)|^2 d\sigma(z)dx < +\infty, \forall\gamma,$$

where $\sigma(z)$ is an element area of the sphere $\{z \in \mathbb{R}^n \big| |z| = 1\}$.

Remark 3.1. Under conditions (i), (ii), (iii) and (iv), the pseudodifferential operator (2) is defined on $C_0^\infty(\mathbb{R}^n)$ and has another form

$$Au(x,q) = (2\pi)^{-n/2} \int\limits_{\mathbb{R}^n_\xi} e^{i\langle x,\xi\rangle} \sigma_A(x,\xi,q)Fu(\xi)d\xi, \tag{3}$$

where

$$\sigma_A(x,\xi,q) = \sum_{|\alpha|+\beta\leq m} q^\beta g_{\alpha\beta}(x,\xi)\xi^\alpha, \tag{4}$$

$$\begin{aligned} g_{\alpha\beta}(x,\xi) &= (2\pi)^{-n/2} \int\limits_{\mathbb{R}^n_y} e^{-i\langle x-y,\xi\rangle} K_{\alpha\beta}(x,x-y)dy \\ &= (2\pi)^{-n/2} \int\limits_{R^n_y} e^{-i\langle z,\xi\rangle} K_{\alpha\beta}(x,z)(-1)^n dz \\ &= g^0_{\alpha\beta}(x,\xi) + g^1_{\alpha\beta}(\xi), \end{aligned} \tag{5}$$

$$g^0_{\alpha\beta}(x,\xi) = (2\pi)^{-n/2}(-1)^n \int\limits_{\mathbb{R}^n_z} e^{-i\langle z,\xi\rangle} k_{\alpha\beta}(x,z)dz, \tag{6}$$

$$g^1_{\alpha\beta}(\xi) = (2\pi)^{-n/2}(-1)^n \int\limits_{\mathbb{R}^n_z} e^{-i\langle z,\xi\rangle} H_{\alpha\beta}(z)dz. \tag{7}$$

Theorem 3.1. *Let ℓ, p be in $\mathbb{R}$, m be in $\mathbb{Z}_+$ such that $1 < p < \infty$. The pseudodifferential operator A of order m of form (2) is a bounded linear operator from $(C_0^\infty(\mathbb{R}^n), |||\cdot|||_{\ell,p})$ to $(H_{\ell-m,p,q}(\mathbb{R}^n), |||\cdot|||_{\ell-m,p})$. This operator can be extended to a bounded linear operator from $H_{\ell,p,q}(\mathbb{R}^n)$ to $H_{\ell-m,p,q}(\mathbb{R}^n)$ acting as an operator of order m of form (3).*

Moreover, the estimate

$$|||Au|||_{\ell-m,p} \leq C|||u|||_{\ell,p}, \quad u \in H_{\ell,p,q}(\mathbb{R}^n), \tag{8}$$

holds true, where C is a constant not depending on u, q.

Proof. In fact, we have to show the estimate (8) for the operator A of form (14). We have

$$Au(x,q) = (2\pi)^{-n/2}\Big[\sum_{|\alpha|+\beta\le m} q^\beta \int_{\mathbb{R}^n_\xi} e^{i\langle x,\xi\rangle} g_{\alpha\beta}(x,\xi)Fu(\xi)d\xi\Big]$$
$$= (2\pi)^{-n/2}\Big[\sum_{|\alpha|+\beta\le m} q^\beta \Big(\int_{\mathbb{R}^n_\xi} e^{i\langle x,\xi\rangle} g^0_{\alpha\beta}(x,\xi)\xi^\alpha Fu(\xi)d\xi$$
$$+ \int_{\mathbb{R}^n_\xi} e^{i\langle x,\xi\rangle} g^1_{\alpha\beta}(\xi)\xi^\alpha Fu(\xi)d\xi\Big)\Big]$$

implies

$$FAu(\eta,q) = (2\pi)^{-\frac{n}{2}}\Big[\sum_{|\alpha|+\beta\le m} q^\beta \int_{\mathbb{R}^n_x}\int_{\mathbb{R}^n_\xi} e^{i\langle x,\xi-\eta\rangle} g^0_{\alpha\beta}(x,\xi)\xi^\alpha Fu(\xi)d\xi du\Big]$$
$$+ \sum_{|\alpha|+\beta\le m} q^\beta \eta^\alpha g^1_{\alpha\beta}(\eta)Fu(\eta),$$

Put

$$I_{\alpha\beta}(\eta,q) = q^\beta \int_{\mathbb{R}^n_x}\int_{\mathbb{R}^n_\xi} e^{i\langle x,\xi-\eta\rangle} g^0_{\alpha\beta}(x,\xi)\xi^\alpha Fu(\xi)d\xi dx$$
$$I(\eta,q) = \sum_{|\alpha|+\beta\le m} q^\beta \eta^\alpha g^1_{\alpha\beta}(\eta)Fu(\eta),$$

then

$$FAu(\eta,q) = (2\pi)^{-\frac{n}{2}} \sum_{|\alpha|+\beta\le m} I_{\alpha\beta}(\eta,q) + I(\eta,q).$$

In order to prove (8), we have to prove that

$$\int_{\mathbb{R}^n_\eta} (1+|\eta|+|q|)^{(\ell-m)p}|I_{\alpha\beta}(\eta,q)|^p d\eta \le C_1 \int_{\mathbb{R}^n_\eta} (1+|\eta|+|q|)^{\ell p}|Fu(\eta)|^p d\eta, \tag{9}$$

$$\int_{\mathbb{R}^n_\eta} (1+|\eta|+|q|)^{(\ell-m)p}|I(\eta,q)|^p d\eta \le C_2 \int_{\mathbb{R}^n_\eta} (1+|\eta|+|q|)^{\ell p}|Fu(\eta)|^p d\eta. \tag{10}$$

First, we prove (10). We have

$$(1+|\eta|+|q|)^{\ell-m}|I(\eta,q)| \le \sum_{|\alpha|+\beta\le m} (1+|\eta|+|q|)^{\ell-m}|q|^{\beta}|q^{\alpha}||g^1_{\alpha\beta}(\eta)||Fu(\eta)|.$$

Since $g^1_{\alpha\beta}(\eta)$ is homogeneous of order 0, we have

$$\sum_{|\alpha|+\beta\le m} (1+|\eta|+|q|)^{\ell-m}|q|^{\beta}|\eta^{\alpha}|\,|g^1_{\alpha\beta}(\eta)|\,|Fu(\eta)| \le C_3(1+|\eta|+|q|)^{\ell}|Fu(\eta)|.$$

This implies that

$$(1+|\eta|+|q|)^{\ell-m}|I(\eta,q)| \le C_3(1+|\eta|+|q|)^{\ell}|Fu(\eta)|$$

Hence, we have (10).

Now, we prove (9). We have

$$\begin{aligned}
&(1+|\eta|+|q|)^{\ell-m}|I_{\alpha\beta}(\eta,q)| = \\
&(1+|\eta|+|q|)^{\ell-m}|q|^{\beta}\Big|\int_{\mathbb{R}^n_\xi} \xi^{\alpha}Fu(\xi)\int_{\mathbb{R}^n_x} e^{i\langle x,\xi-\eta\rangle}g^0_{\alpha\beta}(\ell,\xi)dxd\xi\Big| \\
\le &\int_{\mathbb{R}^n_\xi}(1+|\xi|+|q|)^{\ell}|Fu(\xi)|\Big(\frac{1+|\eta|+|q|}{1+|\xi|+|q|}\Big)^{\ell-m}\Big|\int_{R^n_x} e^{i\langle x,\xi-\eta\rangle}g^0_{\alpha\beta}(x,\xi)dx\Big|d\xi.
\end{aligned} \tag{11}$$

Because of

$$g^0_{\alpha\beta}(x,\xi) = (2\pi)^{-\frac{n}{2}}(-1)^n\int_{\mathbb{R}^n_z} e^{-i\langle z,\xi\rangle}k_{\alpha\beta}(x,z)dz$$

and $D^{\gamma}_x(x,z) \to 0$, when $|x| \to \infty$, for all multi-index γ, so for all multi-index γ,

$$\begin{aligned}
|(\xi-\eta)^{\gamma}|\Big|\int_{\mathbb{R}^n_x} e^{i\langle x,\xi-\eta\rangle}g^0_{\alpha\beta}(x,\xi)dx\Big| &= \Big|\int_{\mathbb{R}^n_x}(D^{\gamma}_x e^{i\langle x,\xi-\eta\rangle})g^0_{\alpha\beta}(x,\xi)dx\Big| \\
&= \Big|\int_{\mathbb{R}^n_x} e^{i\langle x,\xi-\eta\rangle}D^{\gamma}_x g^0_{\alpha\beta}(x,\xi)dx\Big|,
\end{aligned}$$

it implies that for all k in $\mathbb{Z}_+$, we have

$$\begin{aligned}
&(1+|\xi-\eta|)^k\Big|\int\limits_{\mathbb{R}^n_x} e^{i\langle x,\xi-\eta\rangle} g^0_{\alpha\beta}(x,\xi)dx\Big| \\
&\le C_4 \sum_{0\le|\gamma|\le k} \Big|\int\limits_{R^n_x} e^{i\langle x,\xi-\eta\rangle} D^\gamma_x g^0_{\alpha\beta}(x,\xi)dx\Big| \\
&\le C_4 \sum_{0\le|\gamma|\le k} \int\limits_{\mathbb{R}^n_x}\Big|\int\limits_{\mathbb{R}^n_z} e^{-i\langle x,\xi\rangle} D^\gamma_x k_{\alpha\beta}(x,z)dz\Big|dx \\
&\le C_5,
\end{aligned}$$

where C_5 not depending η, ξ, q, so

$$\begin{aligned}
&\Big|\int\limits_{R^n_x} e^{i\langle x,\xi-\eta\rangle} g^0_{\alpha\beta}(x,\xi)dx\Big| \cdot \Big(\frac{1+|\eta|+|q|}{1+|\xi|+|q|}\Big)^{\ell-m} \\
&\le C_5(1+|\xi-\eta|)^{-k}\Big(\frac{1+|\eta|+|q|}{1+|\xi|+|q|}\Big)^{\ell-m} \\
&\le C_5(1+|\xi-\eta|)^{\ell-m-k}. \qquad (12)
\end{aligned}$$

Applying Hölder inequality for $(1+|\xi-\eta|)^{\frac{(\ell-m-k)(p-1)}{p}}$ and $(1+|\xi|+|q|)^\ell|Fu(\xi)|(1+|\xi-\eta|)^{\frac{\ell-m-k}{p}}$, we have

$$\begin{aligned}
&\Big(\int\limits_{\mathbb{R}^n_\xi}(1+|\xi|+|q|)^\ell|Fu(\xi)|(1+|\xi-\eta|)^{\ell-m-k}d\xi\Big)^p \\
&\le \Big(\int\limits_{\mathbb{R}^n_\xi}(1+|\xi-\eta|)^{\ell-m-k}d\xi\Big)^{p-1} \\
&\quad\times \int\limits_{\mathbb{R}^n_\xi}(1+|\xi|+|q|)^{p\ell}|Fu(\xi)|^p(1+|\xi-\eta|)^{\ell-m-k}d\xi
\end{aligned}$$

so from (11), (12) we get

$$\int_{\mathbb{R}^n_\eta} (1+|\eta|+|q|)^{p(\ell-m)} |I_{\alpha\beta}(\eta,q)|^p d\eta$$

$$\leq C_5 \int_{\mathbb{R}^n_\eta} \Big(\int_{\mathbb{R}^n_\xi} (1+|\xi|+|q|)^\ell |Fu(\xi)| (1+|\xi-\eta|)^{\ell-m-k} d\xi \Big)^p d\eta$$

$$\leq C_5 \Big(\int_{\mathbb{R}^n_\xi} (1+|\xi-\eta|)^{\ell-m-k} d\xi \Big)^{p-1}$$

$$\times \int_{\mathbb{R}^n_\eta} \int_{\mathbb{R}^n_\xi} (1+|\xi|+|q|)^{p\ell} |Fu(\xi)|^p (1+|\xi-\eta|)^{\ell-m-k} d\xi d\eta$$

$$\leq C_5 \Big(\int_{\mathbb{R}^n_\xi} (1+|\xi-\eta|)^{\ell-m-k} d\xi \Big)^p \int_{\mathbb{R}^n_\xi} (1+|\xi|+|q|)^{p\ell} |Fu(\xi)|^p d\xi.$$

With large enough k we have (9). □

We now study the pseudodifferential operator which has special form called a homogeneous one with the symbol not depending on x

$$\sigma_A(\xi,q) = \sum_{|\alpha|+\beta=m} q^\beta \xi^\alpha \int_{\mathbb{R}^n_z} e^{-i\langle z,\xi\rangle} K_{\alpha\beta}(z) dz.$$

In this case it is not difficult we obtain the following result.

Theorem 3.2. *Let ℓ, p be in $\mathbb{R}$, m be in $\mathbb{Z}_+$ such that $1 \leq p < \infty$. Suppose that the homogeneous pseudodifferential operator A of order m has symbol $\sigma_A(\xi,q) \neq 0$, when $|\xi|+|q| \neq 0$. Then, with $q \in Q \setminus \{0\}$, A has the inverse operator A^{-1} which is a pseudodifferential operator of order $(-m)$ with symbol $\sigma_A^{-1}(\xi,q)$. Moreover, the estimate*

$$|||A^{-1}f|||_{\ell,p} \leq C|||f|||_{\ell-m,p}, \quad f \in H_{\ell-m,p,q}(\mathbb{R}^n)$$

holds true, where C is a constant not depending on f, q.

4. Pseudodifferential Operator in $\mathbb{R}^n_+$

Definition 4.1. Let m be in $\mathbb{Z}_+$, γ_1, γ_2 be in $\mathbb{R}$ and

$$Q = \{z \in \mathbb{C} | \gamma_1 \leq \arg z \leq \gamma_2\}.$$

The operator defined by

$$A = M\tilde{A}L,$$

where

M is the restricted operator from $\mathbb{R}^n$ to $\mathbb{R}^n_+$,

L is the extended operator from $\mathbb{R}^n_+$ to $\mathbb{R}^n$ (in the Proposition 2.2),

$\tilde{A}$ is the pseudodifferential operator of form (3) of order m in $\mathbb{R}^n$ with symbol

$$\sigma_{\tilde{A}}(x,\xi,q),\ x \in \mathbb{R}^n,\ \xi \in \mathbb{R}^n,\ q \in Q,$$

is called the pseudodifferential operator of order m in $\mathbb{R}^n_+$ with symbol

$$\sigma_A(x,\xi,q) = \sigma_{\tilde{A}}(x,\xi,q), \quad x \in \mathbb{R}^n_+,\ \xi \in \mathbb{R}^n,\ q \in Q.$$

The operator A is said to be homogeneous if the operator $\tilde{A}$ is homogeneous.

The operator A is said to be admissible if its symbol has form

$$\begin{aligned}\sigma_A(x',0,\xi,q) &= \sum_{|\alpha|+\beta \le m} g_{\alpha\beta}(x',0,\xi')\xi^\alpha q^\beta \\ &= \sum_{k=0}^{m} \sigma_{Ak}(x',\xi',q)\xi^k,\end{aligned}$$

where σ_{As} does not depend on ξ'.

We now consider a boundary value problem in $\mathbb{R}^n_+$.

Let $s, m_1, \ldots, m_s$ be in $\mathbb{Z}_+$, γ_1, γ_2 be in $\mathbb{R}$ and

$$Q = \{z \in \mathbb{C} | \gamma_1 \le \arg z \le \gamma_2\}.$$

We now study the boundary value problem

$$Au(x,q) = A(x,D,q)u(x,q) = f(x,q), \quad x_n > 0, \quad (13)$$

$$B_j u(x,q)\big|_{x_n=0} = B_j(x,D,q)u(x,q)\big|_{x_n=0} = g_j(x',q), \quad j = 1,\ldots,s, \quad (14)$$

where $q \in Q$, A, B_j are homogeneous admissible pseudodifferential operators of order (respectively) $2s$, m_j with symbols (respectively) $\sigma_A(x,\xi,q)$, $\sigma_{B_j}(x,\xi,q)$.

Put $\ell_0 = \max\{2s, m_1+1, \ldots, m_s+1\}$,

$$\mathcal{U} = \mathcal{U}(x,D,q) = \left(A, B_1\big|_{x_n=0}, \ldots, B_s\big|_{x_n=0}\right),$$

and when $\ell_0 \le \ell$, $1 \le p < \infty$,

$$H_{\ell,p,q}(\mathbb{R}^n_+, \mathbb{R}^{n-1}) = H_{\ell-2s,p,q}(\mathbb{R}^n_+) \times \prod_{j=1}^{s} H_{\ell-m_j-(1-\frac{1}{p}),p,q}(\mathbb{R}^{n-1})$$

with respect to the norm

$$|||(f,g)|||_{\ell,p} = |||(f,g_1,\ldots,g_s)|||_{\ell,p} = |||f|||_{\ell-2s,p} + \sum_{j=1}^{s} |||g_s|||_{\ell-m_j-(1-\frac{1}{p}),p}.$$

It is easy to prove

Proposition 4.1. *The operator $\mathcal{U}$ is a bounded linear operator from $H_{\ell,p,q}(\mathbb{R}^n_+)$ to $H_{\ell,p,q}(\mathbb{R}^n_+,\mathbb{R}^{n-1})$. Moreover, the estimate*

$$|||\mathcal{U}u|||_{\ell,p} \leq C|||u|||_{\ell,p}, \quad u \in H_{\ell,p}(\mathbb{R}^n_+)$$

holds true, where C is a constant not depending on u, q.

Let us now consider the operators A, B_j possessing symbols $\sigma_A(x,\xi,q)$, $\sigma_{B_j}(x,\xi,q)$ not depending on x, i.e.

$$\sigma_A(x,\xi,q) = \sigma_A(\xi,q), \quad \sigma_{B_j}(x,\xi,q) = \sigma_{B_j}(\xi,q),$$

we have the following useful results.

The problem (13) - (14) is said to satisfy the condition Shapiro-Lopatinski if the problem on the halfline $t \geq 0$

$$\sigma_A(\xi', i\frac{d}{dt}, q)v(t) = 0, \quad t > 0 \tag{15}$$

$$\sigma_{B_j}(\xi', i\frac{d}{dt}, q)v(t)\big|_{t=0} = h_j, \quad j = 1,\ldots,s, \tag{16}$$

for $|\xi'| + |q| \neq 0$, has a unique solution in the space $\mathcal{M}$ of all stable solutions of (15) for arbitrary h_j.

The problem (13) - (14) is said to be elliptic if

- $\sigma_A(\xi,q) \neq 0$, for $|\xi| + |q| \neq 0$,
- the problem satisfies the condition Shapiro - Lopatinski.

Theorem 4.1. *Let p, ℓ be in $\mathbb{R}$ such that $\ell_0 \leq \ell$, $1 \leq p < \infty$, and the problem (13) - (14) is elliptic. Then, the following statements hold true.*

(i) If $q \in Q \setminus \{0\}$ then $\mathcal{U}$ has the inverse operator $\mathcal{U}^{-1}$ is a bounded linear operator from $H_{\ell,p,q}(\mathbb{R}^n_+,\mathbb{R}^{n-1})$ to $H_{\ell,p,q}(\mathbb{R}^n_+)$, not depending on p, ℓ,

(ii) If $q = 0$, there exists a bounded linear operator R from $H_{\ell,p,q}(\mathbb{R}^n_+,\mathbb{R}^{n-1})$ to $H_{\ell,p,q}(\mathbb{R}^n_+)$, not depending on p, ℓ such that

$$\mathcal{U}R = I + T,$$

where, I is the identity operator on $H_{\ell,p}(\mathbb{R}^n_+,\mathbb{R}^{n-1})$ T is a bounded linear operator from $H_{\ell,p,q}(\mathbb{R}^n_+,\mathbb{R}^{n-1})$ to $H_{\ell+1,p,q}(\mathbb{R}^n_+,\mathbb{R}^{n-1})$.

Proof.

(i) When $q \in Q \setminus \{0\}$, the inverse operator $\mathcal{U}^{-1}$ is defined by

$$\mathcal{U}^{-1}(f,g) = MF^{-1}(\sigma_A^{-1}(\xi,q)FLf(\xi,q))$$
$$+\sum_{j=1}^{s} MF^{'-1}(\Omega_j(\xi',x_n,q)F'(g_j - B_j(MF^{-1}(\sigma_A^{-1}(\eta,q)FLf))|_{x_n=0})(\xi',q)).$$

(ii) When $q = 0$, the operator R is defined by

$$R(f,g) = R_0 f + \sum_{j=1}^{s} R_j(g_j - B_j(R_0 f)|_{x_n=0}),$$

where $R_0 f = MF^{-1}(|\xi|^{2s}(1+|\xi|^{2s})^{-1}\sigma_A^{-1}(\xi,0)FLf)$,
$R_j(g_j) = MF^{'-1}(|\xi'|^{m_j+1}(1+|\xi'|^{m_j+1})^{-1}\Omega_j(\xi',x_n,0)F'(g_j)(\xi',0))$. □

5. Pseudodifferential Operator on a Compact Domain Ω

Let Ω be a compact domain in $\mathbb{R}^n$, with an $(n-1)$-dimensional, smooth boundary $\partial\Omega$, and $\{U_j, \varphi_j\}_{j=1}^{N}$ be a partition of identity on Ω.

Definition 5.1. Let m be in $\mathbb{Z}_+$, $\gamma_1, \gamma_2 \in \mathbb{R}$ and

$$Q = \{z \in \mathbb{C} | \gamma_1 \le \arg z \le \gamma_2\}.$$

Let A be the operator of form (3). The operator A is said to be a pseudodifferential operator on Ω of order m if

(i) for each $\varphi \in C^\infty(\Omega)$; φA - $A\varphi$ is an operator of order $(m-1)$ on $H_{\ell,p,q}(\Omega)$, $\ell \ge 0$, $1 < p < \infty$,
(ii) for each φ, $\psi \in C^\infty(\Omega)$ whose supports in U_j
 • if $U_j \cap \Gamma = \emptyset$, in the local coordinates

$$\varphi A[\psi .] = \varphi A_j[\psi .],$$

 where A_j are pseudodifferential operator of order m in $\mathbb{R}^n$ with symbols $\sigma_{A_j}(x,\xi,q)$, $x \in \mathbb{R}^n$, $\xi \in \mathbb{R}^n$, $q \in Q$,
 • if $U_j \cap \Gamma \ne \emptyset$, in the local coordinates

$$\varphi A[\psi .] = \varphi A_j[\psi .],$$

 where A_j are pseudodifferential operators of order m in $\mathbb{R}^n$ with symbols $\delta_{A_j}(x,\xi,q)$, $x \in \mathbb{R}^n_+$, $\xi \in \mathbb{R}^n$, $q \in Q$.

The operator A is said to be admissible if for each φ, $\psi \in C^\infty(\Omega)$ whose supports in U_j, $U_j \cap \Gamma \neq \emptyset$, in the local coordinates

$$\varphi A[\psi .] = \varphi A_j[\psi .],$$

A_j are admissible.

We consider now the boundary value problem on Ω.
Let s, $m_1, \ldots, m_s$ be in $\mathbb{Z}_+$, γ_1, γ_2 be in $\mathbb{R}$ and

$$Q = \big\{ z \in \mathbb{C} \big| \gamma_1 \leq \arg z \leq \gamma_2 \big\}.$$

We now study the boundary value problem

$$Au(x,q) = f(x,q), \quad x \in \Omega, \tag{17}$$

$$B_j u(x,q)\big|_{\partial\Omega} = B_j u(x,q)\big|_{x=x'} = g_j(x',q), \quad j = 1, \ldots, s, \tag{18}$$

where $q \in Q$, A, B_j are admissible pseudodifferential operators of order (respectively) $2s$, m_j.

Put

$$\ell_0 = \max\big\{2s, m_1+1, \ldots, m_s+1\big\},$$

$$\mathcal{U} = \big(A, B_1\big|_{\partial\Omega}, \ldots, B_s\big|_{\partial\Omega}\big),$$

and when $\ell \geq \ell_0$, $1 < p < \infty$,

$$H_{\ell,p,q}(\Omega, \partial\Omega) = H_{\ell-2s,p,q}(\Omega) \times \prod_{j=1}^{s} H_{\ell-m_j-(1-\frac{1}{p}),p,q}(\partial\Omega)$$

with respect to the norm

$$|||(f,g)|||_{\ell,p} = |||(f,g_1,\ldots,g_s)|||_{\ell,p} = |||f|||_{\ell-2s,p} + \sum_{j=1}^{s} |||g_j|||_{\ell-m_j-(1-\frac{1}{p}),p}.$$

It is obvious that

Proposition 5.1. *Let ℓ, p be in $\mathbb{R}$ such that $\ell_0 \leq \ell$, $1 < p < \infty$. Then $\mathcal{U}$ is a bounded linear operator from $H_{\ell,p,q}(\Omega)$ to $H_{\ell,p,q}(\Omega, \partial\Omega)$. Moreover, the estimate*

$$|||\mathcal{U}u|||_{\ell,p} \leq C|||u|||_{\ell,p}, \quad u \in H_{\ell,p}(\Omega),$$

holds true, where C is a constant not depending on u, q.

Ellipticity Condition

Condition 5.1. Let $x_0 \in \Omega$ with a neighbourhood U_j. In the locall coordinate system of x_0, the operator A changes into the pseudodifferential operator A_j of order $2s$ in $\mathbb{R}^n$ having the principal part A_{0j} with the symbol $\sigma_{A_{0j}}(x, \xi, q)$.

The problem (17) - (18) is said to satisfy the Condition 5.1 at x_0 if

$$\sigma_{A_{0j}}(0, \xi, q) \neq 0, \quad \text{for } |\xi| + |q| \neq 0.$$

Condition 5.2. Let $x_0 \in \partial\Omega$ with a neighbourhood U_j. In the locall coordinate system of x_0, the operator A changes into the admissible pseudodifferential operator A_j of order $2s$ in $\mathbb{R}^n_+$ having the principal part A_{0j} with symbol $\sigma_{A_{0j}}(x, \xi, q)$, the operator B_k changes into the admissible pseudodifferential operator of order m_k in $\mathbb{R}^n_+$ having the main part B_{0kj} with symbol $\sigma_{B_{0kj}}(x, \xi, q)$.

The problem (17) - (18) is said to satisfy Condition 5.2 at x_0 if the problem on half-line $t \geq 0$

$$\sigma_{A_{0j}}(0, \xi, -\frac{d}{dt}, q)v(t) = 0, \quad t > 0 \tag{19}$$

$$\sigma_{B_{0kj}}(0, \xi', -i\frac{d}{dt}, q)v(t)\big|_{t=0} = h_k. \quad k = 1, \ldots, s, \tag{20}$$

when $\xi'| + |q| \neq 0$, has a unique solution in the space $\mathcal{M}$ of all stable solutions of (19) for arbitrary h_k.

The problem (17) - (18) is said to be elliptic if it satisfies Condition 1 for every $x_0 \in G$ and Condition 2 for every $x_0 \in \partial\Omega$.

The following proposition is proved without difficulties

Proposition 5.2. *Let ℓ, p be in $\mathbb{R}$ such that $\ell_0 \leq \ell$, $1 \leq p < \infty$, and the problem (17) - (18) be elliptic. Then, the estimate*

$$|||u|||_{\ell,p} \leq C\big(|||\mathcal{U}u|||_{\ell,p} + |||u|||_{0,p}\big), \quad u \in H_{\ell,p}(\Omega)$$

holds true, where C is constant not depending on u, q.

This implies that, when q having large enough $|q|$, for every $(f, g) \in H_{\ell,p}(\Omega, \partial\Omega)$, if the problem (17)-(18) has a solution, then it is unique.

When the problem (17) - (18) is said to be elliptic, we also say the operator $\mathcal{U}$ is elliptic. To obtain the next theorem we need the following easily proved Lemma

Lemma 5.1. *If $\mathcal{U}$ is elliptic, then $\mathcal{U}$ has a regularizer R satisfying*

$$R\mathcal{U} = I_1 + T_1, \quad \mathcal{U}R = I_2 + T_2,$$

where I_1, I_2 are the identity operators (respectively) on $H_{\ell,p,q}(\Omega)$, $H_{\ell,p,q}(\Omega, \partial\Omega)$; T_1 is a bounded linear operator from $H_{\ell,p,q}(\Omega)$ to $H_{\ell+1,p,q}(\Omega)$; T_2 is a bounded linear operator from $H_{\ell,p,q}(\Omega, \partial\Omega)$ to $H_{\ell+1,p,q}(\Omega, \partial\Omega)$.

Lemma 5.2. *Let B_0, B, B' are Banach spaces with respective norm $\|\cdot\|_0$, $\|\cdot\|$, $\|\cdot\|'$. Assume that $B \hookrightarrow B_0$ is a compact imbedding operator and $A : B \to B'$ is a bounded linear operator. Then, the necessary and sufficient condition such that dim(Ker$A$) $< +\infty$ and ImA is closed B', is*

$$\|u\| \leq C\big(\|Au\|' + \|u\|_0\big), \quad u \in B, \tag{21}$$

where C is a constant not depending on u.

By using Proposition 5.1, 5.2, Lemmas 5.1 and 5.2 we obtain

Theorem 5.1. *Let ℓ, p be in $\mathbb{R}$ such that $\ell_0 \leq \ell$, $1 < p < \infty$. If $\mathcal{U}$ is elliptic then for large enough $|q|$ the operator $\mathcal{U}$ is an isomorphism from $H_{\ell,p,q}(\Omega)$ to $H_{\ell,p,q}(\Omega, \partial\Omega)$.*

6. A Linear Non-classical Pseudodifferential Boundary Value Problem

We study now the following non-classical pseudodifferential boundary value problem

$$Au = f(x) \quad \text{in } \Omega \tag{22}$$

$$\mathcal{B}_j u = B_j(D_\nu u) = g_j \text{ on } \partial\Omega, j = 1, 2, \ldots, s \tag{23}$$

where A, B_j are of order $2s, m_j$ respectively in the problem (17)-(18), D_ν is a vector field that can be tangent to $\partial\Omega$ on a smooth manifold $\Gamma_0 \subset \partial\Omega$ of dimension $(n-2)$ and not tangent to Γ_0. We use here the classification of Γ_0 given in Ref. 10.

For the reader's convenience let us recall shortly this classification. Let μ be the inner normal unit vector of the boundary $\partial\Omega$. The real function $\langle \mu, \nu \rangle$ is defined on the boundary $\partial\Omega$,. The submanifold Γ_0 is said to be of the first class if on some neighborhood of Γ_0, the function $\langle \mu, \nu \rangle$ is negative on the negative part (with respect to the direction ν), and positive on the positive one, the second class is positive on the negative part and negative on the positive one, the third class does not change sign when passing Γ_0.

If Γ_0 belongs to the first class, then we add the following conditions

$$D_n^k u(x,q) = U_{0k}(x,q), k = 0,1,\ldots,s-1 \text{ on } \Omega_0. \tag{24}$$

The above problem will be discussed in Sobolev spaces $H_{\ell,p,q}, 1 < p < \infty$.

It is assumed that the problem defined by (A, B_j) is elliptic that is A is an elliptic operator and the Shapiro - Lopatinski condition is satisfied.

Let Γ_d be the intersection of a d-neighborhood of Γ_0 with $\Omega, h = h_d(x) \in C^\infty$ and

$$h_d(x) = \begin{cases} 0, & \text{if } x \text{ is outside of } \Gamma_d \\ 1, & \text{if } x \text{ is in } \Gamma_{d/2}. \end{cases}$$

Denote by $\Pi_{\ell,p,q}(\Omega)$ the space with norm

$$||u||_{\Pi_{\ell,p,q}} = |||u|||_{\ell,p,\Omega} + |||\frac{\partial hu}{\partial \nu}|||_{\ell,p,\Omega} + |||hu|||_{\ell,p,\mathcal{N}},$$

where $D_\nu = \frac{\partial}{\partial \nu}, \mathcal{N} = \mathcal{N}_d = \{x \in G_d | \exists y \in \Gamma_0, \overrightarrow{xy}$ is a normal vector to $\partial\Omega, d(x,\Gamma_0) \le d\}$, if Γ_0 is of the first class and if Γ_0 is not of the first class the term hu in the norm is omitted.

Denote by $G_{\ell,p,q}(\partial\Omega)$ the spaces of functions defined on $\partial\Omega$ with norm

$$||u||_{G_{\ell,p,q}} = |||u|||_{\ell,p,\partial\Omega} + |||hu|||_{\ell+1,p,\partial\Omega}.$$

Denote $\Pi_{\ell,p,q}(\Omega,\partial\Omega) = \Pi_{\ell-2s,p,q}(\Omega) \times \prod_{j=1}^s G_{\ell-j-(1-\frac{1}{p}),p,q}(\partial\Omega)$, for $\ell \ge \ell_1 = \max\{2s, m_j + 2\}$.

Theorem 6.1. *Assume that the operator (A, B_j) is elliptic, $\ell \ge \ell_1$. Then for large enough $|q|$, for each $(f, g_j) \in \Pi_{\ell,p,q}(\Omega,\partial\Omega)$ if Γ_0 belongs to the third class, or additionally $u_0 \in H_{\ell-(1-1/p),p,q}(\Gamma_0)$, if Γ_0 belongs to the first class the problem (22)-(23) or (22)-(23)-(24) has a unique solution $u \in \Pi_{\ell,p,q}(\Omega)$.*

7. A Semilinear Non Classical Pseudodifferential Boundary Value Problem

Let us now consider the semilinear non classical pseudodifferential boundary value problem

$$Au(x,q) = f(x,q,u,\ldots,D^{2m-1}u), \text{in } \Omega, \tag{25}$$

$$\mathcal{B}_j u(x,q) = B_j(D_\nu u(x,q)) = g_j(x,q,u,\ldots,D^{m_j-1}u), \text{on } \partial\Omega, j = 1,\ldots,s \tag{26}$$

where A, B_j are pseudodifferential operators in (22), (23) in Sec. 6. If Γ_0 belongs to the first class then we add the conditions

$$D_n^k u(x,q) = u_{0k}(x,q), k = 0, \ldots, s-1 \text{ on } \Gamma_0. \tag{27}$$

For this problem we get

Theorem 7.1. *Assume that Γ_0 is of the first class, $\ell \geq \ell_1$, (A, B_j) is elliptic, $u_{0,k} \in H_{\ell-k-(1-\frac{1}{p}),p,q}(\Gamma_0)$, $f, g = (g_1, \ldots, g_s)$ satisfy these conditions*

(i) these maps
$(x,q,\vec{u}) \mapsto f(x,q,\vec{u})$, where $\vec{u} = (u_1, \ldots, u_N)$,
$(x,q,\vec{u}) \mapsto g_j(x,q,\vec{u}^j)$, where $\vec{u}^j = (u_1, \ldots, u_{N_j}), j = 1, 2, \ldots, s$
satisfy Carathéodory conditions, i.e. f, g_j are respectively continuous at $\vec{u}, \vec{u}^j$ for almost (x,q), measurable at (x,q) for every $\vec{u}, \vec{u}^j$,

(ii) the map $u(x,q) \mapsto (f(x,q,\vec{u}(x,q)), g_j(x,q,\vec{u}^j(x,q)))$,
is from $\Pi_{\ell,p,q}(\Omega)$ to $\Pi_{\ell,p,q}(\Omega, \partial\Omega)$ transforms each bounded set into a relatively compact set, where

$$\vec{u}(x,q) = (u(x,q), \ldots, D^{2s-1}u(x,q)),$$

$$\vec{u}^j(x,q) = (u(x,q), \ldots, D^{m_j-1}u(x,q)),$$

(iii) $1 > \liminf_{R\to\infty} \limsup_{|q|\to\infty} ||(A, \mathcal{B}_j)^{-1}|| \times ||(f,g)||_{R,q}$, where

$$||(f,g)||_{R,q} = \sup_{||u||_{\Pi_{\ell,p,q}(\Omega)}=R} \frac{||(f(x,q,\vec{u}(x,q)), g_j(x,q,\vec{u}^j(x,q)))||_{\Pi_{\ell,p,q}(\Omega,\partial\Omega)}}{R}.$$

Then, for large enough $|q|$, the problem (25)-(26)-(27) has a solution $u \in \Pi_{\ell,p,q}(\Omega)$.

Proof. According to Theorem 6.1 for $|q|$ large enough, for each $w \in \Pi_{\ell,p,q}(\Omega)$ the linear problem

$$Au(x,q) = f(x,q,w,\ldots,D^{2s-1}w) \text{ in } \Omega, \tag{28}$$

$$B_j u(x,q) = g_j(x,q,w,\ldots,D^{m_j-1}w), \text{ on } \partial\Omega, j = 1, \ldots, s, \tag{29}$$

$$D_n^k u(x,q) = u_{0k}(x,q), k = 0, 1, \ldots, s-1 \text{ on } \Gamma_0 \tag{30}$$

has a unique solution $u \in \Pi_{\ell,p,q}(\Omega)$.
So for $|q|$ large enough, we get the map $J : w \mapsto u$ from $\Pi_{\ell,p,q}(\Omega)$ into itself. By the assumptions $(i) - (ii)$ the map J is compact. By the assumption (iii) there exists a positive R such that the map J maps the sphere $S_R = \{w \in \Pi_{\ell,p,q}(\Omega) | ||w||_{\Pi_{\ell,p,q}(\Omega)} = R\}$ into the ball $B_R = \{w \in$

$\Pi_{\ell,p,q}(\Omega)|\|w\|_{\Pi_{\ell,p,q}(\Omega)} \leq R\}$. Therefore, the map J satisfies the Fixed Theorem of Rothe so it possesses a fixed point which is a solution of the problem (25)-(26)-(27). □

References

1. R. A. Adams, *Sobolev spaces*, (Acad. Press, 1975).
2. S. Agmon, A. Douglis, L. Nirenberg, *Comm. Pure Appl. Math.*, 624 (1959).
3. M. S. Agranovich, *Uspekhi Mat. Nauk* **20**, 3 (1965).
4. M. S. Agranovich, M. I. Vishik, *Uspekhi Mat. Nauk* **19**, 53 (1964).
5. A. V. Bicatze, *Dokl. Acad. Nauk. SSSR* **148**, 749 (1963).
6. R. Borelli, *J. Math. Mech.* **16**, 51 (1966).
7. Nguyen Minh Chuong, Dang Anh Tuan, A boundary value problem for singular integro-differential operators in $H_{\ell,p}, 1 < p < \infty$, *Preprint 2002/28*, (Institute of Mathematics, Hanoi, 2000).
8. Nguyen Minh Chuong, Tran Tri Kiet, On a priori estimates in $L^p(\mathbb{R}^n), p \geq 2$ for nonclassical oblique derivative problems, *Preprint 2000/16*, (Institute of Mathematics, Hanoi, 2000).
9. Yu. V. Egorov, Nguyen Minh Chuong, *Diff. Urav.* **24**, 197 (1969).
10. Yu. V. Egorov, V. A. Kondratiev, *Mat. Sbornik*, **78**, 148 (1969).
11. Yu. V. Egorov, Nguyen Minh Chuong, *Diff. Urav.* **20**, 2163 (1984).
12. Yu. V. Egorov, Nguyen Minh Chuong, *Uspekhi Mat. Nauk* **53**, 249 (1998).
13. Yu. V. Egorov, *Mat. Sbornik* **73**, 356 (1967).
14. L. Hömander, *Ann. of Math.* **83**, 129 (1966).
15. L. Hömander, *The analysis of linear partial differential operators, III, Pseudodifferential operator*, (Springer - Verlag, 1985).
16. J. Leray and J. Schauder, *Ann. Sc. École Norm. Sup.* **51**, 45 (1934).
17. E. H. Lieb, M. Loss, *Analysis*, Vol. 14, (AMS, 1966).
18. G. O. Okikiolu, *Aspects of the theory of bounded integral operators in L^p spaces*, (Academic Press, 1971).
19. Le Quang Trung, *Uspekhi. Mat. Nauk* **44**, 169 (1989).

Advances in Deterministic and Stochastic Analysis
Eds. N. M. Chuong *et al.* (pp. 33–42)

§3. SINGULAR INTEGRAL OPERATORS IN FUNCTIONAL SPACES OF MORREY TYPE

LUBOMIRA SOFTOVA

Postal address: Politecnico di Bari
Dipartimento di Matematica
Via E. Orabona 4,
70125 Bari, Italy
E-mail: luba@dm.uniba.it

Keywords: Generalized Morrey spaces, Singular integrals, Commutators, Calderón-Zygmund kernel, BMO, VMO.
MR(2000) Subject Classification: 42B20

1. Definitions and Preliminary Results

We consider the following integral operator

$$\mathcal{K}f(x) := P.V. \int_{\mathbb{R}^n} k(x; x-y) f(y) dy \tag{1}$$

and its commutator with essentially bounded functions

$$\mathcal{C}[a,k]f(x) := P.V. \int_{\mathbb{R}^n} k(x; x-y)[a(y)-a(x)]f(y)dy = \mathcal{K}(af)(x) - a(x)\mathcal{K}f(x). \tag{2}$$

The generating kernel $k(x;\xi) : \mathbb{R}^n \times \{\mathbb{R}^n \setminus \{0\}\} \to \mathbb{R}$ is the one studied by Fabes and Rivière[8] and is a generalization of the classical kernel of Calderón and Zygmund. Introducing a new metric $\rho(x)$ as a solution of the equation $F(x,\rho) = \sum_{i=1}^n x_i^2 \rho^{-2\alpha_i} = 1$, $\alpha_i \geq 1$, they study (1) in $L^p(\mathbb{R}^n)$. The balls with respect to $\rho(x)$, centered at the origin and of radius r are simply the ellipsoids

$$\mathcal{E}_r(0) = \left\{ x \in \mathbb{R}^n : \quad \frac{x_1^2}{r^{2\alpha_1}} + \ldots + \frac{x_n^2}{r^{2\alpha_n}} < 1 \right\}$$

with Lebesgue measure $|\mathcal{E}_r| = C(n) r^{\alpha}$. It is easy to see that $\mathcal{E}_1(0) \equiv \mathbb{S}^{n-1}$ with respect to the Euclidean metric. This fact allows to impose on the kernel $k(x,\xi)$ the Calderón-Zygmund conditions on the unite sphere, in spite of the lack of "symmetry" of k with respect to the variables ξ_i, $i = 1, \ldots, n$. Let we note, that the standard parabolic metric $\widetilde{\rho} = \sup\{|x|, \sqrt{t}\}$, $x \in \mathbb{R}^n, t \in (0, \infty)$ does not permit to define the mentioned above conditions on kernels having homogeneity of parabolic type. Using the Fourier transform in $L^2(\mathbb{R}^n)$ and the Marcinkiewicz interpolation theorem, Fabes and Rivière obtained that the integral operator (1) is continuous in $L^p(\mathbb{R}^n)$, $p \in (1, \infty)$.

Definition 1.1. The function $k(x; \xi)\colon \mathbb{R}^n \times \{\mathbb{R}^n \setminus \{0\}\} \to \mathbb{R}$ is called a *variable kernel with mixed homogeneity* if:

i) for every fixed x the function $k(x; \cdot)$ is a *constant kernel* satisfying

$i_a)$ $k(x; \cdot) \in C^\infty(\mathbb{R}^n \setminus \{0\})$;
$i_b)$ $k(x; \mu^{\alpha_1}\xi_1, \ldots, \mu^{\alpha_n}\xi_n) = \mu^{-\alpha} k(x; \xi)$, $\forall\ \mu > 0$, $\alpha_i \geq 1$, $\alpha = \sum_{i=1}^n \alpha_i$;
$i_c)$ $\displaystyle\int_{\mathbb{S}^{n-1}} k(x; \xi) d\sigma_\xi = 0 \quad \text{and} \quad \int_{\mathbb{S}^{n-1}} |k(x; \xi)| d\sigma_\xi < \infty$;

ii) for every multiindex β : $\displaystyle\sup_{\xi \in \mathbb{S}^{n-1}} \left| D_\xi^\beta k(x; \xi) \right| \leq C(\beta)$ independently of x.

Let us note that in the special case $\alpha_i = 1$ and thus $\alpha = n$, Definition 29 gives rise to the classical Calderón-Zygmund kernels. One more example is when $\alpha_1 = \ldots = \alpha_{n-1} = 1$, $\alpha_n = \bar{\alpha} > 1$. In this case we obtain the kernels studied by B.F. Jones and discussed in Ref. 8.

Let $\omega\colon \mathbb{R}^n \times \mathbb{R}_+ \to \mathbb{R}_+$ and for any ellipsoid $\mathcal{E}$ we write $\omega(x, r) =: \omega(\mathcal{E})$.

Definition 1.2. A function $f \in L^p_{\mathrm{loc}}(\mathbb{R}^n)$, $p \in (1, \infty)$ belongs to the generalized Morrey space $L^{p,\omega}(\mathbb{R}^n)$ if the following norm is finite

$$\|f\|_{p,\omega} := \left(\sup_{\mathcal{E}} \frac{1}{\omega(\mathcal{E})} \int_{\mathcal{E}} |f(y)|^p dy \right)^{1/p}. \qquad (3)$$

The space $L^{p,\omega}(\Omega)$ and the norm $\|f\|_{p,\omega;\Omega}$ are defined by taking $f \in L^p(\Omega)$ and $\mathcal{E} \cap \Omega$ instead of $\mathcal{E}$ in (3).

For $\omega(x, r) = 1$ we get the Lebesgue space $L^p_{\mathrm{loc}}(\mathbb{R}^n)$ and for $\omega(x, r) = r^\lambda$, $\lambda \in (0, \alpha)$, $L^{p,\omega}(\mathbb{R}^n)$ coincides with the Morrey space $L^{p,\lambda}(\mathbb{R}^n)$. However, there exist weight functions, as $\omega(x, r) = r \ln(r + 2)$ for which $L^{p,\omega}(\mathbb{R}^n)$ does not coincides with any Morrey space.

For a given measurable function $f \in L^1_{\rm loc}(\mathbb{R}^n)$ define the *Hardy–Littlewood maximal operator* Mf and the *sharp maximal operator* $f^\sharp$ as

$$Mf(x) := \sup_{\mathcal{E}(x)} \frac{1}{|\mathcal{E}(x)|} \int_{\mathcal{E}(x)} |f(y)|dy,$$

$$f^\sharp(x) := \sup_{\mathcal{E}(x)} \frac{1}{|\mathcal{E}(x)|} \int_{\mathcal{E}(x)} |f(y) - f_{\mathcal{E}(x)}|dy$$

almost everywhere in $\mathbb{R}^n$ and the supremum is taken over all ellipsoids $\mathcal{E}$ centered at x. Define also the operator $M_s f(x) := (M|f|^s(x))^{1/s}$ for $s \geq 1$.

The next results are weighted variants of the well-known maximal and sharp inequalities (see Refs. 3,5,15).

Lemma 1.1 (Maximal inequality[12]). *Assume that there are constants C_1 and C_2 such that for any $x_0 \in \mathbb{R}^n$ and for any $r > 0$*

$$r \leq t \leq 2r \Longrightarrow C_1 \leq \frac{\omega(x_0, t)}{\omega(x_0, r)} \leq C_2, \tag{4}$$

$$\int_r^\infty \frac{\omega(x_0, t)}{t^{\alpha+1}} dt \leq C \frac{\omega(x_0, r)}{r^\alpha}. \tag{5}$$

For $1 \leq s < p < \infty$, there is a constant $C_{p,s} > 0$ such that for $f \in L^{p,\omega}(\mathbb{R}^n)$

$$\|M_s f\|_{p,\omega} \leq C_{p,s} \|f\|_{p,\omega}.$$

Lemma 1.2 (Sharp inequality). *Let $0 < \sigma \leq 1$ and $\mathcal{E}$ be an ellipsoid centered at $x_0 \in \mathbb{R}^n$ of radius r. Suppose that $\omega(x_0, r)$ satisfies (4) and*

$$\int_r^\infty \frac{\omega(x_0, t)}{t^{\sigma\alpha+1}} dt \leq C \frac{\omega(x_0, r)}{r^{\sigma\alpha}}.$$

Then for $p \in (1, \infty)$ and $f \in L^1_{\rm loc}(\mathbb{R}^n)$ there exists a constant C independent of f such that

$$\|f\|_{p,\omega} \leq C \|f^\sharp\|_{p,\omega}.$$

Proof. Let $\chi_{\mathcal{E}}$ be the characteristic function of the ellipsoid and denote by $2\mathcal{E}$ an ellipsoid centered at x_0 and of radius $2r$. It is easy to verify that $M\chi_{\mathcal{E}}(x) \leq r^\alpha/(\rho(x - x_0) - r)^\alpha \leq 1$, for all $x \in \mathbb{R}^n$. Further, for any $x \in 2^{k+1}\mathcal{E} \setminus 2^k\mathcal{E}$, $k = 1, 2, \ldots$ the maximal function of $\chi_{\mathcal{E}}$ is comparable with $2^{-k\alpha}$ for any x as above. From certain properties of the maximal and sharp functions (see Ref. 9, p. 410) it follows

$$
\begin{aligned}
J &:= \int_{\mathcal{E}} |f(y)|^p dy = \int_{\mathbb{R}^n} |f(y)|^p \chi_{\mathcal{E}}(y) dy \\
&\le \int_{\mathbb{R}^n} |Mf(y)|^p \big(M\chi_{\mathcal{E}}(y)\big)^\sigma dy \le C \int_{\mathbb{R}^n} |f^\#(y)|^p \big(M\chi_{\mathcal{E}}(y)\big)^\sigma dy \\
&\le C\Big\{ \int_{2\mathcal{E}} |f^\#(y)|^p dy + \sum_{k=1}^{\infty} \int_{2^{k+1}\mathcal{E}\setminus 2^k\mathcal{E}} |f^\#(y)|^p \Big(\frac{r}{\rho(y-x_0)-r}\Big)^{\sigma\alpha} dy \Big\} \\
&\le C\Big\{ \omega(2\mathcal{E}) \frac{1}{\omega(2\mathcal{E})} \int_{2\mathcal{E}} |f^\#(y)|^p dy \\
&\quad + r^{\sigma\alpha} \sum_{k=1}^{\infty} \frac{\omega(2^{k+1}\mathcal{E})}{(2^k r)^{\sigma\alpha}} \frac{1}{\omega(2^{k+1}\mathcal{E})} \int_{2^{k+1}\mathcal{E}} |f^\#(y)|^p dy \Big\} \\
&\le C r^{\sigma\alpha} \sum_{k=0}^{\infty} \frac{\omega(2^k\mathcal{E})}{(2^k r)^{\sigma\alpha}} \|f^\#\|_{p,\omega}^p .
\end{aligned}
$$

The assumptions imposed on the weight function give

$$
\frac{\omega(2^k\mathcal{E})}{(2^k r)^{\sigma\alpha}} \sim \int_{2^k r}^{2^{k+1} r} \frac{\omega(x_0,t)}{t^{\sigma\alpha+1}} dt.
$$

Hence

$$
J \le C r^{\sigma\alpha} \int_r^\infty \frac{\omega(x_0,t)}{t^{\sigma\alpha+1}} dt \, \|f^\#\|_{p,\omega}^p \le C\omega(x_0,r) \|f^\#\|_{p,\omega}^p . \qquad \square
$$

One more background we need is that for *spherical harmonics* and their properties (see for details Refs. 4,6,8,13). Recall that the restriction to the unit sphere $\mathbb{S}^{n-1}$ of the homogeneous polynomial $P\colon \mathbb{R}^n \to \mathbb{R}$ of degree m is called *n-dimensional spherical harmonic of degree m*. The space Υ_m of these harmonics is a finite-dimensional linear one with $g_m = \dim \Upsilon_m$ such that $g_0 = 1$, $g_1 = n$ and

$$
g_m = \frac{(m+n-1)!}{(n-1)!m!} - \frac{(m+n-3)!}{(n-1)!(m-2)!} \le C(n) m^{n-2}, \; m \ge 2. \tag{6}
$$

The *orthonormal base* $\{Y_{sm}(x)\}_{s=1}^{g_m}$ of Υ_m and the system $\{Y_{sm}\}_{s=1}^{g_m}{}_{m=0}^{\infty}$ is a *complete orthonormal system* in $L^2(\mathbb{S}^{n-1})$ satisfying

$$
\sup_{x\in\mathbb{S}^{n-1}} \big|D_x^\beta Y_{sm}(x)\big| \le C(n) m^{|\beta|+(n-2)/2}, \quad m = 1,2,\ldots . \tag{7}
$$

If, for instance, $\phi \in C^\infty(\mathbb{S}^{n-1})$ then $\sum_{s,m} b_{sm} Y_{sm}(x)$ is the Fourier series expansion of $\phi(x)$ with respect to $\{Y_{sm}(x)\}$. It converges uniformly to ϕ

(see Ref. 4) and for any integer l

$$|b_{sm}| = \left| \int_{\mathbb{S}^{n-1}} \phi(y) Y_{sm}(y) d\sigma \right| \leq C(n,l) m^{-2l} \sup_{|\beta|=2l} \sup_{y \in \mathbb{S}^{n-1}} \left| D_y^\beta \phi(y) \right|. \quad (8)$$

2. Singular Integral Estimates

Let $k(x;\xi)$ be a kernel in the sense of Definition 29. In order to ensure the existence of the operators (1) and (2) in $L^p(\mathbb{R}^n)$ we restrict our considerations to functions $f \in L^p(\mathbb{R}^n)$, $1 < p < \infty$ for which the norm (3) is finite. For the sake of convenience we still denote these spaces by $L^{p,\omega}(\mathbb{R}^n)$. Note that the last restriction is necessary only to obtain global estimates. If we consider a bounded domain $\Omega \in \mathbb{R}^n$ the operators (1) and (2) with a density $f \in L^{p,\omega}(\Omega)$ will be well defined in $L^p(\Omega)$ without any additional assumptions.

Define the nonsingular operators $\mathcal{K}_\varepsilon$ and $\mathcal{C}_\varepsilon[a,k]$, $a \in BMO$ by

$$\mathcal{K}_\varepsilon f(x) := \int_{\rho(x-y)>\varepsilon} k(x; x-y) f(y) dy,$$

$$\mathcal{C}_\varepsilon[a,k] f(x) := \mathcal{K}_\varepsilon(af)(x) - a(x)\mathcal{K}_\varepsilon f(x).$$

We are going to prove that $\mathcal{K}_\varepsilon$ and $\mathcal{C}_\varepsilon[a,k]$ are bounded and continuous from $L^{p,\omega}(\mathbb{R}^n)$ into itself uniformly in ε. This along with the properties of the kernel will enable to let $\varepsilon \to 0$ obtaining as limits in the $L^{p,\omega}(\mathbb{R}^n)$-topology the *singular* integrals (1) and (2). Moreover, we shall show that the last ones are also continuous in $L^{p,\omega}(\mathbb{R}^n)$.

Let us note that $\mathcal{K}f$ exists in $L^p(\mathbb{R}^n)$ as a limit of $\mathcal{K}_\varepsilon f$ when $\varepsilon \to 0$ in the L^p-norm.[8] Moreover, the operator $\mathcal{K}\colon\ L^p(\mathbb{R}^n) \to L^p(\mathbb{R}^n)$ is continuous and this leads also to continuity in $L^p(\mathbb{R}^n)$ of $\mathcal{C}[a,k]$ if a is an essentially bounded function. As it concerns to the commutator we are going to derive a result similar to that of Coifman-Rochberg-Weiss (Ref. 7, Theorem 1), which asserts: if $\mathcal{K}$ is Calderón-Zygmund operator in $L^p(\mathbb{R}^n)$, $p \in (1,\infty)$ and $a \in BMO$ than the commutator $\mathcal{C}[a,\cdot]$ is a well defined linear continuous operator from $L^p(\mathbb{R}^n)$ into itself. Later, this result has been extended by Bramanti-Cerutti[2] in the framework of homogeneous spaces. Based on this background about Calderón-Zygmund operators, we are going to obtain continuity in $L^{p,\omega}(\mathbb{R}^n)$ and boundedness in terms of $\|a\|_*$ for the commutator (2) having a kernel of more general type.

Theorem 2.1. *Let $k(x;\xi)$ be a variable kernel of mixed homogeneity, $f \in L^{p,\omega}(\mathbb{R}^n)$, $p \in (1,\infty)$, ω satisfies (4) and (5), and $a \in BMO$. Then the integrals $\mathcal{K}f$, $\mathcal{C}[a,k]f \in L^{p,\omega}(\mathbb{R}^n)$ exist as limits of $\mathcal{K}_\varepsilon f$ and $\mathcal{C}_\varepsilon[a,k]f$ when*

$\varepsilon \to 0$ with respect to the $L^{p,\omega}(\mathbb{R}^n)$-norm. The operators $\mathcal{K}$ and $\mathcal{C}[a,k]$ are bounded from $L^{p,\omega}(\mathbb{R}^n)$ into itself and

$$\|\mathcal{K}f\|_{p,\omega} \le C\|f\|_{p,\omega}, \qquad \|\mathcal{C}[a,k]f\|_{p,\omega} \le C\|a\|_*\|f\|_{p,\omega}$$

where the constants depend on n, p, α and k through the constant $C(\beta)$.

Proof. Let $x, y \in \mathbb{R}^n$ and $\overline{y} = \frac{y}{\rho(y)} \in \mathbb{S}^{n-1}$, then

$$k(x; x-y) = \rho(x-y)^{-\alpha} \sum_{s,m} b_{sm}(x) Y_{sm}(\overline{x-y}).$$

This way, the Definition 29 *ii*) and (8) imply

$$\|b_{sm}\|_\infty \le C(n,l,k)m^{-2l} \tag{9}$$

for any integer $l > 1$. Replacing the kernel with its expansion, we get

$$\mathcal{K}_\varepsilon f(x) = \int_{\rho(x-y)>\varepsilon} \sum_{s,m} b_{sm}(x)\mathcal{H}_{sm}(x-y)f(y)dy,$$
$$\mathcal{C}_\varepsilon[a,k]f(x) = \int_{\rho(x-y)>\varepsilon} \sum_{s,m} b_{sm}(x)\mathcal{H}_{sm}(x-y)[a(y)-a(x)]f(y)dy$$

with $\mathcal{H}_{sm}(x-y)$ standing for $Y_{sm}(\overline{x-y})\rho(x-y)^{-\alpha}$ and being a constant kernel in the sense of Definition 29 *i*). In order to get series expansions of $\mathcal{K}_\varepsilon f$ and $\mathcal{C}_\varepsilon[a,k]f$, we let $x \in \mathbb{R}^n$ and $y \in \mathbb{R}^n$ to be such that $\rho(x-y) > \varepsilon$. Then (1), (7) and (9) yield

$$\left|\sum_{m=1}^{N}\sum_{s=1}^{g_m} b_{sm}(x)\frac{Y_{sm}(\overline{x-y})}{\rho(x-y)^\alpha}f(y)\right| \le C(n)\frac{|f(y)|}{\rho(x-y)^\alpha}\sum_{m=1}^{\infty} m^{n-2+(n-2)/2-2l}$$

where $|f(\cdot)|\rho(x-\cdot)^{-\alpha} \in L^1(\mathbb{R}^n)$ for a.a. $x \in \mathbb{R}^n$ and the integer l is preliminary chosen greater than $(3n-4)/4$. Similar inequality holds also for the commutator $\mathcal{C}_\varepsilon[a,k]f$. Thus, by the dominated convergence theorem

$$\mathcal{K}_\varepsilon f(x) = \sum_{s,m} b_{sm}(x)\mathcal{K}_{sm,\varepsilon}(x), \quad \mathcal{C}_\varepsilon[a,k]f(x) = \sum_{s,m} b_{sm}(x)\mathcal{C}_{sm,\varepsilon}[a,k]f(x)$$

with

$$\mathcal{K}_{sm,\varepsilon}(x) := \int_{\rho(x-y)>\varepsilon} \mathcal{H}_{sm}(x-y)f(y)dy,$$
$$\mathcal{C}_{sm,\varepsilon}[a,k]f(x) := \int_{\rho(x-y)>\varepsilon} \mathcal{H}_{sm}(x-y)[a(y)-a(x)]f(y)dy.$$

Thus instead of $\mathcal{K}f$ and $\mathcal{C}[a,k]f$ we shall study the existence in $L^{p,\omega}(\mathbb{R}^n)$ of

$$\mathcal{K}_{sm}f(x) := \text{P.V.}\int_{\mathbb{R}^n} \mathcal{H}_{sm}(x-y)f(y)dy,$$
$$\mathcal{C}_{sm}[a,k]f(x) := \mathcal{K}_{sm}(af)(x) - a(x)\mathcal{K}_{sm}f(x).$$

For what concerns the $\mathcal{K}_{sm}$ we dispose of Ref. 8(Theorem II.1) and this implies, through Ref. 2 (Theorem 2.5), boundedness in $L^p(\mathbb{R}^n)$ of $\mathcal{C}_{sm}[a,k]$ as well. The cited results however require the kernel to have some "integral continuity", called *the Hörmander condition.* It turns out that $\mathcal{H}_{sm}(\cdot)$ satisfies even stronger condition as shows the following lemma.

Lemma 2.1. *(Pointwise Hörmander condition (see Ref. 13, Lemma 2.2))*
Let $\mathcal{E}$ and $2\mathcal{E}$ be ellipsoids centered at x_0 and of radius r and $2r$, respectively. Then

$$|\mathcal{H}_{sm}(x-y) - \mathcal{H}_{sm}(x_0-y)| \le C(n,\alpha)m^{n/2}\frac{\rho(x_0-x)}{\rho(x_0-y)^{\alpha+1}} \tag{10}$$

for each $x \in \mathcal{E}$ and $y \notin 2\mathcal{E}$.

Remark 2.1. *This result ensures that $\mathcal{H}_{sm}$ satisfies also the integral condition*

$$\int_{\{y\in\mathbb{R}^n:\ \rho(y)\ge 4\rho(x)\}} |\mathcal{H}_{sm}(y-x) - \mathcal{H}_{sm}(y)|\, dy \le C$$

with a constant independent of x (see Ref. 8, (1.1)).

In view of the cited above results there exist $\mathcal{K}_{sm}f$, $\mathcal{C}_{sm}[a,k]f \in L^p(\mathbb{R}^n)$ such that

$$\lim_{\varepsilon\to 0}\|\mathcal{K}_{sm,\varepsilon}f - \mathcal{K}_{sm}f\|_{L^p(\mathbb{R}^n)} = \lim_{\varepsilon\to 0}\|\mathcal{C}_{sm,\varepsilon}[a,k]f - \mathcal{C}_{sm}[a,k]f\|_{L^p(\mathbb{R}^n)} = 0.$$

Our goal is to show that this convergence is fulfilled also with respect to the $L^{p,\omega}(\mathbb{R}^n)$-norm. The proof is broken up into several lemmas.

Lemma 2.2. *(see Ref. 13, Lemma 2.4) The singular integrals $\mathcal{K}_{sm}f$ and $\mathcal{C}_{sm}[a,k]f$ satisfy*

$$\begin{aligned}(\mathcal{K}_{sm}f)^\sharp(x) &\le Cm^{n/2}\big(M(|f|^p)(x)\big)^{1/p},\\ (\mathcal{C}_{sm}[a,k]f)^\sharp(x) &\le C\|a\|_*\Big\{\big(M(|\mathcal{K}_{sm}f|^p)(x)\big)^{1/p} + m^{n/2}\big(M(|f|^p)(x)\big)^{1/p}\Big\}\end{aligned} \tag{11}$$

for all $p \in (1,\infty)$ and the constant depends on n, p, α but not on f.

Lemma 2.3. *The operators $\mathcal{K}_{sm}$ and $\mathcal{C}_{sm}[a,k]$ are continuous, acting from $L^{p,\omega}(\mathbb{R}^n)$, $p\in(1,\infty)$, into itself and*

$$\|\mathcal{K}_{sm}f\|_{p,\omega}\le Cm^{n/2}\|f\|_{p,\omega},\quad \|\mathcal{C}_{sm}[a,k]f\|_{p,\omega}\le Cm^{n/2}\|a\|_*\|f\|_{p,\omega}\quad (12)$$

with constants depending on n, p, and α.

Proof. Let $p\in(1,\infty)$. Applying (11) for any $q\in(1,p)$ and the maximal inequality (Lemma 1.1) with $s=1$ we get

$$\begin{aligned}\int_{\mathcal{E}}|(\mathcal{K}_{sm}f)^\sharp(x)|^p dx \le& Cm^{pn/2}\int_{\mathcal{E}}|M(|f|^q)(x)|^{p/q}dx\\ \le& Cm^{pn/2}\omega(\mathcal{E})\|M(|f|^q)\|^{p/q}_{p/q,\omega}\\ \le& Cm^{pn/2}\omega(\mathcal{E})\||f|^q\|^{p/q}_{p/q,\omega}\le Cm^{pn/2}\omega(\mathcal{E})\|f\|^p_{p,\omega}.\end{aligned}$$

Dividing of $\omega(\mathcal{E})$ and taking $\sup_{\mathcal{E}}$, we arrive at

$$\|(\mathcal{K}_{sm}f)^\sharp\|_{p,\omega}\le Cm^{n/2}\|f\|_{p,\omega}$$

which implies the first inequality in (12) through Lemma 1.2. The $L^{p,\omega}$-estimate for the commutator follows in the same manner. □

Lemma 2.4. *The operators $\mathcal{K}_{sm,\varepsilon}$ and $\mathcal{C}_{sm,\varepsilon}[a,k]$ are continuous acting from $L^{p,\omega}(\mathbb{R}^n)$ into itself and satisfy*

$$\begin{aligned}\|\mathcal{K}_{sm,\varepsilon}f\|_{p,\omega}\le& C(n,p,\alpha)m^{n/2}\|f\|_{p,\omega},\\ \|\mathcal{C}_{sm,\varepsilon}[a,k]f\|_{p,\omega}\le& C(n,p,\alpha)m^{n/2}\|a\|_*\|f\|_{p,\omega}.\end{aligned}\quad (13)$$

Proof. Let $\mathcal{E}_\varepsilon$ and $\mathcal{E}_{\varepsilon/2}$ be ellipsoids centered at $x\in\mathbb{R}^n$ and of radius ε and $\varepsilon/2$, respectively. Writing $f=f\chi_{\mathcal{E}_\varepsilon}+f\chi_{(\mathcal{E}_\varepsilon)^c}=f_1+f_2$ we obtain

$$\begin{aligned}\mathcal{K}_{sm,\varepsilon}f(x)\le& \frac{1}{|\mathcal{E}_{\varepsilon/2}|}\int_{\mathcal{E}_{\varepsilon/2}}|\mathcal{K}_{sm,\varepsilon}f(x)|dy\le\frac{1}{|\mathcal{E}_{\varepsilon/2}|}\int_{\mathcal{E}_{\varepsilon/2}}|\mathcal{K}_{sm}f(y)|dy\\ &+\frac{1}{|\mathcal{E}_{\varepsilon/2}|}\int_{\mathcal{E}_{\varepsilon/2}}|\mathcal{K}_{sm,\varepsilon}f(x)-\mathcal{K}_{sm}f(y)|dy\\ \le& \frac{2}{|\mathcal{E}_{\varepsilon/2}|}\int_{\mathcal{E}_{\varepsilon/2}}|\mathcal{K}_{sm}f_1(y)|dy+\frac{1}{|\mathcal{E}_{\varepsilon/2}|}\int_{\mathcal{E}_{\varepsilon/2}}|\mathcal{K}_{sm,\varepsilon}f(x)-\mathcal{K}_{sm}f_2(y)|dy\\ :=& 2I_1(x,\mathcal{E}_{\varepsilon/2})+I_2(x,\mathcal{E}_{\varepsilon/2})\end{aligned}$$

where I_1 and I_2 stand for the terms introduced at the proof of Lemma 2.2, and the same arguments as therein lead to

$$|\mathcal{K}_{sm,\varepsilon}f(x)|\le C(n,p,\alpha)m^{n/2}\left(M(|f|^q)(x)\right)^{1/q}$$

for any $q \in (1, \infty)$. It remains to take the $L^{p,\omega}$-norms of the both sides above for $1 < q < p$ and to apply Lemma 1.1 in order to get (13). The commutator estimate follows analogously. □

Returning to the series expansions (10), we are in a position now to complete the proof of Theorem 2.1. First of all, note that

$$\sum_{m=1}^{\infty}\sum_{s=1}^{g_m} \|b_{sm}\mathcal{K}_{sm,\varepsilon}f\|_{p,\omega} \leq C(n,p,\alpha,k)\|f\|_{p,\omega}\sum_{m=1}^{\infty} m^{-2l+n-2+n/2},$$

$$\sum_{m=1}^{\infty}\sum_{s=1}^{g_m} \|b_{sm}\mathcal{C}_{sm,\varepsilon}[a,k]f\|_{p,\omega} \leq C(n,p,\alpha,k)\|a\|_*\|f\|_{p,\omega}\sum_{m=1}^{\infty} m^{-2l+n-2+n/2}$$

as it follows from (9), (1) and Lemma 16. Choosing $l > (3n-2)/4$ the series in (10) result totally convergent in $L^{p,\omega}(\mathbb{R}^n)$, uniformly in $\varepsilon > 0$, whence

$$\|\mathcal{K}_\varepsilon f\|_{p,\omega} \leq C\|f\|_{p,\omega}, \qquad \|\mathcal{C}_\varepsilon[a,k]f\|_{p,\omega} \leq C\|a\|_*\|f\|_{p,\omega}.$$

Setting

$$\mathcal{K}f(x) := \sum_{s,m} b_{sm}(x)\mathcal{K}_{sm}f(x), \qquad \mathcal{C}[a,k]f(x) := \sum_{s,m} b_{sm}(x)\mathcal{C}_{sm}[a,k]f(x),$$

we obtain as above

$$\|\mathcal{K}f\|_{p,\omega} \leq C\|f\|_{p,\omega}, \qquad \|\mathcal{C}[a,k]f\|_{p,\omega} \leq C\|a\|_*\|f\|_{p,\omega}$$

through (9), (1) and Lemma 2.3. Finally, the total convergence in $L^{p,\omega}(\mathbb{R}^n)$ of (10), uniformly in $\varepsilon > 0$, gives

$$\lim_{\varepsilon\to 0}\mathcal{K}_\varepsilon f(x) = \sum_{s,m} b_{sm}(x)\lim_{\varepsilon\to 0}\mathcal{K}_{sm,\varepsilon}f(x) = \mathcal{K}f(x),$$

$$\lim_{\varepsilon\to 0}\mathcal{C}_\varepsilon[a,k]f(x) = \mathcal{C}[a,k]f(x).$$

It is worth noting that singular integrals like (1) and (2) appear in the representation formulas for the solutions of linear elliptic and parabolic partial differential equations. To make the obtained here results applicable to the study of regularizing properties of these operators we need of some additional local results.

Corollary 2.1. *Let Ω be a bounded domain in $\mathbb{R}^n$ and $k\colon\ \Omega\times\{\mathbb{R}^n\setminus\{0\}\} \to \mathbb{R}$ be a variable kernel of mixed homogeneity, $a \in BMO(\Omega)$, $p \in (1,\infty)$ and ω satisfies* (4) *and* (5)*. Then, for any $f \in L^{p,\omega}(\Omega)$ and almost all $x \in \Omega$,* *the singular integrals*

$$\mathcal{K}f(x) = P.V. \int_\Omega k(x; x-y) f(y) dy$$

$$\mathcal{C}[a,k]f(x) = P.V. \int_\Omega k(x; x-y)[a(y) - a(x)] f(y) dy$$

are well defined in $L^{p,\omega}(\Omega)$ *and*

$$\|\mathcal{K}f\|_{p,\omega;\Omega} \le C\|f\|_{p,\omega;\Omega}, \quad \|\mathcal{C}[a,k]f\|_{p,\omega;\Omega} \le C\|a\|_*\|f\|_{p,\omega;\Omega}$$

with $C = C(n, p, \alpha, \Omega, k)$.

To obtain the above assertion it is sufficient to extend $k(x;\cdot)$ and $f(\cdot)$ as zero outside Ω. One more necessary extension preserving the norm is that of a in $BMO(\mathbb{R}^n)$ and we have it according to the results of Jones[11] and Acquistapace[1] (see Ref. 6 for details).

Another consequence of Theorem 2.1 is the "good behavior" of the commutator for VMO functions a. Thus using the ideas in Ref. 6 (Theorem 2.13) we can prove

Corollary 2.2. *Suppose* $a \in VMO$ *with* VMO*-modulus* γ_a. *Then, for each* $\varepsilon > 0$ *there exists* $r_0 = r_0(\varepsilon, \gamma_a) > 0$ *such that for any* $\varrho \in (0, r_0)$ *and any ellipsoid* $\mathcal{E}_\varrho$ *of radius* ϱ *holds*

$$\|\mathcal{C}[a,k]f\|_{p,\omega;\mathcal{E}_\varrho} \le C\varepsilon\|f\|_{p,\omega;\mathcal{E}_\varrho}. \tag{14}$$

References

1. P. Acquistapace,*Ann. Mat. Pura Appl.* **161**, 231 (1992). **286**, A139 (1978).
2. M. Bramanti, M. C. Cerutti,*Boll. Un. Mat. Ital. B* **10**, 843 (1996).
3. W. Burger ,*C. R. Acad. Sci. Paris* **286**, 139 (1978).
4. *Amer. J. Math.* **79**, 901 (1996).
5. F. Chiarenza, M. Frasca, *Rend. Mat. Appl.* **7**, 273 (1987).
6. F. Chiarenza, M. Frasca, P. Longo, *Ricerche Mat.* **40**, 149 (1991).
7. R. Coifman, R. Rochberg, G. Weiss,*Ann. of Math.* **103**, 611 (1976).
8. E. B. Fabes, N. Rivière, *Studia Math.* **27**, 19 (1966).
9. J. Garcia-Cuerva, J. L. Rubio De Francia, *Weighted Norm Inequalities and Related Topics: North-Holand Math. Studies,* Vol. 116, North-Holand, Amsterdam, 1985.
10. F. John, L. Nirenberg,*Comm. Pure Appl. Math.* **14**, 415 (1961).
11. P.W. Jones, *Indiana Univ. Math. J.* **29**, 41 (1980).
12. E. Nakai, *Math. Nachr.* **166**, 95 (1994).
13. D. Palagachev, L. Softova, *Potent. Anal.* **20**, 237 (2004).
14. D. Sarason, *Trans. Amer. Math. Soc.* **207**, 391 (1975).
15. E. M. Stein, *Singular Integrals and Differentiability Properties of Functions,* (Princeton University Press, Princeton, New Jersey, 1970).
16. L. Softova, *C. R. Acad. Sci., Paris, Ser. I, Math.* **333**, 635 (2001).

Advances in Deterministic and Stochastic Analysis
Eds. N. M. Chuong *et al.* (pp. 43–72)

§4. CLASSIFICATION OF INTEGRAL TRANSFORMS

VU KIM TUAN

Department of Mathematics,
University of West Georgia, Carrollton, GA 30118, USA
E-mail: vu@westga.edu

In this survey paper we give a classification of integral transforms based on their composition structure.

1. Introduction

There is no consensus among mathematicians what should be called an integral transform, and what should not. Some assume that every linear integral operator that appears frequently enough to bear some name can be considered as an integral transform. We accept here another concept. We believe that the most distinguish character that enjoy integral transforms among the vast of linear integral operators is their integral inverses. We call therefore integral transform every linear integral operator whose inverse is also an integral operator of almost the same "complexity". To include Riemann-Liouville fractional operators and Riesz potentials into integral transforms, we should admit some operators of differentiation of finite order into the inverses. So, when we talk about integral transforms we have in mind pairs of direct and inverse integral transforms. Because of the importance of applications and relations of integral transforms with differential equations it is widely believed that new interesting integral transforms can be constructed as eigenfunction expansions of some differential operators. Many classical integral transforms are in fact arisen from self-adjoint Sturm-Liouville differential operators.[31] We will show here another method to construct new integral transforms, based on few well-known integral transforms. Many nonconvolution (index) transforms, constructed by that way, seem impossible to be obtained by eigenfunction expansions, since their kernels do not satisfy any linear differential equations.

The most classical transforms are:

- The pair of Fourier transforms

$$g(x) = (Ff)(x) = \frac{1}{\sqrt{2\pi}} \int_{-\infty}^{\infty} f(y) e^{ixy} dy, \tag{1}$$

$$f(y) = (F^{-1}g)(y) = \frac{1}{\sqrt{2\pi}} \int_{-\infty}^{\infty} g(x) e^{-ixy} dx, \tag{2}$$

- The pair of Laplace transforms

$$g(p) = (Lf)(p) = \int_{0}^{\infty} f(x) e^{-px} dx, \tag{3}$$

$$f(x) = (L^{-1}g)(x) = \frac{1}{2\pi i} \int_{\gamma - i\infty}^{\gamma + i\infty} g(p) e^{px} dp, \tag{4}$$

- The pair of Mellin transforms

$$g(s) = (Mf)(s) = \int_{0}^{\infty} x^{s-1} f(x) dy, \tag{5}$$

$$f(x) = (M^{-1}g)(x) = \frac{1}{2\pi i} \int_{\gamma - i\infty}^{\gamma + i\infty} g(s) x^{-s} ds. \tag{6}$$

The Fourier and Mellin transforms are equivalent subject to some substitution of variables. But each of them has advantages in different areas of applications. If the Fourier transform is considered in the complex domain, then the Laplace transform turns out to be the Fourier transform of functions vanishing on the negative half-line. Otherwise they are distinct. Other well-known transforms include the Hilbert, Hankel, Fourier-sine and -cosine, Y- and H- transforms, Riemann-Liouville and Weyl fractional operators, Kontorovich-Lebedev, Mehler-Fock and Olevskii nonconvolution transforms. We have prepared a draft list of integral transforms coming into our definition. It contained more than 300 integral transforms and was included in final form in Ref. 27.

Of course it is impossible to have some classification of all known and unknown integral transforms. In this work we restrict ourselves on classification of just one-dimensional integral transforms listed in Ref. 27. Similar treatment of multidimensional integral transforms is briefly given in Ref. 3. Our method of classification is based on composition structure of integral transforms. Fortunately (or maybe unfortunately), almost all of these integral transforms have very simple structures. They are build up with the help of very few operators: Fourier transform (equivalent Mellin transform and Laplace transform as a special case are also included), operator of multiplication and substitution of variables.

Throughout the paper, if we say that the integral transform

$$g(x) = \int_{\Omega} k(x,y)f(y)dy$$

is mapping $L_p(\Omega; w)$ into $L_{p'}(\Omega'; w')$, we understand that if $f \in L_p(\Omega; w)$, and Ω_n is any compact subset of Ω, over it $k(x,.)$ is continuous, then the integral

$$g_n(x) = \int_{\Omega_n} k(x,y)f(y)dy$$

converges to a function $g_n \in L_{p'}(\Omega'; w')$, and moreover, if $\Omega_1 \subset \Omega_2 \subset \ldots \Omega_n \subset \ldots$, such that $\bigcup_{n=1}^{\infty} \Omega_n = \overline{\Omega}$, then g_n converges in $L_{p'}(\Omega'; w')$ to g.

All integral transforms can be divided into two classes:

-*Convolution transforms*

and

-*Nonconvolution (index) transforms.*

2. Convolution Transforms

A generalized convolution of functions f and k under three operators K, K_1, K_2, and with some weight function w is a function, denoted by the symbol $k * f$, such that the following factorization property holds

$$K(k * f)(x) = w(x)(K_1 k)(x)(K_2 f)(x).$$

If $K = K_1 = K_2$, we have the classical convolution.

Convolution transforms are called linear integral transforms of the following type

$$f \to g = k * f.$$

Depending on the convolution, convolution transforms can be divided into following types:

2.1. *Fourier convolution transforms*

$$g(x) = \int_{-\infty}^{\infty} k(x-y)f(y)dy. \tag{7}$$

An example is the pair of Hilbert transforms

$$g(x) = (Hf)(x) = \frac{1}{\pi}\int_{-\infty}^{\infty} \frac{f(y)}{y-x}dy, \tag{8}$$

$$f(y) = (H^{-1}g)(y) = \frac{1}{\pi}\int_{-\infty}^{\infty} \frac{g(x)}{y-x}dx. \tag{9}$$

2.2. *Laplace convolution transforms*

$$g(x) = \int_0^x k(x-y)f(y)dy. \tag{10}$$

An example is the pair of Riemann-Liouville fractional integral and differential operators

$$g(x) = (I_{0+}^{\alpha}f)(x) = \int_0^x \frac{(x-y)^{\alpha-1}}{\Gamma(\alpha)}f(y)dy, \tag{11}$$

$$f(y) = (D_{0+}^{\alpha}g)(y) = \frac{d^n}{dy^n}\int_0^y \frac{(y-x)^{n-\alpha-1}}{\Gamma(n-\alpha)}g(x)dx, \quad n > \Re(\alpha) > 0. \tag{12}$$

2.3. *Mellin convolution transforms*

$$g(x) = \int_0^{\infty} k(xy)f(y)dy. \tag{13}$$

An example is the pair of Hankel transforms

$$g(x) = (H_\nu f)(x) = \int_0^{\infty} \sqrt{xy}J_\nu(xy)f(y)dy, \tag{14}$$

$$f(y) = (H_\nu g)(y) = \int_0^{\infty} \sqrt{xy}J_\nu(xy)g(x)dx. \tag{15}$$

where $J_\nu(x)$ is the Bessel function of the first kind.[1]

Similar to the relation between the Fourier and Mellin transforms, transforms of Fourier and Mellin convolution types can be identified by a simple substitution of variables. Transforms of Laplace convolution type are special cases of transforms of Fourier convolution type (if $k(x) = f(x) = 0$ for $x < 0$ in (7)).

Since most of convolution transforms are given in the form of Mellin convolution type[27] we identify the class of Fourier and Laplace convolution transforms with the class of transforms of Mellin convolution type. The class of Mellin convolution transforms is the most representative class among known transforms. They have very simple composition structure

$$(Kf)(x) = (M^{-1}k^*(s)(Mf)(1-s))(x), \tag{16}$$

where $k^*(s) = (Mk)(s)$ - the Mellin transforms of the kernel $k(x)$. Thus, any Mellin convolution transform is a composition of a Mellin transform, a substitution of variable $s \to 1-s$, an operator of multiplication by $k^*(s)$, and an inverse Mellin transform. Although the kernels $k(x)$ in (13) are very distinct in different transforms, their Mellin transforms $k^*(s)$ in general behave very simple on the line $\Re(s) = 1/2$. One can see that in almost cases $k^*(s)$ is bounded and not equal to 0 on the line $\Re(s) = 1/2$. Based on the asymptotic behavior of $k^*(s)$ on the line $\Re(s) = 1/2$ the class of Mellin convolution transforms in the form (13) in its turn can be divided into three subclasses:

2.3.1. *Mellin convolution transforms of Watson type*

These transforms are characterized by the condition[49]

$$C > |k^*(s)| > c > 0, \quad \Re(s) = 1/2. \tag{17}$$

Since the Mellin transform (5) is an homeomorphism from $L_2(0,\infty)$ onto $L_2(1/2-i\infty, 1/2+i\infty)$ and the operator of multiplication by $k^*(s)$ under the condition (17) is an automorphism on $L_2(1/2-i\infty, 1/2+i\infty)$, from (17) it is easy to see that convolution transforms of Watson type are automorphisms on $L_2(0,\infty)$. If, moreover,

$$|k^*(s)| = 1, \quad \Re(s) = 1/2, \tag{18}$$

then convolution transforms of Watson type are unitary on $L_2(0,\infty)$. In the form of Fourier convolution (7) the condition (17) is replaced by

$$C > |(Fk)(x)| > c > 0, \quad x \in R, \tag{19}$$

and transforms satisfying condition (19) are automorphisms on $L_2(-\infty,\infty)$, and unitary there if $|(Fk)(x)| = 1$. Examples of unitary convolution transforms of Watson type are the Hilbert and Hankel transforms, the Fourier-sine transform

$$g(x) = (F_s f)(x) = \sqrt{\frac{2}{\pi}} \int_0^\infty \sin xy\, f(y)dy, \tag{20}$$

$$f(y) = (F_s g)(y) = \sqrt{\frac{2}{\pi}} \int_0^\infty \sin xy\, g(x)dx, \tag{21}$$

and the Fourier cosine transform

$$g(x) =(F_c f)(x) = \sqrt{\frac{2}{\pi}} \int_0^\infty \cos xy\, f(y)dy, \tag{22}$$

$$f(y) =(F_c g)(y) = \sqrt{\frac{2}{\pi}} \int_0^\infty \cos xy\, g(x)dx. \tag{23}$$

The pair of Y- and H- transforms

$$g(x) = \int_0^\infty \sqrt{xy} Y_\nu(xy)\, f(y)dy, \tag{24}$$

$$f(y) = \int_0^\infty \sqrt{xy} \mathbf{H}_\nu(xy)\, g(x)dx, \tag{25}$$

with the Bessel function of the second kind $Y_\nu(x)$ and the Struve function $\mathbf{H}_\nu(x)$ in the kernels[1] are automorphisms, but not unitary in $L_2(0,\infty)$.

Taking a composition of Fourier transforms in the form $Fe^{ix^3/3}Ff$ we obtain the following pair of convolution transforms of Watson type[39,42]

$$g(x) = (Aif)(x) = \int_{-\infty}^\infty Ai(x+y)f(y)dy, \tag{26}$$

$$f(y) = (Aig)(y) = \int_{-\infty}^\infty Ai(x+y)g(x)dx, \tag{27}$$

with the Airy function $Ai(x)$ in the kernel1 that are unitary on $L_2(-\infty,\infty)$.

Taking a composition of Fourier transforms in the form $Fe^{-ia\sinh x}Ff$, $a > 0$, we obtain another pair of convolution transforms of Watson type[41]

$$g(x) =\frac{1}{\pi} \int_{-\infty}^\infty e^{(x+y)\pi/2} K_{i(x+y)}(a)f(y)dy, \tag{28}$$

$$f(y) =\frac{1}{\pi} \int_{-\infty}^\infty e^{(x+y)\pi/2} K_{i(x+y)}(a)g(x)dx, \tag{29}$$

with the Macdonald function $K_\nu(a)$ in the kernels[1] that is unitary on $L_2(-\infty,\infty)$.

Taking a composition of Fourier transforms in the form $Fe^{ia\cosh x}Ff$, $a > 0$, we obtain one more pair of convolution transforms of Watson type[39]

$$g(x) =\frac{1}{2} \int_{-\infty}^\infty e^{(y-x)\pi/2} H^{(1)}_{i(x-y)}(a)f(y)dy, \tag{30}$$

$$f(y) =\frac{1}{2} \int_{-\infty}^\infty e^{(y-x)\pi/2} H^{(2)}_{i(y-x)}(a)g(x)dx, \tag{31}$$

with the Hankel functions $H^{(1)}_\nu(x)$ and $H^{(2)}_\nu(x)$ in the kernels[1] that are unitary on $L_2(-\infty,\infty)$.

2.3.2. *Mellin convolution transforms of Riemann-Liouville type*

The Mellin transform $k^*(s)$ of the kernels of these transforms are characterized by power decay on the line $(1/2 - i\infty, 1/2 + i\infty)$ as $\Im(s) \to \pm\infty$:

$$0 < c < |k^*(s)s^\alpha| < C < \infty, \quad \alpha > 0, \; \Re(s) = 1/2. \tag{32}$$

Many transforms of Laplace convolution type (10) such as Riemann-Liouville fractional integral (11), and Weyl fractional operators

$$g(x) = \int_x^\infty \frac{(y-x)^{\alpha-1}}{\Gamma(\alpha)} f(y)dy, \tag{33}$$

$$f(y) = \left(-\frac{d}{dy}\right)^n \int_y^\infty \frac{(x-y)^{n-\alpha-1}}{\Gamma(n-\alpha)} g(x)dx, \quad n > \Re\alpha > 0, \tag{34}$$

the Riesz potential

$$g(x) = \int_{-\infty}^\infty \frac{f(y)}{|x-y|^{1-\alpha}} dy, \quad 1 > \Re\alpha > 0, \tag{35}$$

$$f(y) = \frac{\alpha}{2\pi} \tan\frac{\alpha\pi}{2} \int_{-\infty}^\infty \frac{g(y) - g(x)}{|x-y|^{1+\alpha}} dx, \tag{36}$$

the Prabhakar transform[27]

$$g(x) = \int_0^x \frac{(x-y)^{c-1}}{\Gamma(c)} {}_1F_1(a; c; \lambda(x-y)) f(y) dy, \tag{37}$$

$$f(y) = \frac{d^n}{dy^n} \int_0^y \frac{(y-x)^{n-c-1}}{\Gamma(n-c)} {}_1F_1(-a; n-c; \lambda(y-x)) g(x)dx, \; 0 < \Re c < n, \tag{38}$$

with the confluent hypergeometric function ${}_1F_1(a; c; x)$ in the kernel, the Saigo transform[28]

$$g(x) = \int_0^x \frac{(x-y)^{c-1}}{\Gamma(c)} {}_2F_1(a, b; c; 1 - \frac{x}{y}) f(y)dy, \quad 0 < \Re c < n, \tag{39}$$

$$f(y) = y^{-a} \frac{d^n}{dy^n} y^a \int_0^y \frac{(y-x)^{n-c-1}}{\Gamma(n-c)} {}_2F_1(-a, n-b; n-c; 1 - \frac{x}{y}) g(x)dx, \tag{40}$$

with the hypergeometric function ${}_2F_1(a,b;c;x)$ in the kernel, the Marichev transform[21]

$$g(x) = \int_0^x \frac{(x-y)^{c-1}}{\Gamma(c)} F_3(a,a',b,b';c;1-x/y,1-y/x)f(y)dy, \tag{41}$$

$$f(y) = y^{n-a-b}\frac{d^n}{dy^n}y^{a+b-n}\int_0^y \frac{(y-x)^{n-c-1}}{\Gamma(n-c)} \tag{42}$$

$$F_3(-a',n-a,-b',n-b;n-c;1-\frac{y}{x},1-\frac{x}{y})g(x)dx, \quad 0<\Re c<n,$$

with the Appell hypergeometric function of two variables $F_3(a,a',b,b';c;x,y)$ in the kernel are convolution transforms of Riemann-Liouville type.

The following pairs of integral transforms[27]

$$g(x) = \int_0^x \frac{(x-y)^{c-1}}{\Gamma(c)} \Xi_1(a,a',b;c;1-\frac{x}{y},\lambda(x-y))f(y)dy, \tag{43}$$

$$f(y) = y^{-b}\frac{d^n}{dy^n}y^{b}\int_0^y \frac{(y-x)^{n-c-1}}{\Gamma(n-c)} \tag{44}$$

$$\Xi_1(n-a,-a',-b;n-c;1-\frac{x}{y},\lambda(y-x))g(x)dx, \quad 0<\Re c<n,$$

and

$$g(x) = \int_0^x \frac{(x-y)^{c-1}}{\Gamma(c)} \Xi_2(a,b;c;1-\frac{x}{y},\lambda(x-y))f(y)dy, \tag{45}$$

$$f(y) = \int_0^y \frac{(y-x)^{n-c-1}}{\Gamma(n-c)} \tag{46}$$

$$\Xi_2(-a,-b;n-c;1-\frac{x}{y},\lambda(x-y))g^{(n)}(x)dx, \quad 0<\Re c<n,$$

with the Humbert hypergeometric functions of two variables Ξ_1 and Ξ_2 in the kernels[8] are nonconvolution transforms, but their properties and composition structure containing only Riemann–Liouville fractional integrals and derivatives are very similar to those considered in this subsection.

Some other special cases such as transforms with orthogonal polynomials in the kernel are compositions of some Riemann-Liouville or Weyl fractional operators with operators of multiplication. Feller and Riesz potentials are, on the other hand, compositions of two fractional operators of different type (Riemann-Liouville and Weyl). Their inverses contain, in general, operators of differentiation of finite order.

2.3.3. *Mellin convolution transforms of Laplace type*

The Mellin transform $k^*(s)$ of the kernels of these transforms is characterized by exponential-power decay on $(1/2 - i\infty, 1/2 + i\infty)$ as $\Im(s) \to \pm\infty$:

$$0 < c < |k^*(s)e^{\gamma|s|}s^{\alpha}| < C < \infty, \quad \gamma > 0, \alpha \in R, \Re(s) = 1/2. \tag{47}$$

Examples are the Laplace transform, the Stieltjes transform

$$g(x) = \int_0^\infty \frac{f(y)}{x+y}dy, \tag{48}$$

$$f(y) = \frac{1}{2\pi i} \lim_{\epsilon\to 0+} [g(-y - i\epsilon) - g(-y + i\epsilon)], \tag{49}$$

the Meijer–Bessel transform

$$g(s) = \int_0^\infty yK_\nu(sy)\, f(y)dy, \tag{50}$$

$$f(y) = \frac{1}{\pi i} \int_{\gamma - i\infty}^{\gamma + i\infty} sI_\nu(sy)\, g(s)ds, \tag{51}$$

with the modified Bessel function $K_\nu(x)$ and $I_\nu(x)$ in the kernels,[1] the Meijer K_ν transform

$$g(s) = \int_0^\infty e^{-sy}K_\nu(sy)\, f(y)dy, \tag{52}$$

$$f(y) = \frac{y}{4\pi\nu i} \int_{\gamma - i\infty}^{\gamma + i\infty} se^{sy} \left[(2\nu + 1)I_{\nu-1}(sy) + 4\nu I_\nu(sy) + (2\nu - 1)I_{\nu+1}(sy)\right] g(s)ds. \tag{53}$$

All of these transforms (Laplace, Stieltjes, Meijer, Meijer–Bessel,...) map integrable functions into functions analytic in some half plane. Their inverses, if considered strictly only on the real line, include some differential operators of infinite order.

For example, the real variable inverse of the Laplace transform has the form

$$f(y) = \lim_{n\to\infty} \frac{(-1)^n}{n!} \left(\frac{n}{y}\right)^{n+1} g^{(n)}\left(\frac{n}{y}\right), \tag{54}$$

while the real variable inverse of the Stietjes transform has the form

$$f(y) = \lim_{n\to\infty} \frac{(-1)^n}{2\pi} \left(\frac{e}{n}\right)^{2n} \left[y^{2n}g^{(n)}(y)\right]^{(n)}. \tag{55}$$

Most of the Mellin convolution transforms are special cases of the following **convolution G-transform**

$$g(x) = \int_0^\infty G^{m,n}_{p,q}\left(xy \,\middle|\, {(a_p) \atop (b_q)}\right) f(y)dy, \tag{56}$$

with the Meijer G-function in the kernel of Ref. 8. Kesarwani[16–18] has found the condition when this transform is of Watson type

$$c^* := p+q-2(m+n) = 0, \quad \gamma^* := \Re\left(\sum_{i=1}^{p} a_i - \sum_{j=1}^{q} b_j\right) = 0, \quad p \neq q. \tag{57}$$

In this case the inverse has the form

$$f(y) = \int_0^\infty G^{q-m,\, p-n}_{p,\, q}\left(xy \left| \begin{matrix} -(a_p^{n+1}), -(a_n) \\ -(b_q^{m+1}), -(b_m) \end{matrix} \right.\right) g(x)dx. \tag{58}$$

One can find conditions when this transform is of Riemann–Liouville type

$$c^* = p+q-2(m+n) = 0, \quad \gamma^* = \Re\left(\sum_{i=1}^{p} a_i - \sum_{j=1}^{q} b_j\right) > 0, \tag{59}$$

and the inverse formula in this case has the form

$$f(y) = \frac{d^k}{dy^k} y^k \int_0^\infty G^{q-m,\, p-n+1}_{p+1,\, q+1}\left(xy \left| \begin{matrix} 0, -(a_p^{n+1}), -(a_n) \\ -(b_q^{m+1}), -(b_m), -k \end{matrix} \right.\right) g(x)dx, \quad k > \gamma^*, \tag{60}$$

or of Laplace type

$$c^* = p+q-2(m+n) > 0. \tag{61}$$

with the real variable inverse of the form

$$f(y) = (2\pi)^{\frac{1-r}{2}} r^{r-\frac{1}{2}} \lim_{k\to\infty} \frac{k^k y^{\frac{k}{r}}}{(k-1)!} \left(r y^{\frac{r+1}{r}} \frac{d}{dy}\right)^k \tag{62}$$

$$\int_0^\infty G^{q-m+r,\, p-n}_{p,\, q+r}\left(k^r xy \left| \begin{matrix} -(a_p^{n+1}), -(a_n) \\ 0, \frac{1}{r}, \ldots, \frac{r-1}{r}, -(b_q^{m+1}), -(b_m) \end{matrix} \right.\right) g(r^r x)dx, \quad r > 2c^*$$

(see Ref. 37 for more details).

As it was said, a Mellin convolution transform can be decomposed as composition of direct and inverse Mellin transforms and an operator of multiplication

$$(Kf)(x) = M^{-1}\left\{k^*(s) M\{f(y); 1-s\}; x\right\}. \tag{63}$$

As a special case of the Mellin convolution transform, the convolution G–transform (56) enjoys more interesting composition structure. The Mellin transform of the kernel, the Meijer G-function $G^{m,\,n}_{p,\,q}\left(x \left| \begin{matrix} (a_p) \\ (b_q) \end{matrix} \right.\right)$, is a quotient of products of Gamma functions ($m+n$ in the numerator and $p+q-m-n$ in the denominator). Each Gamma function in the numerator corresponds to a modified Laplace transform, and each Gamma function in the denominator corresponds to a modified inverse Laplace transform.[2] Therefore, the convolution G-transform (56) can be decomposed as composition of $m+n$ direct Laplace transforms and $p+q-m-n$ inverse Laplace transforms. If we do not want to include the inverse Laplace transform, that contains integration in the complex plane, into the composition, then in the composition we have to use some Riemann–Liouville, Weyl fractional operators[27] and Hankel transforms (the Mellin transforms of their kernels are quotients of one Gamma function over one Gamma function). The convolution G–transform is of Riemann–Liouville or Watson type if and only if it can be decomposed into composition of only Riemann–Liouville, Weyl fractional operators and the Hankel transforms (without the Laplace and inverse Laplace transforms).

Meijer G-function $G^{m,n}_{p,q}\left(x\,\middle|\,\begin{smallmatrix}(a_p)\\(b_q)\end{smallmatrix}\right)$ contains as special cases many classical special functions (orthogonal polynomials, Bessel functions, confluent hypergeometric functions, hypergeometric functions, to name just a few). Therefore it has a very complicated asymptotic behavior near 0 and infinity,[8] and studying existence and mapping properties of convolution G-transform (56) in classical L_p spaces is a very challenging problem. However, its Mellin transform $k^*(s)$ is a quotient of products of Gamma functions, and therefore has a very simple asymptotic behavior at infinity, power-exponential decay. We introduce now a space of functions $\mathcal{M}^{-1}_{c,\gamma}(L_p)$ based on the power-exponential asymptotic behavior of $k^*(s)$ on $\Re s = 1/2$. For simplicity we assume $p = 2$. The general case p is considered in.[37] This space turns out to be very convenient in studying convolution G-transform (56).

Definition 2.1. Let $2\,\mathrm{sign}c + \mathrm{sign}\gamma \geq 0$. By $\mathcal{M}^{-1}_{c,\gamma}(L_2)$ we denote the set of functions f on $R_+ = (0, \infty)$, which can be represented as the inverse Mellin transform

$$f(x) = \frac{1}{2\pi i}\int_{1/2-i\infty}^{1/2+i\infty} f^*(s)x^{-s}ds$$

of $f^*(s)$ from $L_2(1/2 - i\infty, 1/2 + i\infty)$ with weight $|s|^\gamma e^{\pi c|s|}$. If the norm of f in $\mathcal{M}^{-1}_{c,\gamma}(L_2)$ is defined through the norm of $f^*(s)$, then $\mathcal{M}^{-1}_{c,\gamma}(L_2)$ is a Banach space.

If $c = \gamma = 0$, then $\mathcal{M}^{-1}_{0,0}(L_2) = L_2(R_+)$. If $c = 0, \gamma > 0$, then $f \in \mathcal{M}^{-1}_{0,\gamma}(L_2)$ if and only if $e^{x/2}f(e^x)$ belongs to the space of Bessel potentials $L^\gamma_2(R)$. If $c > 0$, then $f \in \mathcal{M}^{-1}_{c,\gamma}(L_2)$ if and only if f is infinitely differentiable a.e., and moreover,

$$\sum_{n=0}^{\infty} \frac{(2\pi c)^{2n}}{(2n)!} \left\| \left(x\frac{d}{dx} \right)^n x^{\gamma} I_{0+}^{-\gamma} f(x) \right\|_{L_2(R_+)}^2 < \infty.$$

The importance of the space $\mathcal{M}_{c,\gamma}^{-1}(L_p)$ can be seen from the following fundamental result

Theorem 2.1.[37] *All convolution G–transforms, whether they are of Watson, Riemann-Liouville, or Laplace types, are automorphisms from* $\mathcal{M}_{c,\gamma}^{-1}(L_2)$ *onto* $\mathcal{M}_{c+c^*,\gamma+\gamma*}^{-1}(L_2)$.

2.4. *Fourier cosine and sine convolution transforms*

There exist some different convolutions related to the Fourier cosine and Fourier sine transforms.

The first convolution has the form[29]

$$(f{*}k)(x) = \frac{1}{\sqrt{2\pi}} \int_0^{\infty} [k(\mid x-y \mid) + k(x+y)] f(y) dy, \tag{64}$$

and it satisfies the convolution property

$$F_c(f{*}g)(x) = (F_c f)(x)\,(F_c k)(x). \tag{65}$$

Another convolution was first introduced by Churchill[7]

$$(f{*}k)(x) = \frac{1}{\sqrt{2\pi}} \int_0^{\infty} [k(\mid x-y \mid) - k(x+y)] f(y) dy, \tag{66}$$

and the respective convolution property for (66) has the form

$$F_s(f{*}k)(x) = (F_s f)(x)\,(F_c k)(x). \tag{67}$$

Recently, the third convolution for the Fourier cosine and sine transforms has been discovered[25]

$$(f{*}k)(x) = \frac{1}{\sqrt{2\pi}} \int_0^{\infty} [\mathrm{sign}(y-x) k(\mid y-x \mid) + k(y+x)] f(y) dy. \tag{68}$$

It was shown there that

$$F_c(f{*}k)(x) = (F_s f)(x)\,(F_s k)(x), \qquad x \epsilon R_+ \ . \tag{69}$$

Corresponding to these three convolutions we can construct three classes of integral transforms of convolution type. For example, for the convolution transform of the type (64) it has been proved[43] that the condition

$$\int_0^{\infty} (k_1(x+y) + k_1(|x-y|))\, \overline{k_1(y)} dy = \frac{1}{4}(x+1)e^{-x}, \quad x > 0, \tag{70}$$

is necessary and sufficient for the transform

$$g(x) = \left(1 - \frac{d^2}{dx^2}\right) \int_0^\infty \left(k_1(x+y) + k_1(|x-y|)\right) f(y)\, dy, \quad x \in R_+ \ , \tag{71}$$

to be unitary in $L_2(R_+)$ with the symmetric reciprocal formula

$$f(x) = \left(1 - \frac{d^2}{dx^2}\right) \int_0^\infty \left(\overline{k_1(x+y)} + \overline{k_1(|x-y|)}\right) g(y)\, dy, \quad x \in R_+ \ . \tag{72}$$

If moreover, k_1 is twice differentiable, and $k(x) = k_1(x) - k_1''(x)$ is locally bounded on R_+, then the integral transform

$$g(x) = \int_0^\infty \left(k(x+y) + k(|x-y|)\right) f(y) dy, \quad x \in R_+ \ , \tag{73}$$

satisfies the Plancherel formula

$$\|g\|_{L_2(R_+)} = \|f\|_{L_2(R_+)}, \tag{74}$$

and reciprocally,

$$f(x) = \int_0^\infty \left(\overline{k(x+y)} + \overline{k(|x-y|)}\right) g(y) dy, \quad x \in R_+ \ . \tag{75}$$

As examples we can take

$$k(x) = \frac{1}{2}(Ai(x) + Ai(-x) + iGi(x) + iGi(-x)), \tag{76}$$

and therefore,

$$\overline{k(x)} = \frac{1}{2}(Ai(x) + Ai(-x) - iGi(x) - iGi(-x)). \tag{77}$$

Since $Ai(x)$ and $Gi(x)$, the Airy functions, are bounded on the whole real line,[1] the kernel $k(x)$ is bounded, therefore, transform (73) with the kernel (18) is a bounded operator from $L_p(R_+)$, $1 \le p \le 2$, into $L_q(R_+)$, $p^{-1} + q^{-1} = 1$. Moreover, it is unitary on $L_2(R_+)$ and $\overline{k(x)}$ is the kernel of the inverse operator.

Another example is

$$k(x) = \frac{1}{2} e^{-\frac{\pi x}{2}} H_{ix}^{(1)}(a), \tag{78}$$

and

$$\overline{k(x)} = \frac{1}{2} e^{\frac{\pi x}{2}} H_{ix}^{(2)}(a), \tag{79}$$

where $H_\nu^{(1)}(a)$ and $H_\nu^{(2)}(a)$ are the Bessel functions of the third kind.[1] Transform (73) with kernel (78) is a bounded operator from $L_p(R_+)$, $1 \leq p \leq 2$, into $L_q(R_+)$, $p^{-1} + q^{-1} = 1$, and moreover, it is unitary on $L_2(R_+)$ and (79) gives the kernel of the inverse operator.

Let

$$k(x) = \frac{1}{\pi}\sqrt{2a \sinh \pi a}\; x^{ia-1/2} K_{1/2-ia}(x), \quad x \in R_+ \ . \tag{80}$$

Then

$$\overline{k(x)} = \frac{1}{\pi}\sqrt{2a \sinh \pi a}\; x^{-ia-1/2} K_{1/2+ia}(x), \quad x \in R_+ \ . \tag{81}$$

Transform (73) with kernel (80) is unitary on $L_2(R_+)$ and its inverse has (81) as the kernel.

For the convolution (66) it was shown that if $k_1 \in L_2(R_+)$, then the transformation

$$g(x) = \left(1 - \frac{d^2}{dx^2}\right) \int_0^\infty \left[k_1(|x-y|) - k_1(x+y)\right] f(y)dy, \quad x \in R_+,$$

is unitary on $L_2(R_+)$, and the reciprocal formula

$$f(x) = \left(1 - \frac{d^2}{dx^2}\right) \int_0^\infty \left[\overline{k_1}(|x-y|) - \overline{k_1}(x+y)\right] g(y)dy, \quad x \in R_+,$$

holds almost everywhere, if, and only if, the kernel satisfies the condition (18).

Suppose that the kernel k_1 is twice differentiable, and $k = k_1 - k_1''$ is locally bounded on R_+, then the integral transform

$$g(x) = \int_0^\infty \left(k(|x-y| - k(x+y|)\right) f(y)dy, \quad x \in R_+ \ . \tag{82}$$

satisfies the Plancherel formula

$$\|g\|_{L_2(R_+)} = \|f\|_{L_2(R_+)}, \tag{83}$$

and reciprocally,

$$f(x) = \int_0^\infty \left(\overline{k(|x-y|)} - \overline{k(x+y)}\right) g(y)dy, \quad x \in R_+ \ . \tag{84}$$

As examples we can take

$$k(x) = -\frac{i}{2\sinh \pi x}\left[\mathbf{J}_{ix}(ia) - \mathbf{J}_{-ix}(ia) + e^{\frac{\pi x}{2}} I_{-ix}(a) - e^{-\frac{\pi x}{2}} I_{ix}(a)\right],$$

where $\mathbf{J}_p(z)$ is the Anger-Weber function.[1] For the convolution (68) it was shown that if $k_1 \in L_2(R_+)$, then the transformation

$$g(x) = \left(1 - \frac{d^2}{dx^2}\right) \int_0^\infty [\text{sign}(y-x)k_1(|x-y|) + k_1(x+y)]\ f(y)dy, \quad x \in R_+,$$

is unitary on $L_2(R_+)$, and the reciprocal formula

$$f(x) = \left(1 - \frac{d^2}{dx^2}\right) \int_0^\infty [g(|x-y|) - g(x+y)]\ \overline{k}(y)dy, \quad x \in R_+,$$

holds almost everywhere, if, and only if, the kernel satisfies the condition

$$\int_0^\infty [\text{sign}(y-x)k_1(|x-y|) + k_1(x+y)]\ \overline{k_1}(y)dy = \frac{1}{4}(x+1)e^{-x}, \quad x \in R_+.$$

Similar results are obtained when the kernel k_1 is twice differentiable, and $k = k_1 - k_1''$ is locally bounded on R_+.

As examples we can take

$$k(x) = \frac{1}{2}(Gi(x) - Gi(-x) + iAi(-x) - iAi(x)),$$

and

$$k(x) = \frac{i}{2\sinh \pi x}\left[\mathbf{J}_{ix}(ia) + \mathbf{J}_{-ix}(ia) - e^{\frac{\pi x}{2}} I_{-ix}(a) - e^{-\frac{\pi x}{2}} I_{ix}(a)\right].$$

3. Nonconvolution Transforms

3.1. *Wimp nonconvolution G-transform*

The only class of nonconvolution transforms known until the eighties is the *Wimp nonconvolution G–transform*[51,52]

$$g(x) = \int_0^\infty G^{m,\,n+2}_{p+2,\,q}\left(y \,\middle|\, {}^{\nu+ix,\nu-ix,(a_p)}_{(b_q)}\right) f(y)dy. \tag{85}$$

Special cases of this class of transforms are the Kontorovich–Lebedev transform[19]

$$g(x) = \int_0^\infty K_{ix}(y) f(y)dy, \tag{86}$$

$$f(y) = \frac{2}{\pi^2 y}\int_0^\infty x \sinh \pi x K_{ix}(y) g(x)dx, \tag{87}$$

the Mehler–Fock transform[10]

$$g(x) = \int_1^\infty P_{ix-1/2}(y) f(y)dy, \tag{88}$$

$$f(y) = \int_0^\infty x \tanh \pi x P_{ix-1/2}(y) g(x)dx, \tag{89}$$

with the Legendre function of the first kind $P_\nu(x)$[1] in the kernel, the Olevskii transform[26]

$$g(x) = \int_0^\infty {}_2F_1(a+ix, a-ix; c; -y) f(y)dy, \tag{90}$$

$$f(y) = \frac{y^{c-1}(1+y)^{2a-c}}{\pi^2\Gamma^2(c)} \int_0^\infty x\sinh 2\pi x\, |\Gamma(a+ix)\Gamma(c-a+ix)|^2 \; {}_2F_1(a+ix, a-ix; c; -y) g(x)dx, \tag{91}$$

with the Gaussian hypergeometric function ${}_2F_1$ in the kernel,[1] and the Lebedev and Lebedev–Skalskaia transforms.[27] The conventional wisdom is that this is the only class of nonconvolution transforms. You will see soon how wrong it is. Analyzing the composition structure of the Wimp nonconvolution G–transform, Yakubovich[52,53] has found that the Wimp nonconvolution G–transform can be decomposed into the composition of a Kontorovich–Lebedev transform and a convolution G–transform (56). But the Kontorovich–Lebedev transform is a composition of two cosine Fourier transforms and an operator of substitution, and the convolution G–transform (56) is a composition of Mellin and inverse Mellin transforms and an operator of multiplication, then the Wimp nonconvolution G–transform is a composition of just Fourier transforms (Mellin and Fourier transforms are the same, up to a change of variables), an operator of multiplication and substitution of variables.

Using this composition structure we have proved a Plancherel theorem for the Wimp nonconvolution G–transform as well as for its special cases[37,45]

Theorem 3.1. *The Wimp integral transform*

$$g(x) = \int_0^\infty G^{m,\, n+2}_{p+2,\, q}\left(y \,\middle|\, {}^{1/2+ix, 1/2-ix, (a_p)}_{(b_q)}\right) f(y)dy, \tag{92}$$

where $c^* = \gamma^* = 0$, $p \neq q$, *and the integral is understood in* $L_2(R_+; x\sinh 2\pi x)$ *is a homeomorphism from* $L_2(R_+)$ *onto* $L_2(R_+; x\sinh 2\pi x)$ *and the inverse transform has the form*

$$f(y) = \frac{1}{\pi^2}\int_0^\infty G^{q-m,\, p-n+2}_{p+2,\, q}\left(y \,\middle|\, {}^{1/2+ix, 1/2-ix, -(a_p^{n+1}), -(a_n)}_{-(b_q^{m+1}), -(b_m)}\right) f(y)dy. \tag{93}$$

If, moreover, $p = 2n$, $q = 2m$, $m \neq n$; $a_{n+i} = -a_i$, $a_i \in R$, $i = 1, \ldots, n$; $b_{m+j} = -b_j$, $b_j \in R$, $j = 1, \ldots, m$, then we have the Parseval identity

$$\frac{1}{\pi^2}\int_0^\infty x\sinh 2\pi x |g(x)|^2 dx = \int_0^\infty |f(y)|^2 dy. \tag{94}$$

In particular, put $m = 1, n = p = 0, q = 2, b_1 = a - 1/2, b_2 = 1/2 - a$, we obtain the Olevskii transform with $c = a$. Hence, we have

Theorem 3.2. *Let $c = a$. Then the pair of Olevskii transforms (90), (91), where the integrals are understood in $L_2(R_+; x^2(x+1)^{4a-3})$ and $L_2(R_+; y^{1-2a})$ norms, respectively, is a homeomorphism between $L_2(R_+; y^{1-2a})$ and $L_2(R_+; x^2(x+1)^{4a-3})$. Moreover, the Parseval identity holds*

$$\int_0^\infty y^{1-2a}|f(y)|^2 dy = \frac{1}{\pi^2\Gamma^2(a)} \int_0^\infty x \sinh 2\pi x \, |\Gamma^2(a+ix)g(x)|^2 dx. \quad (95)$$

In particular, taking $a = 1/2$ we get the L_2– theorem for the Mehler-Fock transform established before by Lebedev.[53]

3.2. *Transforms of Kontorovich-Lebedev type*

Another generalization of the Kontorovich-Lebedev transform comes directly from its composition structure.[37]

Let $\varphi(t)$ be a strictly increasing on R, $\varphi(-\infty) = -\infty, \varphi(\infty) = \infty$, so that the improper integral

$$k(x,y) = \frac{1}{2\pi} \int_{-\infty}^\infty e^{-ixt+iy\varphi(t)} dt \quad (96)$$

converges uniformly on any closed interval of $R\backslash\{0\}$. For example, if $\varphi(t)$ is odd and twice differentiable function, $\varphi'(\infty) = \infty$, $\varphi''(t) > c > 0$ on $[N, \infty)$ and $\varphi''(t)/(\varphi'(t))^2 \in L(N, \infty)$, then the integral (96) converges uniformly on any closed interval of $R\backslash\{0\}$. Then

$$g(x) = \int_{-\infty}^\infty k(x,y) f(y) dy, \quad (97)$$

$$f(y) = \frac{1}{y} \int_{-\infty}^\infty k(-x,-y) x g(x) dx, \quad (98)$$

is a pair of integral transforms.

– If $\varphi(t) = a \sinh t$, $a > 0$, we obtain the following pair of integral transforms

$$g(x) = \frac{1}{\pi} \int_{-\infty}^\infty e^{\frac{\pi x}{2} \operatorname{sign} y} K_{ix}(|y|) f(y) dy, \quad (99)$$

$$f(y) = \frac{1}{\pi y} \int_{-\infty}^\infty e^{\frac{\pi x}{2} \operatorname{sign} y} K_{ix}(|y|) x g(x) dx. \quad (100)$$

If f is an odd function, (99) and (100) become the pair of Kontorovich – Lebedev transforms.

– If $\varphi(t) = \sinh t + at$, $a > -1$, we obtain the following pair of integral transforms

$$g(x) = \frac{1}{\pi} \int_{-\infty}^{\infty} e^{\frac{\pi}{2}(x-ay)\,\mathrm{sign}\, y} K_{i(x-ay)}(|y|) f(y) dy, \tag{101}$$

$$f(y) = \frac{1}{\pi y} \int_{-\infty}^{\infty} e^{\frac{\pi}{2}(x-ay)\,\mathrm{sign}\, y} K_{i(x-ay)}(|y|) x g(x) dx. \tag{102}$$

– If $\varphi(t) = t^3/3 + at$, we obtain the following pair of integral transforms

$$g(x) = \int_{-\infty}^{\infty} y^{-1/3} Ai(y^{-1/3}(ay - x)) f(y) dy \tag{103}$$

$$f(y) = -\frac{1}{y} \int_{-\infty}^{\infty} y^{-1/3} Ai(y^{-1/3}(ay - x)) x g(x) dx. \tag{104}$$

3.3. *The Buchholz nonconvolution G-transform*

The following "odd" transform was introduced by Buchholz[4]

$$g(x) = \int_0^{\infty} M_{ix,\nu}(iy) f(y) dy \tag{105}$$

$$f(y) = \frac{1}{2\pi y \Gamma^2(2\nu + 1)} \tag{106}$$

$$\times \int_{-\infty}^{\infty} e^{-\pi x} \Gamma(\frac{1}{2} + \nu + ix) \Gamma(\frac{1}{2} + \nu - ix) M_{ix,\nu}(-iy) g(x) dx,$$

with the Whittaker function $M_{\mu,\nu}(x)$ in the kernel.[1] It is a composition of a Mellin transform, a multiplication operator, and a Hankel transform. In its composition structure replace the Hankel transform by a more general convolution G-transform (56) we could construct a nonconvolution G–transform including the Buchholz transform[37]

$$g(x) = \int_0^{\infty} G_{p+r,\,q}^{m,\,n+r} \left(e^{-\pi r i/2} y \middle| \begin{matrix} 1/r, 2/r, \ldots, 1, (a_p) + 1/2 + ix \\ (b_q) + 1/2 + ix \end{matrix} \right) y^{-ix} f(y) dy, \tag{107}$$

as a special case.

Theorem 3.3. *The G– transform (107) is a homeomorphism from $L_2(R_+)$ onto $L_2(R)$ and has the inverse*

$$f(y) = (2\pi)^{-r} r y^{-\frac{1}{2}} \tag{108}$$

$$\times \int_{-\infty}^{\infty} G_{p+r,\,q}^{q-m,\,p-n+r} \left(e^{\pi r i/2} y \middle| \begin{matrix} \frac{1}{r}, \frac{2}{r}, \ldots, 1, \frac{1}{2} - (a_p^{n+1}) - ix, \frac{1}{2} - (a_n) - ix \\ \frac{1}{2} - (b_q^{m+1}) - ix, \frac{1}{2} - (b_m) - ix \end{matrix} \right) y^{ix - \frac{1}{2}} g(x) dx.$$

If, moreover, $p = 2n, q = 2n, a_{n+j} = -a_j \in R, j = 1, ..., n; b_{m+j} = -b_j \in R, j = 1, ..., m$, then we have the Parseval formula

$$\frac{r}{(2\pi)^r} \int_{-\infty}^{\infty} |g(x)|^2 dx = \int_0^{\infty} y|f(y)|^2 \, dy.$$

The G-transform (107) is a composition of a Mellin transform, multiplication by $e^{-irx^{1/r}}$, and a convolution G-transform.

Put in formulas (107) and (108) $m = 1, n = p = 0, q = 2, b_1 = \nu, b_2 = -\nu, r = a = 1$, we obtained the pair of Buchholz integral transforms. Hence, we have

Theorem 3.4. *The pair of Buchholz transforms (105) and (106) are homeomorphisms between* $L_2(R_+; y)$ *and* $L_2\left(R; (x^2+1)^{\Re(\nu)} e^{\pi(x-|x|)}\right)$. *Moreover, if* $\nu \in R$ *then the following Parseval equation holds*

$$2\pi\Gamma^2(1+2\nu) \int_0^{\infty} y|f(y)|^2 dy = \int_{-\infty}^{\infty} e^{\pi x} |\Gamma(1/2+\nu+ix) g(x)|^2 dx. \quad (109)$$

In particular, if we take the Fourier cosine (sine) transform as the inner convolution G-transform then we obtain the pair of transforms

$$g(x) = \int_0^{\infty} \left[(a+y)^{-ix-1/2} \pm (a-y)^{-ix-1/2}\right] f(y) dy \quad (110)$$

$$f(y) = \frac{1}{4\pi} \int_{-\infty}^{\infty} \left[(a+y)^{ix-1/2} \pm (a-y)^{ix-1/2}\right] \frac{e^{\pi x}}{\cosh \pi x} g(x) dx. \quad (111)$$

The Parseval equality

$$\frac{1}{4\pi} \int_{-\infty}^{\infty} \frac{e^{\pi x}}{\cosh \pi x} |g(x)|^2 dx = \int_0^{\infty} |f(y)|^2 \, dy$$

also holds.

Here is the composite structure

$$g(x) = \frac{\sqrt{2\pi} \exp(-\pi x/2 \pm \pi i/4)}{\Gamma(1/2+ix)} M \left\{\exp(-iat) F_{\binom{c}{s}} \left\{f(y); t\right\}; ix + 1/2\right\}.$$

If we replace the Fourier cosine transform by the Hankel transform we obtain an integral transform with the Legendre function in the kernel.[38]

3.4. *The cherry nonconvolution G-transform*

In 1949 Cherry[6] considered an index transformation with $D_\nu(z)$ as its kernel and established that, if $f \in L_1(\mathbf{R})$ is of bounded variation and continuous at x, then

$$\begin{aligned} f(x) =& \frac{1}{4\pi}\int_{-\infty}^{\infty}\frac{e^{\frac{\lambda\pi}{2}}}{\cosh\lambda\pi}d\lambda\int_{-\infty}^{\infty}\Big[D_{-i\lambda-\frac{1}{2}}(e^{\frac{\pi i}{4}}x)D_{i\lambda-\frac{1}{2}}(e^{-\frac{\pi i}{4}}y) \\ &+ D_{-i\lambda-\frac{1}{2}}(-e^{\frac{\pi i}{4}}x)D_{i\lambda-\frac{1}{2}}(-e^{-\frac{\pi i}{4}}y)\Big]\, f(y)dy. \end{aligned} \tag{112}$$

In fact, Cherry's expansion (112) is still valid when the integrability condition on f is replaced by some condition on the asymptotic behavior of f at infinity such as

$$f(x) = e^{-\frac{ix^2}{4}}\left(\frac{c_1}{x^\alpha}+\frac{c_2}{x^{\alpha+1}}\right)\{1+O(x^{-1})\},$$

where $\alpha > \frac{1}{2}$, c_1 and c_2 are constants, which may be different for $x \to +\infty$ and $x \to -\infty$.

To derive composition structure and a Plancherel-type theorem for the Cherry transform, we rewrite the Cherry expansion (112) in the following form:

$$f(x) = (D_+^{-1}D_+f)(x) + (D_-^{-1}D_-f)(x), \tag{113}$$

where operators D_+, D_-, D_+^{-1} and D_-^{-1} are defined by[1]

$$(D_+f)(x) = \int_{-\infty}^{\infty} D_{ix-\frac{1}{2}}(-e^{-\frac{\pi i}{4}}y)f(y)dy, \tag{114}$$

$$(D_-f)(x) = \int_{-\infty}^{\infty} D_{ix-\frac{1}{2}}(e^{-\frac{\pi i}{4}}y)f(y)dy, \tag{115}$$

$$(D_+^{-1}f)(y) = \frac{1}{4\pi}\int_{-\infty}^{\infty}\frac{e^{\frac{\pi x}{2}}}{\cosh\pi x}D_{-ix-\frac{1}{2}}(-e^{\frac{\pi i}{4}}y)f(x)dx \tag{116}$$

and

$$(D_-^{-1}f)(y) = \frac{1}{4\pi}\int_{-\infty}^{\infty}\frac{e^{\frac{\pi x}{2}}}{\cosh\pi x}D_{-ix-\frac{1}{2}}(e^{\frac{\pi i}{4}}y)f(x)dx, \tag{117}$$

respectively. Substituting the integral representation (2.3.15.3) from Ref. 9 for the parabolic cylinder function $D_\nu(z)$, that is,

$$D_{ix-\frac{1}{2}}(\pm e^{-\frac{\pi i}{4}}t)e^{-\frac{it^2}{4}} = \frac{e^{-\frac{\pi x}{4}-\frac{\pi i}{8}}}{\Gamma(\frac{1}{2}-ix)}\int_0^\infty e^{\frac{iy^2}{2}\pm iyt}y^{-ix-\frac{1}{2}}dy \tag{118}$$

and

$$D_{-ix-\frac{1}{2}}(\pm e^{\frac{\pi i}{4}}t)e^{\frac{it^2}{4}} = \frac{e^{-\frac{\pi x}{4}+\frac{\pi i}{8}}}{\Gamma(\frac{1}{2}+ix)}\int_0^\infty e^{-\frac{iy^2}{2}\mp iyt}y^{ix-\frac{1}{2}}dy, \tag{119}$$

into the formulas (7) and (8), and interchanging the order of integration that is permissible if $f(y) \in \mathcal{S}(\mathbf{R})$, the Schwartz space of all infinitely differentiable functions which rapidly vanish at infinity together with all derivatives, we arrive at the following composite representations for the transforms D_+, D_-, D_+^{-1} and D_-^{-1}:

$$(D_+f)(x) = \frac{\sqrt{2\pi}e^{-\frac{\pi x}{4}-\frac{\pi i}{8}}}{\Gamma(\frac{1}{2}-ix)} M\left\{e^{\frac{iy^2}{2}}F\{f(t)e^{\frac{it^2}{4}};-y\};-ix+\frac{1}{2}\right\}, \quad (120)$$

$$(D_-f)(x) = \frac{\sqrt{2\pi}e^{-\frac{\pi x}{4}-\frac{\pi i}{8}}}{\Gamma(\frac{1}{2}-ix)} M\left\{e^{\frac{iy^2}{2}}F\{f(t)e^{\frac{it^2}{4}};y\};-ix+\frac{1}{2}\right\}, \quad (121)$$

$$(D_+^{-1}f)(t) = e^{-\frac{it^2}{4}}F^{-1}\left\{e^{-\frac{iy^2}{2}}M^{-1}\left\{f(x)\frac{\Gamma(\frac{1}{2}-ix)}{\sqrt{2\pi}e^{-\frac{\pi x}{4}-\frac{\pi i}{8}}};y\right\};-t\right\} \quad (122)$$

and

$$(D_-^{-1}f)(t) = e^{-\frac{it^2}{4}}F^{-1}\left\{e^{-\frac{iy^2}{2}}M^{-1}\left\{f(x)\frac{\Gamma(\frac{1}{2}-ix)}{\sqrt{2\pi}e^{-\frac{\pi x}{4}-\frac{\pi i}{8}}};y\right\};t\right\}. \quad (123)$$

Here M and M^{-1} are the Mellin transform and its inverse, and F and F^{-1} are the Fourier transform and its inverse. The composite representations (13) and (14) of the transforms D_+ and D_- are key tools in studying the Cherry expansion (6).

By $L_2^+(R;e^{\frac{iy^2}{4}})$ (similarly, by $L_2^-(R;e^{\frac{iy^2}{4}})$) we denote a subspace of $L_2(R)$ consisting of all square-integrable functions $f(y)$ such that the Fourier transforms of $f(y)e^{\frac{iy^2}{4}}$ vanish on the negative (positive) part of the real axis. These spaces are Hilbert spaces and their elements can be analytically continued onto the whole upper (lower) complex half-plane. Any function $f \in L_2(R)$ can be uniquely factorized by

$$f(y) = f^+(y) + f^-(y),$$

where

$$f^+ \in L_2^+(R;e^{\frac{iy^2}{4}}) \quad \text{and} \quad f^- \in L_2^-(R;e^{\frac{iy^2}{4}}).$$

Applying the Parseval equalities for the Mellin and Fourier transforms to the composite representations (13) and (14), we obtain the following Parseval equalities for the transforms D_+ and D_-:

$$\frac{1}{4\pi}\int_{-\infty}^{\infty}\frac{e^{\frac{\pi x}{2}}}{\cosh \pi x}|(D_+f^+)(x)|^2dx = \int_{-\infty}^{\infty}|f^+(t)|^2dt, \quad (124)$$

and

$$\frac{1}{4\pi}\int_{-\infty}^{\infty}\frac{e^{\frac{\pi x}{2}}}{\cosh \pi x}|(D_- f^-)(x)|^2 dx = \int_{-\infty}^{\infty}|f^-(t)|^2 dt. \qquad (125)$$

From the composite representations (13) and (14), the Parseval equalities, and the mapping properties of the Fourier and Mellin transforms, by standard procedure we can expand these identities to the whole spaces $L_2^+(R; e^{\frac{iy^2}{4}})$ and $L_2^-(R; e^{\frac{iy^2}{4}})$. Thus D_+ and D_- are isomorphisms from $L_2^+(R; e^{\frac{iy^2}{4}})$ and $L_2^-(R; e^{\frac{iy^2}{4}})$, respectively, onto the space $L_2(R; e^{\frac{\pi x}{2}}(\cosh \pi x)^{-1})$, and the integrals (7) and (8) converge in the latter Hilbert space.

Comparing the composite representations (13), (14), (15) and (123), it is easy to see that (15) is the inverse of (13), and (123) is the inverse of (14). Hence the operator D_+^{-1} is the inverse of D_+, and D_-^{-1} is the inverse operator for D_-, and the integrals (9) and (117) converge in $L_2(\mathbf{R})$. Thus

$$D_+^{-1}D_+ f^+ = f^+ \qquad \text{and} \qquad D_-^{-1}D_- f^- = f^-.$$

By the definitions of f^+ and f^-, we have

$$F\{f^+(y)e^{\frac{iy^2}{4}}; -x\} = 0 \qquad \text{and} \qquad F\{f^-(y)e^{\frac{iy^2}{4}}; x\} = 0 \quad (x > 0).$$

Consequently, from (13) and (14) it is easy to see that

$$(D_+ f^-)(x) = 0 \qquad \text{and} \qquad (D_- f^+)(x) = 0.$$

Thus, for any $f \in L_2(\mathbf{R})$, we have

$$D_+^{-1}D_+ f + D_-^{-1}D_- f = D_+^{-1}D_+ f^+ + D_-^{-1}D_- f^- = f^+ + f^- = f.$$

We sum up our results in the following Plancherel-type theorem:

Theorem 3.5. *The Cherry transforms* (7) *and* (8) *isomorphically map the spaces* $L_2^+(R; e^{\frac{it^2}{4}})$ *and* $L_2^-(R; e^{\frac{it^2}{4}})$ *onto the space* $L_2(R; e^{\frac{\pi x}{2}}(\cosh \pi x)^{-1})$, *admit the inversion formulas* (9) *and* (117), *and satisfy the Parseval equalities* (124) *and* (125), *respectively. Furthermore, the expansion formula* (6) *holds true in* $L_2(R)$.

We will show now that the Cherry transform is related to a singular Sturm-Liouville problem on the whole line. As a consequence, another Plancherel-type theorem for the Cherry transform is established. For more details see.[30]

Let us consider the following singular Sturm-Liouville problem:

$$Ly = \frac{d^2 y}{dx^2} - q(x)y = -\lambda y \qquad (-\infty < x < \infty), \qquad (126)$$

$$|y(\pm\infty)| < \infty,$$

with $q(x)$ being even on R such that the spectrum of the singular Sturm-Liouville problem (1) is the whole R. Let $\phi(x,\lambda)$ and $\theta(x,\lambda)$ be the solutions of the equation (1) such that

$$\phi(0,\lambda) = 0, \quad \phi'(0,\lambda) = -1, \tag{127}$$

$$\theta(0,\lambda) = 1, \quad \theta'(0,\lambda) = 0. \tag{128}$$

There exist two functions $m_1(\lambda)$ and $m_2(\lambda)$ analytic in the upper half-plane such that, as a function of x,

$$\psi_1(x,\lambda) = \theta(x,\lambda) + m_1(\lambda)\phi(x,\lambda)$$

is in $L_2(-\infty, 0)$, and

$$\psi_2(x,\lambda) = \theta(x,\lambda) + m_2(\lambda)\phi(x,\lambda)$$

is in $L_2(0,\infty)$, for non-real λ. Moreover, since $q(x)$ is even, we have $m_1(\lambda) = -m_2(\lambda)$. In addition, there exist two non-decreasing functions $\xi(\lambda)$ and $\zeta(\lambda)$ such that

$$\lim_{\delta\to 0+} \int_0^\lambda \Im\left[\frac{1}{2m_2(u+i\delta)}\right] du = \xi(\lambda)$$

and

$$\lim_{\delta\to 0+} \int_0^\lambda \Im\left[\frac{m_2(u+i\delta)}{2}\right] du = -\zeta(\lambda),$$

where the symbol $\Im$ means the imaginary part of a complex-valued function.

It is a known fact[31] that if $f(x) \in L_2(R)$, then

$$F(\lambda) = \int_{-\infty}^{\infty} f(x)\phi(x,\lambda)\,dx \tag{129}$$

belongs to $L_2(R, d\zeta)$, and

$$E(\lambda) = \int_{-\infty}^{\infty} f(x)\theta(x,\lambda)\,dx \tag{130}$$

belongs to $L_2(R, d\xi)$. The following inversion formula

$$f(x) = \frac{1}{\pi}\int_{-\infty}^{\infty} E(\lambda)\theta(x,\lambda)d\xi(\lambda) + \frac{1}{\pi}\int_{-\infty}^{\infty} F(\lambda)\phi(x,\lambda)d\zeta(\lambda), \tag{131}$$

holds true; and in addition, the Parseval equality:

$$\int_{-\infty}^{\infty} |f(x)|^2 dx = \frac{1}{\pi}\int_{-\infty}^{\infty} |E(\lambda)|^2\, d\xi(\lambda) + \frac{1}{\pi}\int_{-\infty}^{\infty} |F(\lambda)|^2\, d\zeta(\lambda) \tag{132}$$

is valid.

Now we are ready to consider the following singular Sturm-Liouville problem

$$\frac{d^2y}{dx^2}+\frac{x^2}{4}y=-\lambda y \qquad (-\infty<x<\infty), \tag{133}$$

$$|y(\pm\infty)|<\infty.$$

Functions

$$D_{-i\lambda-\frac{1}{2}}\left(e^{\frac{\pi i}{4}}x\right),\quad D_{-i\lambda-\frac{1}{2}}\left(-e^{\frac{\pi i}{4}}x\right),$$
$$D_{i\lambda-\frac{1}{2}}\left(e^{-\frac{\pi i}{4}}x\right),\quad D_{i\lambda-\frac{1}{2}}\left(-e^{-\frac{\pi i}{4}}x\right)$$

are solutions of the equation (133). Since $q(x)=-\frac{x^2}{4}$ is even, $q(x)\to-\infty$ as $x\to\pm\infty$, and the integral $\int_0^\infty |q(x)|^{-\frac{1}{2}}dx$ is divergent, there is a continuous spectrum over $(-\infty,\infty)$,[31] and the formulas (4), (5), (8) and (132) hold true. It is easy to see that

$$\theta(x,\lambda)=\frac{\Gamma\left(\frac{3}{4}+\frac{i\lambda}{2}\right)}{2^{-\frac{i\lambda}{2}+\frac{3}{4}}\sqrt{\pi}}\left[D_{-i\lambda-\frac{1}{2}}\left(e^{\frac{\pi i}{4}}x\right)+D_{-i\lambda-\frac{1}{2}}\left(-e^{\frac{\pi i}{4}}x\right)\right] \tag{134}$$

and

$$\phi(x,\lambda)=\frac{e^{-\frac{\pi i}{4}}\Gamma\left(\frac{1}{4}+\frac{i\lambda}{2}\right)}{2^{-\frac{i\lambda}{2}+\frac{5}{4}}\sqrt{\pi}}\left[D_{-i\lambda-\frac{1}{2}}\left(e^{\frac{\pi i}{4}}x\right)-D_{-i\lambda-\frac{1}{2}}\left(-e^{\frac{\pi i}{4}}x\right)\right] \tag{135}$$

satisfy the initial conditions (2) and (3). As $x\to\infty$, $e^{\frac{\pi i}{4}}x$ has the argument $\frac{\pi}{4}$, and $-e^{\frac{\pi i}{4}}x$ has the argument $-\frac{3\pi}{4}$. Consequently, the following asymptotic formulas for the parabolic cylinder function[9] can be used:

$$\begin{aligned}D_\nu(z)=&z^\nu e^{-\frac{z^2}{4}}\left[\sum_{n=0}^{N}\frac{(-\frac{\nu}{2})_n(\frac{1-\nu}{2})_n}{n!}\left(-\frac{2}{z^2}\right)^n+O\left(|z|^{-2N-2}\right)\right]\\&\left(-\frac{3\pi}{4}<\arg(z)<\frac{3\pi}{4}\right);\end{aligned} \tag{136}$$

$$\begin{aligned}D_\nu(z)=&z^\nu e^{-\frac{z^2}{4}}\left[\sum_{n=0}^{N}\frac{(-\frac{\nu}{2})_n(\frac{1-\nu}{2})_n}{n!}\left(-\frac{2}{z^2}\right)^n+O\left(|z|^{-2N-2}\right)\right]\\&-\frac{\sqrt{2\pi}}{\Gamma(-\nu)}e^{-\nu\pi i}z^{-\nu-1}e^{\frac{z^2}{4}}\left[\sum_{n=0}^{N}\frac{(\frac{\nu}{2})_n(\frac{1-\nu}{2})_n}{n!}\left(\frac{2}{z^2}\right)^n+O\left(|z|^{-2N-2}\right)\right]\\&\left(-\frac{5\pi}{4}<\arg(z)<-\frac{\pi}{4}\right).\end{aligned} \tag{137}$$

Letting $N = 0$ in (136) and (137), we obtain

$$D_{-i\lambda-\frac{1}{2}}\left(e^{\frac{\pi i}{4}}x\right) = e^{\frac{\lambda\pi}{4}-\frac{\pi i}{8}-\frac{ix^2}{4}}x^{-i\lambda-\frac{1}{2}}\left[1+O(x^{-2})\right]\ (x\to\infty);$$

$$D_{-i\lambda-\frac{1}{2}}\left(-e^{\frac{\pi i}{4}}x\right) = e^{-\frac{3\lambda\pi}{4}+\frac{3\pi i}{8}-\frac{ix^2}{4}}x^{-i\lambda-\frac{1}{2}}\left[1+O(x^{-2})\right]$$
$$-\frac{\sqrt{2\pi}}{\Gamma(\frac{1}{2}+i\lambda)}e^{-\frac{\lambda\pi}{4}+\frac{7\pi i}{8}+\frac{ix^2}{4}}x^{i\lambda-\frac{1}{2}}\left[1+O(x^{-2})\right]$$
$$(x\to\infty).$$

If λ is in the upper half-plane ($\lambda = \tau + i\delta$; $\delta > 0$), then

$$x^{i\lambda-\frac{1}{2}} = x^{i\tau-\delta-\frac{1}{2}} \in L_2(1,\infty).$$

Consequently, we observe that

$$\theta(x,\lambda) + m_2(\lambda)\phi(x,\lambda) \in L_2(0,\infty)$$

for all λ in the upper half-plane if and only if the term $x^{-i\lambda-\frac{1}{2}}e^{-\frac{ix^2}{4}}$ vanishes, that is,

$$\frac{\Gamma\left(\frac{3}{4}+\frac{i\lambda}{2}\right)}{2^{-\frac{i\lambda}{2}+\frac{3}{4}}\sqrt{\pi}}e^{\frac{\lambda\pi}{4}-\frac{\pi i}{8}}(1+ie^{-\lambda\pi})$$
$$+m_2(\lambda)\frac{\Gamma\left(\frac{1}{4}+\frac{i\lambda}{2}\right)}{2^{-\frac{i\lambda}{2}+\frac{5}{4}}\sqrt{\pi}}e^{\frac{\lambda\pi}{4}-\frac{3\pi i}{8}}(1-ie^{-\lambda\pi}) = 0.$$

Hence

$$m_2(\lambda) = \frac{2^{\frac{3}{2}}\pi e^{\frac{\lambda\pi}{2}}}{\left|\Gamma\left(\frac{1}{4}+\frac{i\lambda}{2}\right)\right|^2(1+ie^{\lambda\pi})},$$

and therefore, we have

$$\xi'(\lambda) = \Im\left[\frac{1}{2m_2(\lambda)}\right] = \Im\left[\frac{1+ie^{\lambda\pi}}{2^{\frac{5}{2}}\pi e^{\frac{\lambda\pi}{2}}}\left|\Gamma\left(\frac{1}{4}+\frac{i\lambda}{2}\right)\right|^2\right]$$
$$= \frac{e^{\frac{\lambda\pi}{2}}}{2^{\frac{5}{2}}\pi}\left|\Gamma\left(\frac{1}{4}+\frac{i\lambda}{2}\right)\right|^2 \tag{138}$$

and

$$\zeta'(\lambda) = \frac{\sqrt{2}\pi e^{\frac{3\lambda\pi}{2}}}{1+e^{2\lambda\pi}}\frac{1}{\left|\Gamma\left(\frac{1}{4}+\frac{i\lambda}{2}\right)\right|^2}. \tag{139}$$

So, finally, the formulas (4), (5), (8) and (132) take the forms:

$$F(\lambda) = \int_{-\infty}^{\infty}\frac{e^{-\frac{\pi i}{4}}\Gamma\left(\frac{1}{4}+\frac{i\lambda}{2}\right)}{2^{-\frac{i\lambda}{2}+\frac{5}{4}}\sqrt{\pi}}\left[D_{-i\lambda-\frac{1}{2}}\left(e^{\frac{\pi i}{4}}x\right) - D_{-i\lambda-\frac{1}{2}}\left(-e^{\frac{\pi i}{4}}x\right)\right]f(x)\,dx, \tag{140}$$

$$E(\lambda) = \int_{-\infty}^{\infty} \frac{\Gamma\left(\frac{3}{4}+\frac{i\lambda}{2}\right)}{2^{-\frac{i\lambda}{2}+\frac{3}{4}}\sqrt{\pi}} \left[D_{-i\lambda-\frac{1}{2}}\left(e^{\frac{\pi i}{4}}x\right) + D_{-i\lambda-\frac{1}{2}}\left(-e^{\frac{\pi i}{4}}x\right)\right] f(x)\,dx, \tag{141}$$

$$\begin{aligned} f(x) =& \frac{1}{\pi}\int_{-\infty}^{\infty} \frac{\Gamma\left(\frac{3}{4}+\frac{i\lambda}{2}\right)}{2^{-\frac{i\lambda}{2}+\frac{3}{4}}\sqrt{\pi}} \frac{e^{\frac{\lambda\pi}{2}}}{2^{\frac{5}{2}}\pi} \left|\Gamma\left(\frac{1}{4}+\frac{i\lambda}{2}\right)\right|^2 \\ & \left[D_{-i\lambda-\frac{1}{2}}\left(e^{\frac{\pi i}{4}}x\right) + D_{-i\lambda-\frac{1}{2}}\left(-e^{\frac{\pi i}{4}}x\right)\right] E(\lambda)\,d\lambda \\ & + \frac{1}{\pi}\int_{-\infty}^{\infty} \frac{e^{-\frac{\pi i}{4}}\Gamma\left(\frac{1}{4}+\frac{i\lambda}{2}\right)}{2^{-\frac{i\lambda}{2}+\frac{5}{4}}\sqrt{\pi}} \frac{\sqrt{2}\pi e^{\frac{3\lambda\pi}{2}}}{1+e^{2\lambda\pi}} \frac{1}{\left|\Gamma\left(\frac{1}{4}+\frac{i\lambda}{2}\right)\right|^2} \\ & \left[D_{-i\lambda-\frac{1}{2}}\left(e^{\frac{\pi i}{4}}x\right) - D_{-i\lambda-\frac{1}{2}}\left(-e^{\frac{\pi i}{4}}x\right)\right] F(\lambda)\,d\lambda \end{aligned} \tag{142}$$

and

$$\begin{aligned} \int_{-\infty}^{\infty} |f(x)|^2\,dx =& \frac{1}{\pi}\int_{-\infty}^{\infty} |E(\lambda)|^2 \frac{e^{\frac{\lambda\pi}{2}}}{2^{\frac{5}{2}}\pi} \left|\Gamma\left(\frac{1}{4}+\frac{i\lambda}{2}\right)\right|^2 d\lambda \\ & + \frac{1}{\pi}\int_{-\infty}^{\infty} |F(\lambda)|^2 \frac{\sqrt{2}\pi e^{\frac{3\lambda\pi}{2}}}{1+e^{2\lambda\pi}} \frac{1}{\left|\Gamma\left(\frac{1}{4}+\frac{i\lambda}{2}\right)\right|^2} d\lambda, \end{aligned} \tag{143}$$

respectively.

If we put

$$\tilde{F}(\lambda) = \frac{2^{-\frac{i\lambda}{2}+\frac{5}{4}}\sqrt{\pi}}{e^{-\frac{\pi i}{4}}\Gamma\left(\frac{1}{4}+\frac{i\lambda}{2}\right)} F(\lambda) \qquad \text{and} \qquad \tilde{E}(\lambda) = \frac{2^{-\frac{i\lambda}{2}+\frac{3}{4}}\sqrt{\pi}}{\Gamma\left(\frac{3}{4}+\frac{i\lambda}{2}\right)} E(\lambda),$$

then the formulas (140), (141), (142) and (143) become

$$\tilde{F}(\lambda) = \int_{-\infty}^{\infty} \left[D_{-i\lambda-\frac{1}{2}}\left(e^{\frac{\pi i}{4}}x\right) - D_{-i\lambda-\frac{1}{2}}\left(-e^{\frac{\pi i}{4}}x\right)\right] f(x)\,dx, \tag{144}$$

$$\tilde{E}(\lambda) = \int_{-\infty}^{\infty} \left[D_{-i\lambda-\frac{1}{2}}\left(e^{\frac{\pi i}{4}}x\right) + D_{-i\lambda-\frac{1}{2}}\left(-e^{\frac{\pi i}{4}}x\right)\right] f(x)\,dx, \tag{145}$$

$$\begin{aligned} f(x) =& \int_{-\infty}^{\infty} \frac{e^{\frac{\lambda\pi}{2}}\cosh\left(\frac{\lambda\pi}{2}+\frac{i\pi}{4}\right)\Gamma\left(\frac{1}{2}+i\lambda\right)}{4\pi\sqrt{2\pi}\cosh\lambda\pi} \\ & \left[D_{-i\lambda-\frac{1}{2}}\left(e^{\frac{\pi i}{4}}x\right) + D_{-i\lambda-\frac{1}{2}}\left(-e^{\frac{\pi i}{4}}x\right)\right] \tilde{E}(\lambda)\,d\lambda \\ & - \int_{-\infty}^{\infty} \frac{e^{\frac{\lambda\pi}{2}}\sinh\left(\frac{\lambda\pi}{2}+\frac{i\pi}{4}\right)\Gamma\left(\frac{1}{2}+i\lambda\right)}{4\pi\sqrt{2\pi}\cosh\lambda\pi} \\ & \left[D_{-i\lambda-\frac{1}{2}}\left(e^{\frac{\pi i}{4}}x\right) - D_{-i\lambda-\frac{1}{2}}\left(-e^{\frac{\pi i}{4}}x\right)\right] \tilde{F}(\lambda)\,d\lambda \end{aligned} \tag{146}$$

and

$$\int\limits_{-\infty}^{\infty} |f(x)|^2\, dx = \frac{1}{4\sqrt{2\pi}} \int_{-\infty}^{\infty} |\tilde{E}(\lambda)|^2 \frac{e^{\frac{\lambda\pi}{2}} \left|\cosh\left(\frac{\lambda\pi}{2} + \frac{i\pi}{4}\right)\right|}{(\cosh\lambda\pi)^{\frac{3}{2}}} d\lambda + \frac{1}{4\sqrt{2\pi}} \int_{-\infty}^{\infty} |\tilde{F}(\lambda)|^2 \frac{e^{\frac{\lambda\pi}{2}} \left|\sinh\left(\frac{\lambda\pi}{2} + \frac{i\pi}{4}\right)\right|}{(\cosh\lambda\pi)^{\frac{3}{2}}} d\lambda \tag{147}$$

respectively.

Hence we can sum up our results in the following Plancherel-type theorem.

Theorem 3.6. *The integral transforms* (144) *and* (145) *are isomorphisms from* $L_2(R)$ *onto* $L_2(R; e^{\frac{\lambda\pi}{2}}(\cosh\lambda\pi)^{-1})$. *Moreover, the inversion formula has the form* (146) *and the Parseval equality* (147) *holds true.*

One can show that the transforms (144), (145) and (146) are equivalent to the Cherry expansion (112).

The composition structure of Cherry's transform gives rise to a new Cherry nonconvolution G-transform.[37,45]

Let $p = 2n$, $q = 2m$, $m \neq n$; $a_{n+i} = -a_i$, $a_i \in R$, $i = 1, \ldots, n$; $b_{m+j} = -b_j$, $b_j \in R$, $j = 1, \ldots, m$. Then Cherry G-transform

$$(G_+f)(x) = g(x) = \int_{-\infty}^{\infty} G_{p+1,q}^{m,n+1}\left(-\frac{ix}{y}\middle|\begin{matrix}0,(a_p)\\(b_q)\end{matrix}\right)\frac{f(y)}{y}dy \tag{148}$$

is an isomorphism from $L_2^+(R)$ onto $L_2(R_+)$ and the following inverse formula holds

$$(G_+^{-1}g)(y) = f(y) = \frac{1}{2\pi y}\int_0^{\infty} G_{p+1,q}^{q-m,p-n+1}\left(\frac{ix}{y}\middle|\begin{matrix}0,-(a_p^{n+1}),-(a_n)\\-(b_q^{m+1}),-(b_m)\end{matrix}\right)g(x)dx. \tag{149}$$

Similarly, Cherry G-transform

$$(G_+f)(x) = g(x) = \int_{-\infty}^{\infty} G_{p+1,q}^{m,n+1}\left(\frac{ix}{y}\middle|\begin{matrix}0,(a_p)\\(b_q)\end{matrix}\right)\frac{f(y)}{y}dy \tag{150}$$

is an isomorphism from $L_2^-(R)$ onto $L_2(R_+)$ and the following inverse formula holds

$$(G_+^{-1}g)(y) = f(y) = \frac{1}{2\pi y}\int_0^{\infty} G_{p+1,q}^{q-m,p-n+1}\left(-\frac{ix}{y}\middle|\begin{matrix}0,-(a_p^{n+1}),-(a_n)\\-(b_q^{m+1}),-(b_m)\end{matrix}\right)g(x)dx. \tag{151}$$

The following expansion holds in $L_2(R)$

$$f(x) = \left(G_+^{-1}G_+f\right)(x) + \left(G_-^{-1}G_-f\right)(x).$$

3.5. *Other nonconvolution G-transforms*

Let $\lambda, \gamma \notin (-\infty, 0]$. Then the integral transform[38]

$$g(x) = \int_0^\infty \left[y^{-ix-\frac{1}{2}} + (\lambda - \gamma)\gamma^{-ix-\frac{1}{2}} \frac{\prod_{j=1}^q \Gamma(\frac{1}{2} + b_j + ix) \prod_{j=1}^n \Gamma(\frac{1}{2} - d_j - ix)}{\prod_{j=1}^p \Gamma(\frac{1}{2} - a_j - ix) \prod_{j=1}^m \Gamma(\frac{1}{2} + c_j + ix)} \right.$$

$$\left. G^{m+1,\, p+1}_{p+q+1,\, m+n+1} \left(\gamma y \left| \begin{matrix} \frac{1}{2}-ix,(a_p),(b_q) \\ \frac{1}{2}-ix,(c_m),(d_n) \end{matrix} \right. \right) \right] f(y)dy, \tag{152}$$

is a bounded operator from $L_2(0, \infty)$ onto $L_2(-\infty, \infty)$ and has the inverse

$$f(y) = \frac{1}{2\pi} \int_{-\infty}^\infty \left[y^{ix-\frac{1}{2}} + (\gamma - \lambda)\lambda^{ix-\frac{1}{2}} \frac{\prod_{j=1}^p \Gamma(\frac{1}{2} - a_j - ix) \prod_{j=1}^m \Gamma(\frac{1}{2} + c_j + ix)}{\prod_{j=1}^q \Gamma(\frac{1}{2} + b_j + ix) \prod_{j=1}^n \Gamma(\frac{1}{2} - d_j - ix)} \right.$$

$$\left. G^{n+1,\, q+1}_{p+q+1,\, m+n+1} \left(\lambda y \left| \begin{matrix} \frac{1}{2}+ix,(-b_q),(-a_p) \\ \frac{1}{2}+ix,(-d_n),(-c_m) \end{matrix} \right. \right) \right] g(x)dx. \tag{153}$$

If, moreover, $m = n, p = q, b_j = -a_j, j = 1, \ldots, p, d_j = -c_j, j = 1, \ldots, m, \lambda = -\gamma = i\delta, \delta \in R$, then the Plancherel equality holds

$$\frac{1}{2\pi} \int_{-\infty}^\infty |g(x)|^2 dx = \int_0^\infty |f(y)|^2 dy.$$

G-transforms (107) and (152) look completely different, but they have very similar composition structure. In fact (152) can be obtained from (107) by replacing the multiplicator $e^{-irx^{1/r}}$ by $\frac{x+\lambda}{x+\gamma}$.

Playing with only the Fourier transforms, operators of multiplications and substitutions of variables we have found some more new classes of nonconvolution G–transforms and prove L_2 theorems for them. As special cases about 60 new transforms with L_2 theorems are added to the list of around 300 known transforms.[27]

References

1. M. Abramowitz and I. A. Stegun *Handbook of Mathematical Functions with Formulas, Graphs, and Mathematical Tables.* Reprint of the 1972 edition. (Dover Publications, Inc., New York, 1992), pp. xiv+1046
2. Yu. A. Brychkov, H.-J. Glaeske and O.I. Marichev *Itogi Nauki i Tekhniki. Mat. Anal., VINITI AN SSSR* **21**, 3 (1983).
3. Yu. A. Brychkov, H.-J. Glaeske, A.P. Prudnikov and Vu Kim Tuan, *Multidimentional Integral Transformations*, Gordon and Breach, (New York - Philadelphia - London - Paris - Montreux - Tokyo - Melbourne - Singapore, 1992).
4. H. Buchholz, *The Confluent Hypergeometric Function*, (Springer Verlag, Berlin - New York, 1969), pp. 240
5. I. W. Busbridge, *J. London Math. Soc.* **9**, 179 (1934).

6. T. M. Cherry, *Proc. Edinburgh Math. Soc. (2)* **8**, 50-65 (1948).
7. R. V. Churchill, *Operational Mathematics.* 2nd Ed. (McGraw-Hill Book Co., Inc., New York - Toronto - London, 1958), pp. ix+337
8. A. Erdélyi, W. Magnus, F. Oberhettinger, and F.G. Tricomi, *Higher Transcendental Functions*, Vol. I. Based, in part, on notes left by Harry Bateman, (McGraw-Hill Book Company, Inc., New York - Toronto - London, 1953), pp. xxvi+302
9. A. Erdélyi, W. Magnus, F. Oberhettinger, and F.G. Tricomi, *Higher Transcendental Functions*, Vol. II. Based, in part, on notes left by Harry Bateman, (McGraw-Hill Book Company, Inc., New York - Toronto - London, 1953), pp. xxvi+396
10. A. Fock, *Dokl. Akad. Nauk SSSR* **39**, 279 (1943).
11. H.-J. Glaeske and Vu Kim Tuan, *Serdica Bulgar. Mat. Publ.* **16**, 143 (1990), MR 91k: 44009.
12. H.-J. Glaeske and Vu Kim Tuan, *Boundary Value and Initial Value Problems in Complex Analysis: Studies in Complex Analysis and Its Applications to Partial Differential Equations 1*, (Halle, 1988), pp. 209-220, Pitman Res. Notes Math. Ser. N **256**, 1991, Longman Sci. Tech., Harlow., MR 93h: 42010.
13. H.-J. Glaeske and Vu Kim Tuan, *Math. Nachr.* **152**, 179(1991).
14. G. H. Hardy and E. C. Titchmarsh, *Proc. London Math. Soc.* **35**, 116 (1933).
15. Jr. Hirschman and D. V. Widder *The Convolution Transform*, (Princeton Univ. Press, Princeton, 1955).
16. N. Kesarwani, *Proc. Amer. Math. Soc.* **13**, 950 (1962).
17. N. Kesarwani, *Proc. Amer. Math. Soc.* **14**, 18 (1963).
18. N. Kesarwani, *Proc. Amer. Math. Soc.* **14**, 271 (1963).
19. M. I. Kontorovich and N. N. Lebedev, *Zhurn. Eksperim. i Teoret. Fis.* **8**, 1192 (1938).
20. E. R. Love, *Proc. Edinburgh Math. Soc.* **15**, 169 (1967).
21. O. I. Marichev, *Soviet Math. Dokl.* **13**, 703 (1972).
22. O. I. Marichev and Vu Kim Tuan, Some Volterra equations with the Appell function F_1 in the kernel (Russian). *Scientific Works of the Jubilee Seminar on Boundary Value Questions*, (Collect. Artic., Minsk, 1985.) pp. 167-172.
23. O. I. Marichev and Vu Kim Tuan, *Reports of the Extended Sessions of a Seminar of the I. N. Vekua Institute of Applied Mathematics (Tbilisi, 1985)*, (Tbilisi Gos. Univ., Tbilisi, 1985), pp. 139-142.
24. O. I. Marichev and Vu Kim Tuan, *Complex Analysis and Applications '85* (Varna, 1985), pp. 418-433, (Bulgar. Acad. Sci., Sofia, 1986).
25. Nguyen Xuan Thao, V. A. Kakichev, and Vu Kim Tuan, *East–West J. Math.* **1**, 85 (1998).
26. M. N. Olevskii, *Dokl. Akad. Nauk SSSR* **69**, 11 (1949).
27. A. P. Prudnikov, Yu. A. Brychkov and O. I. Marichev, *Integrals and Series. Vol. 5. Inverse Laplace Transforms*, (Gordon and Breach Science Publishers, New York, 1992), pp. xx+595.
28. M. Saigo, *Math. Rep. Kyushu Univ.* **11**, 135 (1977/78).
29. I. N. Sneddon *Fourier Transforms*, Reprint of the 1951 original, (Dover Publications, Inc., New York, 1995), pp. xii+542.
30. H. M. Srivastava, Vu Kim Tuan, and S. B. Yakubovich, *Quart. J. Math. Oxford Ser.* **51**, 371 (2000).

31. E. C. Titchmarsh, *Eigenfunction Expansions Associated with Second-Order Differential Equations. Part I*, Second Edition, (Clarendon Press, Oxford 1962), pp. vi+203.
32. Vu Kim Tuan, *Dokl. Akad. Nauk BSSR* (Russian) **29**, 584 (1985).
33. Vu Kim Tuan, *Dokl. Akad. Nauk. Armyan SSR* (Russian) **83**, 7 (1986).
34. Vu Kim Tuan, *Dokl. Akad. Nauk SSSR* (Russian) **286**, 521 (1986), translation in *Soviet Math. Dokl*, **33**, 103 (1986).
35. Vu Kim Tuan, *Dokl. Akad. Nauk BSSR* (Russian) **30**, 689 (1986).
36. Vu Kim Tuan, Generalized integral transformations of convolution type in some space of functions. *Complex Analysis and Applications '85* (Varna, 1985), (Bulgar. Acad. Sci., Sofia, 1986) pp. 720-735.
37. Vu Kim Tuan, *Integral Transforms and Their Composition Structure* (Russian), Dr. Sci. Dissertation, (Belarusian State University, Minsk, 1987), p. 275.
38. Vu Kim Tuan, *Dokl. Akad. Nauk SSSR* (Russian) **299**, (1988), translation in *Soviet Math. Dokl.* **37**, 317 (1988).
39. Vu Kim Tuan, *Dokl. Akad. Nauk SSSR* (Russian) **300**, 521 (1988), translation in *Soviet Math. Dokl.* **37**, 669 (1988).
40. Vu Kim Tuan, Some integral transformations with a Macdonald function in the kernel (Russian) *Current Analysis and Its Applications*, (Naukova Dumka, Kiev, 1989), pp. 16-22.
41. Vu Kim Tuan, *Ukrain. Mat. Zh.* (Russian) **42**, 990 (1990), translation in *Ukr. Math. J.* **42**, 880 (1991).
42. Vu Kim Tuan, Airy integral transform and the Paley–Wiener theorem, in *Proceedings of the 2nd International Workshop on Transform Methods and Special Functions, held in Varna, Bulgaria, August 24–29, 1996*, Editors P. Rusev, I. Dimovski and V. Kiryakova, (Institute of Mathematics, Sophia, 1998), pp. 523-531.
43. Vu Kim Tuan, *J. Math. Anal. Appl.* **229**, 519 (1999).
44. Vu Kim Tuan, *Dokl. Akad. Nauk SSSR* (Russian) **313**, 1299 (1990), translation in *Soviet Math. Dokl.* **42**, 150 (1991), MR 91k: 44006.
45. Vu Kim Tuan, O. I. Marichev and S. B. Yakubovich, *Dokl. Akad. Nauk SSSR* (Russian) **286**,786 (1986), translation in *Soviet Math. Dokl.* **33**, 166 (1986).
46. Vu Kim Tuan and Nguyen Thanh Hai, *Dokl. Akad. Nauk SSSR* (Russian) **317**, 797 (1991), translation in *Soviet Math. Dokl.* **43**, 508 (1991).
47. Vu Kim Tuan and S. B. Yakubovich, *Dokl. Akad. Nauk BSSR* (Russian)**29**, 11 (1985), translation in *Amer. Math. Soc. Transl.* **137**, 61 (1987).
48. Vu Kim Tuan and S. B. Yakubovich, *Ukrain. Math. Zh.* (Russian) **44**, 697 (1992), translation in *Ukrainian Math. J.* **44**, 630 (1993).
49. G. N. Watson, *Proc. London Math. Soc. (2)* **35**, 156 (1932).
50. D. V. Widder, *The Laplace Transform*, (Princeton Univ. Press, Princeton, 1946).
51. J. Wimp, *Proc. Edinburgh Math. Soc. (2)* **14**, 33 (1964).
52. S. B. Yakubovich, *Izv. Vyssh. Uchevn. Zaved Mat.* (6), 77-79 (1986).
53. S. B. Yakubovich, *Index Transforms*, (World Scientific Publishing Co., Inc., River Edge, NJ, 1996), pp. xiv+248.
54. S. B. Yakubovich, Vu Kim Tuan., O. I. Marichev , and S. L. Kalla, *Rev. Tecn. Fac. Ingr. Univ. Zulia, Edicion Especial* **10**, 105 (1987).

Chapter II

PARTIAL DIFFERENTIAL EQUATIONS

Advances in Deterministic and Stochastic Analysis
Eds. N. M. Chuong *et al.* (pp. 75–89)

§5. UNIFIED MINIMAX METHODS

MARTIN SCHECHTER

Department of Mathematics University of California, Irvine
Irvine, CA 92697-3875
E-mail: mschecht@math.uci.edu

The variational approach to solving nonlinear problems eventually leads to the search for critical points of related functionals. In case of semibounded functionals, one can look for extrema. Otherwise, one is forced to use minimax methods. There are several approaches to such methods. In this paper we unify these approaches providing one theory that works for all of them. The usual approach has used Palais Smale sequences. We show that all of them lead to Cerami sequences as well. Applications are given.

1. Introduction

Many problems arising in science and engineering call for the solving of the Euler equations of functionals, i.e., equations of the form

$$G'(u) = 0,$$

where G(u) is a C^1 functional (usually representing the energy) arising from the given data. The classical approach was to look for maxima or minima. If one is looking for a minimum, it is not sufficient to know that the functional G(u) is bounded from below. However, if it is bounded from below, one can show that there is a sequence satisfying

$$G(u_k) \to a, \quad G'(u_k) \to 0 \tag{1}$$

for $a = \inf G$. If the sequence has a convergent subsequence, this will produce a minimum.

A functional G is said to satisfy the PS condition if every sequence of the form (3) has a convergent subsequence and consequently produces a critical point.

AMS Subject Classification: Primary 35J65, 58E05, 49B27.
Key words and phrases: Critical point theory, variational methods, saddle point theory, semilinear differential equations.

Actually, one can do better. If the functional G(u) is bounded from below, then there exists a sequence satisfying

$$G(u_k) \to a, \quad (1 + \|u_k\|)G'(u_k) \to 0 \tag{2}$$

for $a = \inf G$. Such a sequence is called a Cerami sequence. It will have a convergent subsequence even in cases when a PS sequence does not.

However, when the functional is not semi-bounded, there is no organized procedure for producing critical points. One approach is to search for circumstances other than semibound- edness which will produce PS sequences (3). A successful method has been to find subsets A and B having the property that when they separate a functional, a PS sequence (3) results. Specifically, we want the subsets to have the property that for every $G \in C^1(E, \mathbb{R})$ bounded on bounded sets and satisfying

$$a_0 := \sup_A G < b_0 := \inf_B G, \tag{3}$$

there is a sequence $\{u_k\} \subset E$ and a constant a such that

$$b_0 \leq a < \infty \tag{4}$$

and (3) holds. When this is true, we say that A links B. Fortunately, such pairs exist and help solve many problems.

There have been several approaches to linking;[2,6,7,13,14] we describe them below. Here we shall present an approach which includes all of them, is more general and is easy to apply.

The idea is as follows. For each $A \subset E$ one wishes to find a collection $\mathcal{K}$ of sets K with the following properties.

1. If $G \in C^1(E, \mathbb{R})$ satisfies

$$a_0 := \sup_A G < a := \inf_{K \in \mathcal{K}} \sup_K G, \tag{5}$$

then there is a sequence $\{u_k\} \subset E$ such that (3) holds.

2. If

$$a_0 \leq \sup_{K \setminus A} G, \quad K \in \mathcal{K}, \tag{6}$$

then there is a sequence $\{u_k\} \subset E$ such that (3) holds.

3. If A, $B \subset E$ satisfy

$$A \cap B = \phi, \quad B \cap K = \phi, \quad K \in \mathcal{K} \tag{7}$$

and

$$a_0 := \sup_A G \leq b_0 := \inf_B G, \tag{8}$$

then there is a sequence $\{u_k\} \subset E$ such that (3) holds.

4. If

$$g_K = \{v \in K \backslash A : G(v) \geq a_0\} \neq \phi, K \in \mathcal{K}, \tag{9}$$

then there is a sequence $\{u_k\} \subset E$ such that (3) holds.

In the next two sections we determine collections $\mathcal{K}$ that have these properties. Moreover, we shall show that all of these collections produce not only PS sequences (3), but also Cerami sequences (4). In Sections 4–6 we describe known methods of linking and find the sets $\mathcal{K}$ in each case. In Section 7 we give some applications.

2. Definitions and Theorems

In this section we begin with a Banach space E.

Definition 2.1. We shall say that a map φ : E $\to$ E is of class Λ, if it is a homeomorphism onto E, and both φ, $\varphi^{-1}$are bounded on bounded sets.

Definition 2.2. For A $\subset$ E, we define

$$\Lambda(A) = \{\varphi \in \Lambda : \varphi(u) = u, \quad u \in A\}.$$

Definition 2.3. For a nonempty set $A \subset E$, we define a nonempty collection $\mathcal{K} = \mathcal{K}(A)$ of subsets $\mathcal{K} \subset E$ to be a *minimax system* for A if it has the following property:

$$\varphi(K) \in \mathcal{K}, \quad \varphi \in \Lambda(A), K \in \mathcal{K}.$$

Every nonempty set has a minimax system. We have

Theorem 2.1. *Let K be a minimax system for a nonempty subset A of E, and let $G(u)$ be a C^1 functional on E. Define*

$$a := \inf_{K \in \mathcal{K}} \sup_K G, \tag{10}$$

and assume that a is finite and satisfies

$$a > a_0 := \sup_A G. \tag{11}$$

Let $\psi(t)$ be a positive, locally Lipschitz continuous function on $[0, \infty)$ such that

$$\int_0^\infty \psi(r)dr = \infty. \tag{12}$$

Then there is a sequence $\{u_k\} \subset E$ such that

$$G(u_k) \to a, G^0(u_k)/\psi(\|u_k\|) \to 0. \tag{13}$$

Definition 2.4. We shall call $\sigma(t) \in C([0,T] \times E, E)$, $T > 0$, a *flow* if $\sigma(0) = I$ and for each $t \in [0,T]$, $\sigma(t)$ is a homeomorphism of E onto itself.

Theorem 2.2. *Let K be a minimax system for a nonempty set A and let $G(u)$ be a C^1 functional on E such that*

$$g_K = \{v \in K \backslash A : G(v) \geq a_0\} \neq \phi, K \in \mathcal{K}, \tag{14}$$

and

$$a = \inf_{\mathcal{K}} \sup_{K} G < \infty. \tag{15}$$

Assume that for any $b > a$, $K \in \mathcal{K}$ and flow $\sigma(t)$ satisfying

$$G(\sigma(t)v) < b, v \in K, t > 0, \tag{16}$$

and

$$G(\sigma(t)u) < a, u \in A, t > 0, \tag{17}$$

there is a $\tilde{K} \in \mathcal{K}$ such that there is a $\tilde{K} \in \mathcal{K}$ such that

$$\tilde{K} \subset \bigcup_{t\in[0,t]} \sigma(t)A \cup \sigma(T)[E_b \cup K], \tag{18}$$

where

$$E_\alpha = \{u \in E : G(u) \leq \alpha\}. \tag{19}$$

Then the conclusions of Theorem 2.1 hold.

3. Linking Subsets

We now show how linking can play a major role in finding critical points.

Definition 3.1. We shall say that a set A in E *links a set* $B \subset E$ *relative to a minimax* system $\mathcal{K}$ for A if

$$A \cap B = \phi \tag{20}$$

and

$$B \cap K \neq \phi, \quad K \in K. \tag{21}$$

We shall say that A *links* B *[mm]* if there is a minimax system $\mathcal{K}$ for A such that A links B relative to $\mathcal{K}$.

Theorem 3.1. *Let K be a minimax system for a nonempty set A, and assume that there is a subset B of E such that A links B relative to $\mathcal{K}$. Assume that $G \in C^1(E,\mathbb{R})$ satisfies*

$$a_0 := \sup_A G < b_0 := \inf_B G, \tag{22}$$

and that the quantity a given by (1) is finite. Then for each positive, locally Lipschitz continuous function $\psi(t)$ on $[0,\infty)$ satisfying (12), there is a sequence $\{u_k\} \subset E$ such that (13) holds.

Definition 3.2. A subset A of a Banach space E *links* a subset B of E if for every $G \in C^1(E,\mathbb{R})$ bounded on bounded sets and satisfying (22) there is a sequence $\{u_k\} \subset E$ and a constant a such that

$$b_0 \leq a < \infty \tag{23}$$

and (3) holds.

Definition 3.3. A subset A of a Banach space E links a subset B of E strongly if for every $G \in C^1(E,\mathbb{R})$ bounded on bounded sets and satisfying (22) and each positive, locally Lipschitz continuous function $\psi(t)$ on $[0,\infty)$ satisfying (12), there is a sequence $\{u_k\} \subset E$ such that (10) and (13) hold.

Theorem 3.2. *If A links B [mm], then it links B strongly.*

Theorem 3.3. *Let $\mathcal{K}$ be a minimax system for a nonempty set A, and assume that A links a subset B of E relative to $\mathcal{K}$. Let G be a C^1 functional satisfying (14), (15) and (24)*

$$a_0 := \sup_A G \leq b_0 := \inf_B G. \tag{24}$$

Assume that for any $b > a$, $K \in \mathcal{K}$ and flow $\sigma(t)$ satisfying (16), (17) and such that

$$\sigma(t)A \cap B = \phi, t \in [0,T], \tag{25}$$

there is a $\tilde{K} \in \mathcal{K}$ such that (18) holds. Then the conclusions of Theorem 2.1 hold.

Theorem 3.4. *Let $\mathcal{K}$ be a minimax system for a nonempty set A satsfying*

$$K \backslash A = \phi, \quad K \in \mathcal{K}. \tag{26}$$

Let $G \in C^1$ satisfy (14) and (15). Assume that for any $K \in \mathcal{K}$ and flow $\sigma(t)$ satisfying (16), (17), one has $S(K) \in \mathcal{K}$, where

$$S(u) = \sigma(d(u,A))u, u \in E. \tag{27}$$

Assume also that for each $K \in \mathcal{K}$ there is a $\rho > 0$ such that $G(u)$ is Lipschitz continuous on

$$K_\rho = \{u \in K : d(u, A) \leq \rho\}. \tag{28}$$

Then the conclusions of Theorem 2.1 hold.

Theorem 3.5. *Let $\mathcal{K}$ be a minimax system for a nonempty set A, and assume that there is a subset B of E such that A links B relative to $\mathcal{K}$. Assume for any $K \in \mathcal{K}$ and flow $\sigma(t)$ satisfying (16), (17), one has $S(K) \in \mathcal{K}$, where $S(u)$ is given by (27). Let G be a C^1 functional satisfying (15) and(24). Assume also that for each $K \in \mathcal{K}$ there is a $\rho > 0$ such that $G(u)$ is Lipschitz continuous on K_ρ given by (28). Then the conclusions of Theorem 2.1 hold.*

Corollary 3.1. *Let $\mathcal{K}$ be a minimax system for a nonempty set A, and let $G \in C^1$ satisfy (14) and (15). Assume that for any $b > a$, $K \in \mathcal{K}$ and flow $\sigma(t)$ satisfying (16),(17), and (25) for*

$$B = \{u \in E \backslash A : G(u) \geq a_0\}, \tag{29}$$

there is a $\tilde{K} \in \mathcal{K}$ satisfying $\tilde{K} \subset \sigma(T)E_b$. Then the conclusions of Theorem 2.1 hold.

We present some linking methods which are special cases of linking with respect to minimax systems.

4. A Method Using Homeomorphisms

One linking method can be described as follows. Let Ebe a Banach space and let Φ be the set of all continuous maps $\Gamma = \Gamma(t)$ from $E \times [0,1]$ to E such that

1. $\Gamma(0) = I$, the identity map.
2. For each $t \in [0,1), \Gamma(t)$ is a homeomorphism of Eonto $E$and $\Gamma^{-1}(t) \in C$ $(E \times [0,1), E)$.
3. $\Gamma(1)E$is a single point in $E$and $\Gamma(t)A$converges uniformly to $\Gamma(1)E$as $t \to 1$ for each bounded set $A \subset E$.
4. For each $t_0 \in [0,1)$ and each bounded set $A \subset E$

$$\sup_{\substack{0 \leq t \leq t_0 \\ u \in A}} \{\|\Gamma(t)u\| + \|\Gamma^{-1}(t)u\|\} < \infty. \tag{30}$$

We have

Theorem 4.1. *Let G be a C^1-functional on E, and let A be a bounded subset of E. Assume that*

$$a_0 := \sup G < a := \inf_{\Gamma \in \Phi} \sup_{\substack{0 \le s \le 1 \\ u \in A}} G(\Gamma(s)u) < \infty. \tag{31}$$

Let $\psi(t)$ be a positive, locally Lipschitz continuous function on $[0, \infty)$ satisfying (12). Then there is a sequence $\{u_k\} \subset E$ such that (13) holds. In particular, there is a Cerami sequence satisfying (4).

Proof. In this case, we let $\mathcal{K}$ be the collection of sets of the form

$$K = \{\Gamma(t)u : t \in [0, 1], u \in A, \Gamma \in \Phi\}. \tag{32}$$

To show that this is a minimax system, let φ be a mapping in $\Lambda(A)$, and let $\Gamma(t)$ be in Φ. Define the mapping

$$\Gamma_1(t)u = \varphi \circ \Gamma(t) \circ \varphi^{-1}u, \quad u \in E.$$

Then $\Gamma_1(t) \in \Phi$. Moreover,

$$\Gamma_1(t)u = \varphi \circ \Gamma(t)u, \quad u \in A.$$

Consequently,

$$\Gamma_1(t)A = \varphi(\{\Gamma(t)u : u \in A\}),$$

showing that the collection $K$is a minimax system. Since (31) now becomes (3), the result follows. □

Definition 4.1. We say that A links $B[st]$ if A, B are subsets of E such that $A \cap B = \phi$ and, for each $\Gamma(t) \in \Phi$, there is a $t \in (0, 1]$ such that $\Gamma(t)A \cap B \neq \phi$.

We have

Corollary 4.1. *If A links $B[st]$, then it links $B[mm]$.*

We also have

Theorem 4.2. *Let A, B be subsets of E such that A links $B[st]$, and let G be a C^1 -functional on E, satisfying*

$$a_0 := \sup_A G \le b_0 := \inf_B G. \tag{33}$$

Assume that

$$a := \inf_{\Gamma \in \Phi} \sup_{\substack{0 \le s \le 1 \\ u \in A}} G(\Gamma(s)u) \tag{34}$$

is finite. Let $\psi(t)$ be a positive, locally Lipschitz continuous function on $[0, \infty)$ satisfying (12). Then there is a sequence $\{u_k\} \subset E$ such that (13) holds. In particular, there is a Cerami sequence satisfying (4).

Proof. The theorem follows from Theorem 3.3. If

$$K = \{\Gamma(t)u : t \in [0,1], u \in A\} \tag{35}$$

for some $\Gamma \in \Phi$, and $\sigma(t)$ is any flow, let

$$\begin{aligned} \tilde{\Gamma}(s) &= \sigma(2Ts), \quad 0 \le s \le \frac{1}{2}, \\ &= \sigma(T)\Gamma(2s-1), \quad \frac{1}{2} < s \le 1. \end{aligned}$$

It is easily checked that $\tilde{\Gamma} \in \Phi$. Morever,

$$\tilde{K} \subset \bigcup_{t \in [0,T]} \sigma(t)A \cup \sigma(T)K. \tag{36}$$

Hence, (2.9) is satisfied. □

5. A Method Using Metric Spaces

Another general procedure is described in Mawhin-Willem[7] and Brezis-Nirenberg[2] as follows. One finds a compact metric space Σ and selects a closed subset Σ^* of Σ such that $\Sigma^* \ne \phi$, $\Sigma^* \ne \Sigma$. One then picks a map $p^* \in C(\Sigma, E)$ and defines

$$A = \{p \in C(\Sigma, E) : p = p^* \text{ on } \Sigma^*\}$$

$$a = \inf_{p \in \mathcal{A}} \max_{\xi \in \Sigma} G(p(\xi)).$$

They assume

(A) For each $p \in \mathcal{A}$, $\max_{\xi \in \Sigma} G(p(\xi))$ is attained at a point in $\Sigma \backslash \Sigma^*$.

They then prove

Theorem 5.1. *Under the above hypotheses there is a sequence satisfying (1.1).*

We shall prove

Theorem 5.2. *Under the same hypotheses, let $\psi(t)$ be a positive, locally Lipschitz continuous function on $[0, \infty)$ satisfying (12). Then there is a sequence $\{u_k\} \subset E$ such that (13) holds. In particular, there is a Cerami sequence satisfying (4).*

Proof. If $a > a_0$, this follows from Theorem 2.1. In fact, we can take $A = p^*(\Sigma^*)$ and $\mathcal{K}$ as the collection of sets

$$K = \{p(\xi) : \xi \in \Sigma, \quad p \in A\}.$$

If $\varphi \in \Lambda(A)$, then

$$\varphi(p(\xi)) = \varphi(p^*(\xi)) = p^*(\xi), \quad \xi \in \Sigma^*.$$

Thus, $\varphi(p(\xi)) \in \mathcal{K}$. Hence, $\mathcal{K}$ is a minimax system for A, and we can apply Theorem 2.1 to come to the desired conclusion.

Now assume $a_0 = a$. Since each $K \in \mathcal{K}$ is compact, the same is true of K_ρ given by (28) for each $\rho > 0$. Moreover, if $\sigma(t) \in C(R^+ \times E, E)$ is such that $\sigma(0) = I$, one has $S(K) \in \mathcal{K}$, where

$$S(p(\xi)) = \sigma(d(p(\xi), A))p(\xi), \quad \xi \in \Sigma.$$

The result follows from Theorem 3.5. □

6. A Method Using Homotopy Stable Families

Another approach is found in Ghoussoub.[6] The author condiders a closed subset A of E, and a collection $\mathcal{K}$ of compact subsets of E having the property that if $K \in \mathcal{K}$ and $\psi \in C([0,1] \times E, E)$ satisfies

$$\psi(0)u = u, \quad u \in E,$$

and

$$\psi(t)u = u, \quad t \in [0,1], \quad u \in A,$$

then $\psi(1)K \in \mathcal{K}$. He calls such a collection a homotopy-stable family with extended boundary A. The author proves

Theorem 6.1. *If $\mathcal{K}$ is a homotopy-stable family with extended closed boundary A, B is a closed subset of E satisfying (9) and G is a C^1 functional which satisfies*

$$\sup_A G \le a \le \inf_B G,$$

where the quantity a is given by (1), then there is a PS sequence satisfying (3).

We shall prove

Theorem 6.2. *Under the same assumptions, for each function $\psi(t)$ satisfying the hypotheses of Theorem 2.1 there is a sequence satisfying (13). In particular, there is a Cerami sequence satisfying (4).*

Proof. It is easy to show that such a collection $\mathcal{K}$ is indeed a minimax system. Indeed, if $K \in \mathcal{K}$, and $\varphi \in \Lambda(A)$, then

$$\psi(t)u = t\varphi(u) + (1-t)u$$

satisfies the stipulations above, and consequently $\varphi(K) \in \mathcal{K}$. Since each of the members of $\mathcal{K}$ are compact, it follows that every C^1 functional is Lipschitz continuous on some set K_ρ defined by (28), provided $\rho > 0$ is sufficiently small. Moreover, if $\sigma(t) \in C(R^+ \times E, E)$ is such that $\sigma(0) = I$, one has $S(K) \in \mathcal{K}$, where

$$S(u) = \sigma(d(u,A))u, \quad u \in E.$$

This follows from the fact that $S(t)u = \sigma(td(u,A))u$ is in $\in C([0,1] \times E, E)$ and satisfies

$$S(0)u = u, \quad u \in E,$$

and

$$S(t)u = u, \quad t \in [0,1], \quad u \in A.$$

Consequently, $S(K) = S(1)K \in \mathcal{K}$ by hypothesis. It now follows that the conclusion of Theorem 3.5 holds. □

Note: It is not required to have a satisfy

$$a \le \inf_B G.$$

If it does, one obtains additional information as described in Ref. 6.

Definition 6.1. We say that A links $B[gh]$ if A, B are subsets of E satisfying the hypotheses of Theorem 6.1.

7. Some Applications

We now apply the theorems of the preceding sections to various geometries in Banach space. As before, we assume that $G \in C^1(E, \mathbb{R})$ and that satisfies the hypotheses of Theorem 2.1.

Theorem 7.1.

$$G(0) \leq \alpha \leq G(u), u \in \partial B_\delta \tag{37}$$

and that there is a $\varphi_0 \in \partial B_1$ *such that*

$$G(R\varphi_0) \leq \gamma, \quad R > 0. \tag{38}$$

Then for each function $\psi(t)$ *satisfying the hypotheses of Theorem 2.1, there is a sequence* $\{u_k\} \subset E$ *such that*

$$G(u_k) \to c, \quad \alpha \leq c \leq \gamma, \quad G'(u_k)/\psi(\|u_k\|) \to 0. \tag{39}$$

Many elliptic semi-linear problems can be described in the following way. Let Ω be a domain in $\mathbb{R}^n$, and let A be a self-adjoint operator on $L^2(\Omega)$. We assume that $A \geq \lambda_0 > 0$ and that

$$C_0^\infty(\Omega) \subset D := D(A^{1/2}) \subset H^{m,2}(\Omega) \tag{40}$$

for some $m > 0$, where $C^\infty(\Omega)$ denotes the set of test functions in Ω (i.e., infinitely differentiable functions with compact supports in Ω), and $H^{m,2}(\Omega)$ denotes the Sobolev space. If m is an integer, the norm in $H^{m,2}(\Omega)$ is given by

$$\|u\|_{m,2} := \left(\sum_{|\mu| \leq m} \|D^\mu u\|^2 \right)^{1/2}.$$

Here D^μ represents the generic derivative of order $|\mu|$ and the norm on the right hand side of (7) is that of $L^2(\Omega)$. We shall not assume that m is an integer.

Let q be any number satisfying

$$\begin{array}{ll} 2 \leq\leq 2n/(n-2m), & 2m < n \\ 2 \leq q < \infty, & n \leq 2m, \end{array} \tag{41}$$

and let $f(x,t)$ be a Carathéodory function on $\Omega \times \boldsymbol{R}$. This means that $f(x,t)$ is continuous in t for a.e. $x \in \Omega$ and measurable in x for every $t \in \boldsymbol{R}$. Throught this section we make the following assumptions:

(A) The function $f(x,t)$ satisfies

$$|f(x,t)| \leq V_0(x)^q |t^{q-1} + V_0(x) W_0(x) \tag{42}$$

and

$$f(x,t)/V_0(x)^q = o(|t|^{q-1}) \text{ as } |t| \to \infty, \tag{43}$$

where $V_0(x) > 0$ is a function such that

$$\|V_0 u\|_q \le C\|u\|_D, \quad u \in D \tag{44}$$

and W_0 is a function in $L^{q^0}(\Omega)$. Here

$$\|u\|_q := (|u(x)|^q dx)^{1/q}, \tag{45}$$

$$\|u\|_D := \|A^{1/2}u\| \tag{46}$$

and $q' = q/(q-1)$. If Ω and $V_0(x)$ are bounded, then (44) will hold automatically by the Sobolev inequality. However, there are functions $V_0(x)$ which are unbounded and such that (44) holds even on unbounded regions Ω. With the norm (46), D becomes a Hilbert space. Define

$$F(x,t) := \int_0^t f(x,s)ds \tag{47}$$

and

$$G(u) := \|u\|_D^2 - 2\int_\Omega F(x,u)dx. \tag{48}$$

It follows that G is a continuously differentiable functional on the whole of D (cf., e.g., Ref. 11).

For our first result, we assume further that

$$H(x,t) = 2F(x,t) - tf(x,t) \ge -W_1(x) \in L^1(\Omega), \quad x \in \Omega, \quad t \in \mathbb{R} \tag{49}$$

and

$$H(x,t) \to \infty \text{ a.e. as } |t| \to \infty. \tag{50}$$

Moreover, we assume that there are functions $V(x), W(x) \in L^2(\Omega)$ such that multiplication by $V(x)$ is a compact operator from D to $L^2(\Omega)$ and

$$F(x,t) \le C(V(x)^2|t|^2 + V(x)W(x)|t|) \tag{51}$$

We wish to obtain a solution of

$$Au = f(x,u), \quad u \in D. \tag{52}$$

By a solution of (52) we shall mean a function $u \in D$ such that

$$(u,v)_D = (f(\cdot,u),v), \quad v \in D. \tag{53}$$

If $f(x,u)$ is in $L^2(\Omega)$, then a solution of (53) is in $D(A)$ and solves (52) in the classical sense. Otherwise we call it a weak (or semi-strong) solution. We have

Theorem 7.2. *Assume that λ_0 is an eigenvalue of A with eigenfunction φ_0. Assume also*

$$2F(x,t) \leq \lambda_0 t^2, \quad |t| \leq \delta \text{ for some } \delta > 0 \tag{54}$$

and

$$2F(x,t) \geq \lambda_0 t^2 - W_1(x), \quad t > 0, x \in \Omega, \tag{55}$$

where $W_1 \in L^1(\Omega)$. Then (52) has a solution $u = 0$.

Proof. Under the hypotheses of the theorem, it was shown in Theorem 3.2.1 of Ref. 11 that the following alternative holds:

Either

(a) there is an infinite number of $y(x) \in D(A)\backslash\{0\}$ such that

$$Ay = f(x,y) = \lambda_0 y \tag{56}$$

or

(b) for each $\rho > 0$ sufficiently small, there is an $\varepsilon > 0$ such that

$$G(u) \geq \varepsilon, \quad \|u\|_D = \rho. \tag{57}$$

We may assume that option (*b*) holds, for otherwise we are done. By (55) we have

$$G(R\varphi_0) \leq R^2(\|\varphi_0\|_D^2 - \lambda_0\|\varphi_0\|) + \int_\Omega W_1(x)dx = \int_\Omega W_1(x)dx = B_1.$$

Thus (38) holds. By Theorem 7.1, there is a sequence satisfying (39). Taking $\psi(r) = 1/(r+1)$, we conclude that there is a sequence $\{u_k\} \subset D$ such that

$$G(u_k) \to c, \quad \varepsilon \leq c \leq B_1, \quad (1 + \|u_k\|_D)G^0(u_k) \to 0. \tag{58}$$

In particular, we have

$$\|u\|_k\|_D^2 - 2\int_\Omega F(x,u_k)dx \to c \tag{59}$$

and

$$\|u_k\|_D^2 - (f(\cdot, x_k), u_k) \to 0. \tag{60}$$

Consequently

$$\int_\Omega H(x,u_k)dx \to -c. \tag{61}$$

These imply

$$\int_\Omega H(x, u_k)dx \le K. \tag{62}$$

If $\rho_k = \|u_k\|_D \to \infty$, let $\tilde{u}_k = u_k/\rho_k$. Then $\|\tilde{u}_k\|_D = 1$. Consequently there is a renamed subsequence such that $\tilde{u}_k \to \tilde{u}$ weakly in D, strongly in $L^2(\Omega)$ and a.e. in Ω. We have by (51)

$$1 \le (m_1 + \delta)/\rho_k^2 + 2C \int_\Omega \{V(x)^2 \tilde{u}_k^2 + V(x)W(x)|\tilde{u}_k|\rho_k^{-1}\}dx.$$

Consequently

$$1 \le 2C \int_\Omega V(x)^2 \tilde{u}_k^2 dx. \tag{63}$$

This shows that $\tilde{u} \not\equiv 0$. Let Ω_0 be the subset of Ω on which $\tilde{u} \neq 0$. Then

$$|u_k(x)| = \rho_k |\tilde{u}_k(x)| \to \infty, \quad x \in \Omega_0. \tag{64}$$

If $\Omega_1 = \Omega \backslash \Omega_0$, then we have

$$\int_\Omega H(x, u_k)dx = \int_{\Omega_0} + \int_{\Omega_1} \ge \int_{\Omega_0} H(x, u_k)dx - \int_{\Omega_1} W_1(x)dx \to \infty. \tag{65}$$

This contradicts (62), and we see that $\rho_k = \|u_k\|_D$ is bounded. Once we know that the ρ_k are bounded we can apply Theorem 3.4.1 of Ref. 11 to obtain the desired conclusion. □

Remark 7.1. It should be noted that the crucial element in the proof of Theorem 7.2 was (60). If we had been dealing with an ordinary PalaisSmale sequence, we could only conclude that

$$\|u_k\|_D^2 - (f(, u_k), u_k) = o(\rho_k),$$

which would imply only

$$H(x, u_k)dx = o(\rho_k).$$

This would not contradict (65), and the argument would not go through.

References

1. P. Bartolo, V. Benci and D. Fortunato, *Nonlinear Analysis TMA*, **7**, 981 (1983).
2. H. Brezis and L. Nirenberg, *Comm. Pure Appl. Math.* **44**, 939 (1991).
3. V. Benci and P. H. Rabinowitz, *Invent. Math.* **52**, 241 (1979).
4. K. C. Chang, *Infinite dimensional Morse theory and multiple solution problems*, (Birkhauser, Boston, 1993).

5. C. Corduneanu, *Principles of Differential and Integral Equations*, (Chelsea, N.Y., 1977).
6. N. Ghoussoub, *Duality and Perturbation Methods in Critical Point Theory*, (Combridge Tracts in Mathemtics, 107,1993).
7. J. Mawhin and M. Willem, *Critical Point Theory and Hamiltonian Systems*, (Springer-Verlag, 1989).
8. L. Nirenberg, *Bull. Amer. Math. Soc.* **4** 267 (1981).
9. P. H. Rabinowitz, Minimax methods in critical point theory with applications to differential equations, *Conf. Board of Math. Sci. Reg. Conf. Ser. in Math. No. 65, Amer. Math. Soc. 1986.*
10. M. Schechter, *Pacific J. Math.* **171**, 529 (1995).
11. M. Schechter, *Rend. Sem. Mat. Univ. Padova* **99**, 255 (1998).
12. M. Schechter, *Linking Methods in Critical Point Theory*, (Birkhauser Boston, 1999).
13. E. A. de B. e. Silva, *Nonlinear Analysis TMA* **16**, 455 (1991).
14. M. Schechter and K. Tintarev, *Bull. Soc. Math. Belg.* **44**, 249 (1992).[15] K. Tintarev, Isotopic linking and critical points of functionals, in *Proceedings of the Second World Congress of Nonlinear Analysts*, Part 7 (Athens, 1996), *Nonlinear Anal.* **30**, 4145 (1997).
15. M. Willem, *Minimax Theorems*, (Birkhauser, 1996).

Advances in Deterministic and Stochastic Analysis
Eds. N. M. Chuong *et al.* (pp. 91–102)

§6. SOME REMARKS ON SINGLE CONSERVATION LAWS

MIKIO TSUJI

Department of Mathematics, Kyoto Sangyo University,
Kita-Ku, Kyoto 603-8555, Japan
mtsuji@cc.kyoto-su.ac.jp

PETER WAGNER

Institut für Technische Mathematik,
Geometrie und Bauinformatik, Universität Innsbruck,
Technikerstr. 13, A-6020 Innsbruck, Austria

First we consider the Cauchy problem for Burgers equation whose initial function is x^2, and study the existence domain of a piecewise smooth weak solution of the Cauchy problem. Then the existence domain of its real-valued solution is different from that of the complex-valued solution. Next we consider the Cauchy problem for a single conservation law which does not satisfy the convexity condition, and study the regularity of weak solutions of the Cauchy problem. Our question is whether the weak solutions become piecewise smooth, and whether "rarefaction waves" and "contact discontinuities" would really appear in the weak solutions. In this talk we solve an example as a tentative trial and construct its piecewise smooth weak solution by the geometric method.

1. Introduction

We consider the Cauchy problem for an equation of conservation law in one space dimension as follows:

$$\frac{\partial u}{\partial t} + \frac{\partial}{\partial x} f(u) = 0 \quad \text{in} \quad \{t > 0, x \in \mathbf{R}^1\}, \tag{1}$$

$$u(0, x) = \phi(x) \quad \text{on} \quad \{t = 0, x \in \mathbf{R}^1\} \tag{2}$$

where $f = f(u)$ is of class C^∞. We assume that the initial functions $\phi = \phi(x)$ are also sufficiently smooth. The global existence of weak solutions of (1)-(2) has been well studied. As there exist too many papers on this subject, we do not mention anything on references. In this talk we will study the regularity of weak solutions of (1)-(2) and their domains of existence. We think that the most typical phenomenon in nonlinear hyperbolic equations is the appearance of "shock wave". Moreover the shock is defined only for the piecewise smooth weak solution. This is the reason why we pay much attention on the existence of "piecewise smooth weak solution". Before considering the Cauchy problem (1)-(2), we state our program for nonlinear hyperbolic equations and systems. Our first step is to lift the equation into higher dimensional space so that the singularities of solution would disappear. Next we construct a global solution of the lifted equation in the higher dimensional space. We call this as a "geometric solution" of the original problem. Finally we project the geometric solution to the base space and construct a single valued weak solution which satisfies certain additional condition called as the "entropy condition". For single first order partial differential equations, we have already proved in Refs. 9,10 and 11 that, if the equations are convex, our program is correct. But we do not yet succeed in the realization of our program for non-convex first order partial equations. For second order nonlinear hyperbolic equations and p-systems, we could not construct weak solutions by our method.[14,15] By these investigations, we have now some questions on the mathematical formulation for nonlinear hyperbolic equations and systems. Our final aim is to show that our program is true for all kinds of hyperbolic problems.

In this talk we will first consider the Cauchy problem where an initial function is analytic and not bounded, and study an existence domain of solution. In analytic case, the complex method has been effective especially for linear partial differential equations. Therefore we will study, in Sec. 2, what would happen if we might accept a complex-valued solution. for linear partial differential equations. Therefore we will study, in Sec. 2, what would happen if we might accept a complex-valued solution.

Next we will construct piecewise smooth weak solutions of (1)-(2). In the case where $f = f(u)$ is convex, we think that we could understand the mechanics of the "propagation of singularities". See Refs. 9–11, and 12. Therefore we will consider the case where $f''(u)$ changes the sign. This problem has been studied by J. Guckenheimer,[1] G. Jennings,[3] S. Izumiya and G. T. Kossioris,[2] etc. In these papers they have used a "rarefaction wave" and a "contact discontinuity", because they have reduced the starting problem to the Cauchy problem whose initial functions have jump discontinuities. Our question is on this reduction. If our program written in the above might be accepted, we would not need to introduce the rarefaction waves and contact discontinuities for the construction of weak solutions of (1)-(2), because a family of characteristic curves covers the whole space. As we do not yet arrive at the final conclusion on this problem, we have tried to solve a specific problem in an exact form as a tentative trial. This is the subject of the second part of this talk.

2. Existence Domains of Solutions for Burgers Equation

In this section we consider the following Cauchy problem:

$$\begin{cases} \dfrac{\partial u}{\partial t} + \dfrac{\partial}{\partial x}(\dfrac{1}{2}u^2) = 0, \ (t,x) \in \mathbf{R}^2 \\ \\ u(0,x) = x^2, \qquad\qquad x \in \mathbf{R}^1 \end{cases}, \tag{3}$$

and study an existence domain of a piecewise smooth weak solution of (3). What we would like to show in this section is the following: If we may accept only a real-valued solution, then the solution does not exist in the whole space $\mathbf{R}^2$ and its existence domain is surrounded by an envelope of a family of characteristic curves. But, if we may accept a complex-valued solution, then the Cauchy problem (3) admits a piecewise smooth weak solution defined in the whole space $\mathbf{R}^2$. First we solve the Cauchy problem (3) by the characteristic method. As a system of characteristic differential equations is written by

$$\dot{x} = u, \qquad \dot{u} = 0$$

whose initial condition is

$$x(0) = \xi, \qquad u(0) = \xi^2,$$

a family of characteristic curves is given by

$$x = \xi + \xi^2 t, \qquad u = \xi^2, \qquad \xi \in \mathbf{R}^1. \tag{4}$$

Here we denote $D_\pm = \{(t,x); 4tx + 1 \gtrless 0\}$ and $C = \{(t,x); 4tx + 1 = 0\}$. Then we can easily see that $D_+ \cup C = \cup_{\xi \in \mathbf{R}^1}\{(t,x); x = \xi + \xi^2 t\}$, and that the curve C is obtained as an envelope of the characteristic curves $\{(t,x); x = \xi + \xi^2 t\}$ where ξ is a parameter moving in $\mathbf{R}^1$. Solving the first equation of (4) with respect to ξ, we get the solution of (3) by

$$u(t,x) = 4x^2/(1 + \sqrt{4tx+1})^2 \tag{5}$$

in a neighbourhood of the initial line $\{t = 0\}$. Though the right hand side of (5) is a continuous function defined in the whole space $\mathbf{R}^2$, it is not differentiable along the curve C and it takes complex values in the domain D_-. Therefore our problem is how to extend the solution (5) beyond the curve C as a real-valued piecewise smooth weak solution of (3).

Proposition 2.1. *Assume that the solution (5), written by $u = u(t,x)$, can be extended as a piecewise smooth real-valued weak solution of (3) beyond the curve C. Then $u(t,x)$ does not have a jump discontinuity along the curve C.*

Proof. Assume that $u = u(t,x)$ has a jump discontinuity along the curve C. Here we write the curve C as $x = \gamma(t) = -1/4t$. Then it holds $[u]\dot{\gamma} = [u^2/2]$ where $[u] = u(t,\gamma(t)+0) - u(t,\gamma(t)-0)$. As $u(t,\gamma(t)+0) = 1/4t^2$ and $\dot{\gamma} = 1/4t^2$, we get $u(t,\gamma(t)-0) = 1/4t^2$, that is to say, $u(t,\gamma(t)+0) = u(t,\gamma(t)-0)$. Hence $u = u(t,x)$ must be continuous along the curve C. □

Next we consider the following Cauchy problem:

$$\begin{cases} \dfrac{\partial u}{\partial t} + \dfrac{\partial}{\partial x}(\dfrac{1}{2}u^2) = 0, \ (t,x) \in D_- \\ \\ u(t,x) = 4x^2, \qquad \text{on } C \end{cases} . \tag{6}$$

Proposition 2.2. *The Cauchy problem (6) can not have a solution in $C^1(D_-) \cap C^0(\bar{D_-})$ where $\bar{D_-}$ is the closure of D_-.*

Proof. Assume that (6) has a solution in $C^1(D_-) \cap C^0(\bar{D_-})$. Take any point P on C and a point $Q = (t_0, x_0)$ in D_- satisfying $u(t_0,x_0) \neq 0$. Then a characteristic curve passing through the point Q is given by

$$x = x_0 + u(t_0,x_0)(t - t_0). \tag{7}$$

As $u(t_0,x_0) \neq 0$, the characteristic line (7) meets with the curve C at two different points A and B. Obviously it holds $u|_A \neq u|_B$. But, as the solution is constant along any characteristic curve, this is a contradiction. □

We can localize the above proposition. Take any point P on the curve C and any open neighbourhoon U of the point P, and consider the following Cauchy problem:

$$\begin{cases} \dfrac{\partial u}{\partial t} + \dfrac{\partial}{\partial x}(\dfrac{1}{2}u^2) = 0, \ (t,x) \in D_- \cap U \\ \\ u(t,x) = 4x^2, \qquad \text{on } C \cap U \end{cases} . \tag{8}$$

Proposition 2.3. *The Cauchy problem (8) can not have a solution in $C^1(D_- \cap U) \cap C^0(\bar{D_-} \cap U)$.*

The proof is almost the same as that of Proposition 2.2. Summing up the above results, we get the following

Theorem 2.1. *The Cauchy problem (3) can not admit a real-valued piecewise smooth weak solution defined in the whole space* $\mathbf{R}^2$ *and the boundary of the existence domain of the solution is the envelope of the family of the characteristic curves coresponding to the Cauchy problem (3).*

Remark 2.1. If we may accept a complex-valued solution, the solution (5) is a piecewise smooth and continuous weak solution of (3) defined in the whole space $\mathbf{R}^2$.

3. A Single Conservation Law without Convexity Condition

In the following sections we will study the Cauchy problem for a single conservation law which does not satisfy the convexity condition. Assume that the graphs of $f = f(u)$ and $\phi = \phi(x)$ are drawn as in Fig. 1 (i) and Fig. 2 (i) respectively. The graphs of their derivatives are Fig. 1 (ii)-(iii) and Fig. 2 (ii). In the above case, we assume that $f(a_i) = f'(a_i) = 0$ $(i = 1, 2)$, $f''(b_i) = 0$ $(i = 1, 2)$ and $\phi'(c) = \phi'(d) = 0$. Moreover we assume $\max_x \phi(x) > b_2$. For the Cauchy problem (1)-(2), the characteristic curves are written by

$$x = x(t, y) = y + tf'(\phi(y)), \quad v = v(t, y) = \phi(y). \tag{9}$$

Therefore it follows that

$$\frac{\partial x}{\partial y}(t, y) = 1 + tf''(\phi(y))\phi'(y). \tag{10}$$

We define $h(y) \stackrel{\text{def}}{=} f''(\phi(y))\phi'(y)$, and assume that the graph of $h = h(y)$ is drawn as in Fig. 3. Next, denote $A_i = (y_i, h(y_i))(i = 1, 2)$ where $h(y_i) < 0$ and $h'(y_i) = 0$. Then we have at A_1

$$\phi'(y_1) > 0 \quad \text{and} \quad f''(\phi(y_1)) < 0,$$

and at A_2

$$\phi'(y_2) < 0 \quad \text{and} \quad f''(\phi(y_2)) > 0.$$

We now put $t_i = -1/h(y_i)$ and $X_i = x(t_i, y_i)$ $(i = 1, 2)$. Then we can see the appearance of a shock whose starting point is (t_i, X_i) $(i = 1, 2)$, and we denote it by S_i $(i = 1, 2)$. We will briefly explain the construction of shocks

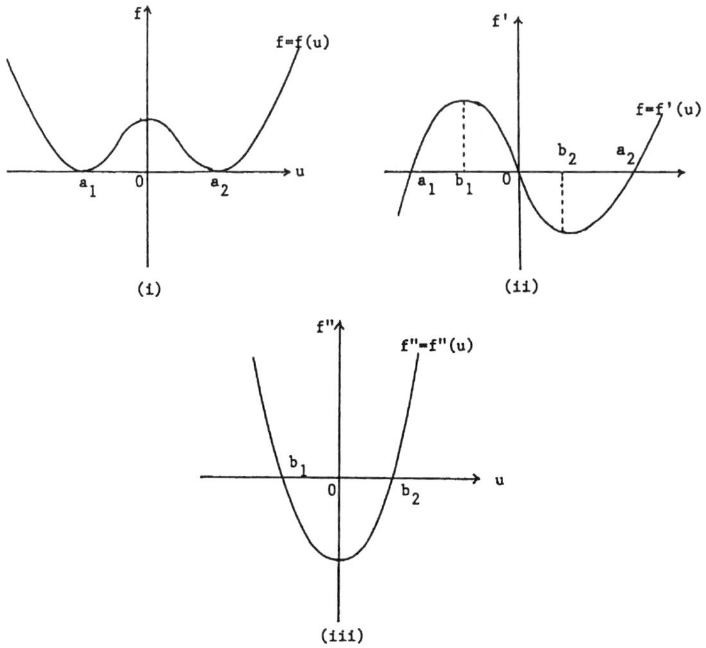
f
f=f(u)
u
a1
0
a2
(i)
f'
f=f'(u)
b2
a2
a1
b1
0
(ii)
f"
f"=f"(u)
b1
0
b2
u
(iii)

Fig. 1.

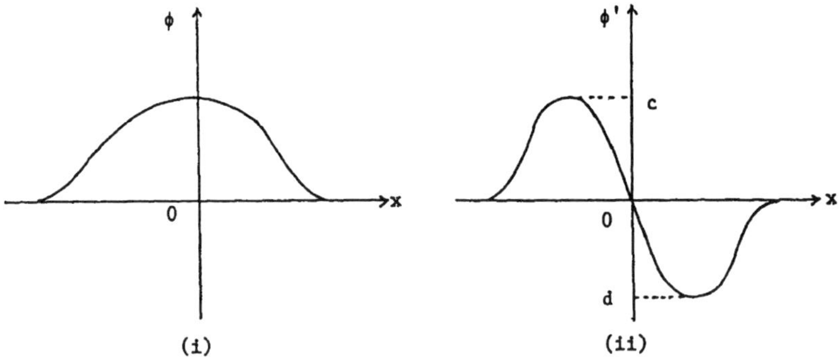
ϕ
x
0
(i)
ϕ'
c
x
0
d
(ii)

Fig. 2.

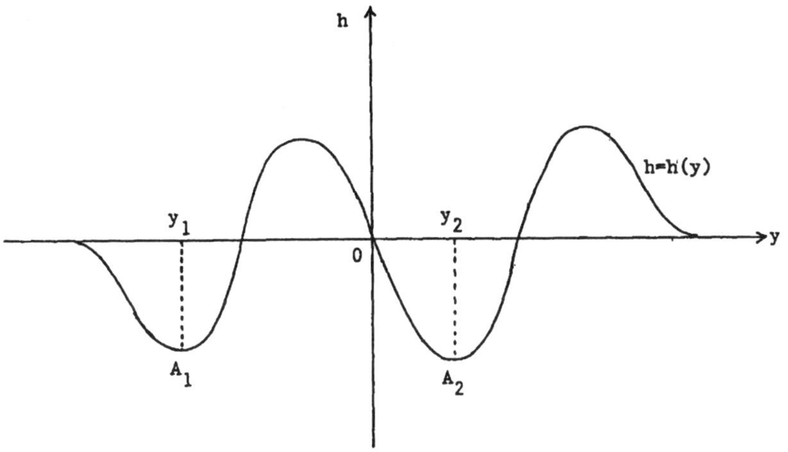

Fig. 3.

at the beginning of Sec. 4. For the detailed explanation.[10,11] Our problem is how to extend each shock S_i $(i = 1, 2)$ for large t.

We give some comment on the above example. Let us consider the Riemann problem for (1), that is to say, we assume that $\phi(x) = c_+$ for $x > 0$ and $\phi(x) = c_-$ for $x < 0$ where $c_\pm$ are constant. If $f'(c_-) < f'(c_+)$, the region $\{(t, x); f'(c_-)t < x < f'(c_+)t\}$ is not covered by the family of characteristic curves (9). Therefore the rarefaction wave was introduced as a solution in this region. On the other hand, as we assume that the initial function (2) is in $C_0^\infty(\mathbf{R}^1)$, the family of characteristic curves covers the whole space $\mathbf{R}^2$. This is the principal reason why we doubt the necessity to use rarefaction waves in the process of construction of weak solutions.

4. Behavior of the Shock S_1

In this section we will extend the shock S_1 for large t. To explain the situation, we repeat briefly how we have constructed the shock S_1. The graph of $x = x(t, y)$ for $t > t_i$ is drawn as in Fig. 4. We explain the meanings of the notations used in Fig. 4. Let $y = \xi_i(t)$ $(\xi_1(t) < \xi_2(t), t > t_1)$ be the solutions of $(\partial x/\partial y)(t, y) = 0$ with respect to y in a neighbourhood of $y = y_1$ and denote $x_i(t) = x(t, \xi_i(t)(i = 1, 2)$. Then we see that $x_1(t) > x_2(t)$ for $t > t_1$. Solving the equation $x = x(t, y)$ with respect to y for $x \in (x_2(t), x_1(t))$, we get three solutions $y = g_i(t, x)$ $(i = 1, 2, 3)$ satisfying $g_1(t, x) < g_2(t, x) < g_3(t, x)$. As $v(t, y) = \phi(y)$ for all $(t, y) \in \mathbf{R}^2$, we define

$u_i(t,x) = \phi(g_i(t,x))$ $(i = 1,2,3)$. Then, as $\phi'(y_1) > 0$, we have

$$u_1(t,x) < u_2(t,x) < u_3(t,x).$$

As we are looking for a single-valued weak solution, we jump from the first branch $\{u = u_1(t,x)\}$ to the third one $\{u = u_3(t,x)\}$. A shock curve is determined by the Rankine-Hugoniot jump condition as follows:

$$\frac{dx}{dt} = \frac{f(u_1(t,x)) - f(u_3(t,x))}{u_1(t,x) - u_3(t,x)}, x(t_1) = X_1. \tag{11}$$

Though the right hand term is not Lipschitz continuous at the starting point, we have proved in Ref. 12 that the Cauchy problem (4.1) has a unique solution $x = \gamma_1(t)$. This is the shock curve S_1. To extend the shock S_1, we consider the behavior of $u_i(t,\gamma_1(t)) = \phi(g_i(t,\gamma_1(t)))$ $(i = 1,3)$. We put $u_\pm(t) = u(t,\gamma_1(t) \pm 0)$. Then $u_+(t) = u_3(t,\gamma_1(t))$ and $u_-(t) = u_1(t,\gamma_1(t))$.

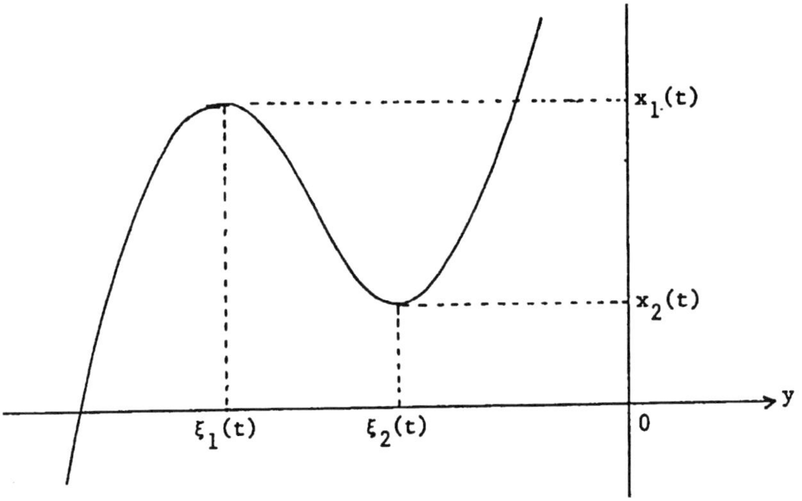

Fig. 4.

Lemma 4.1. *As long as* S_1 *satisfies the entropy condition, the function* $g_1(t,\gamma_1(t))$ *is decreasing and* $g_3(t,\gamma_1(t))$ *is increasing.*

Proof. By the definition of $y = g_i(t,x)$, we have

$$x = g_i(t,x) + tf'(\phi(g_i(t,x))) \quad (i = 1,2,3).$$

Taking the derivatives with respect to t and x, we have

$$0 = \frac{\partial g_i}{\partial t} + f'(\phi(g_i)) + tf''(\phi(g_i))\phi'(g_i)\frac{\partial g_i}{\partial t}$$

and

$$1 = \frac{\partial g_i}{\partial x}\{1 + tf''(\phi(g_i))\phi'(g_i)\}.$$

Hence it holds that

$$\frac{\partial g_i}{\partial t}(t,x) = -f'(u_i(t,x))\frac{\partial g_i}{\partial x}(t,x) \quad (i = 1,2,3) \tag{12}$$

which leads us to

$$\frac{d}{dt}g_i(t,\gamma_1(t)) = \{\frac{d\gamma_1}{dt} - f'(u_i(t,\gamma_1(t)))\}\frac{\partial g_i}{\partial x}(t,\gamma_1(t)) \quad (i = 1,2,3).$$

Here we recall

$$u_+(t) = u_3(t,\gamma_1(t)) > u_1(t,\gamma_1(t)) = u_-(t).$$

Then the entropy condition means

$$\frac{d\gamma_1}{dt}(t) \leq f'(u_1(t,\gamma_1(t))) = f'(u_-(t))$$

and

$$\frac{d\gamma_1}{dt}(t) \geq f'(u_3(t,\gamma_1(t))) = f'(u_+(t)).$$

Since $(\partial g_i/\partial x)(t,x) > 0$ for $i = 1$ and $i = 3$, we obtain

$$\frac{d}{dt}g_1(t,\gamma_1(t)) \leq 0 \quad\text{and}\quad \frac{d}{dt}g_3(t,\gamma_1(t)) \geq 0.$$

Hence the proof of the lemma is complete. □

The initial function $\phi = \phi(x)$ has the properties as drawn in Fig. 2. As long as the entropy condition is satisfied, we see by Lemma 4.1 that $\phi(g_1(t,\gamma_1(t))) = u_-(t)$ is decreasing, and that $\phi(g_3(t,\gamma_1(t))) = u_+(t)$ is increasing since $g_3(t,x) < 0$. Hence $\phi(g_1(t,\gamma_1(t)))$ advances to 0, and $\phi(g_3(t,\gamma_1(t)))$ goes to its maximum. We write here $P_+(t) = (u_+(t), f(u_+(t)))$ and $P_-(t) = (u_-(t), f(u_-(t)))$. The shock S_1 satisfies the entropy condition for $t > t_1$ where $t - t_1$ is small. This means that the graph of $f = f(u)$ lies entirely on the upper side of the chord $\overline{P_+P_-}$ joining the two points $P_+(t)$ and $P_-(t)$. If we extend the shock S_1 further, then the chord $\overline{P_+P_-}$ may become tangent to the curve $\{(u, f(u)) \; : u \in \mathbf{R}^1\}$ in finite time. We assume that $\overline{P_+P_-}$ is tangent to the curve $f = f(u)$ at $t = T$. Since $0 < u_-(t) < \phi(y_1) < b_2$ and thus $f"(u_-(t) < 0$, we see that

$\overline{P_+P_-}$ is tangent to the curve $f = f(u)$ at the point $P_+(T)$, but not at the point $P_-(T)$. Therefore it holds that

$$\frac{d\gamma_1}{dt}(T) = f'(u_+(T)). \tag{13}$$

Lemma 4.2. *i)* $\dfrac{d^2\gamma_1}{dt^2}(T) > 0,$ *ii)* $\dfrac{d}{dt}f'(u_+(t))|_{t=T} = 0.$

Proof. i) Using the definition (4.1) of $(d\gamma_1/dt)(t)$, we have

$$\frac{d^2\gamma_1}{dt^2}(t) = \frac{d}{dt}(\frac{d\gamma_1}{dt}(t)) = \left[\frac{f'(u_+)}{u_+ - u_-} - \frac{f(u_+) - f(u_-)}{(u_+ - u_-)^2}\right]\frac{du_+}{dt}(t)$$

$$+ \left[\frac{f(u_+) - f(u_-)}{(u_+ - u_-)^2} - \frac{f'(u_-)}{u_+ - u_-}\right]\frac{du_-}{dt}(t).$$

As $u_+(t) = u_3(t, \gamma_1(t))$, we get by (18)

$$\frac{d}{dt}u_+(t)|_{t=T} = \left[\frac{\partial u_3}{\partial t} + \frac{\partial u_3}{\partial x}\frac{d\gamma_1}{dt}\right]|_{t=T} = \left[\frac{d\gamma_1}{dt} - f'(u_3(t,\gamma_1))\right]\frac{\partial u_3}{\partial x}|_{t=T} = 0. \tag{14}$$

Put $Y_1 = g_1(T, \gamma_1(T))$. As $u_-(t) = u_1(t, \gamma_1(t)) = \phi(g_1(t, \gamma_1(t)))$, we obtain by (17)

$$\frac{d}{dt}u_-(t) = \frac{d}{dt}\phi(g_1(t,\gamma_1(t))) = \{\frac{d\gamma_1}{dt}(t) - f'(u_-(t))\}\phi'(g_1(t,x))\frac{\partial g_1}{\partial x}(t,x)|_{x=\gamma_1(t)}.$$

This leads us to

$$\frac{d^2\gamma_1}{dt^2}(T) = \frac{1}{u_+(T) - u_-(T)}\left[\frac{d\gamma_1}{dt}(T) - f'(u_-(T))\right]^2\phi'(Y_1)\frac{\partial g_1}{\partial x}(T, \gamma_1(T)).$$

Here we recall that the starting point of the shock S_1 is the point (t_1, X_1). Therefore it follows that $\phi'(Y_1) > 0$, $u_+(t) - u_-(t) > 0$ and $(\partial g_1/\partial x)(t,x) > 0$. Hence we get $(d^2\gamma_1/dt^2)(T) > 0$.

Part (ii) is easily obtained from (19). □

By Lemma 4.2, we have

$$\frac{d\gamma_1}{dt}(t) > f'(u_+(t)), \quad t > T.$$

This implies that the entropy condition is satisfied for $t > T$. Summing up the above results, we have the following

Theorem 4.1. *The shock S_1 starting from the point (t_1, X_1) always satisfies the entropy condition.*

5. Behavior of the Shock S_2

In this section we extend the shock S_2 whose starting point is (t_2, X_2). The graph of $x = x(t, y)$ for $t > t_2$ can be drawn as in Fig. 4. Though $x = x(t, y)$ in this section is different from $x = x(t, y)$ in the previous section, we use the same notations introduced there. As in Sec. 3, we solve the Cauchy problem for (16) whose initial condition is $x(t_2) = X_2$, and denote the solution by $x = \gamma_2(t)$. We write here also $u_\pm(t) = u(t, \gamma_2(t) \pm 0)$; that is to say, $u_+(t) = \phi(g_3(t, \gamma_2(t)))$ and $u_-(t) = \phi(g_1(t, \gamma_2(t)))$. Put $P_\pm(t) = (u_\pm(t), f(u_\pm(t)))$. The shock S_2 satisfies the entropy condition for $t > t_2$ where $t - t_2$ is small. In this case, as $f''(\phi(y_2)) > 0$, the entropy condition says that the graph of $f = f(u)$ lies entirely on the lower side of the chord $\overline{P_+P_-}$.

Lemma 5.1. *As long as the entropy condition for S_2 is satisfied, $g_1(t, \gamma_2(t))$ is decreasing and $g_3(t, \gamma_2(t))$ is increasing.*

The proof of this lemma is almost the same as that of Lemma 4.1. When t gets larger, $\phi(g_3(t, \gamma_2(t)))$ tends to 0 and $\phi(g_1(t, \gamma_2(t)))$ tends to the maximum of $\phi = \phi(x)$. Therefore we assume that the entropy condition is satisfied for $t < T$, and that $\overline{P_+P_-}$ becomes tangent to the curve $\{(u, f(u)); u \in \mathbf{R}^1\}$ at $t = T$. Then it follows that

$$\frac{d\gamma_2}{dt}(T) = f'(\phi(g_3(t, \gamma_2(t))))|_{t=T} = f'(u_+(T)).$$

Lemma 5.2.

i) $\dfrac{d^2\gamma_2}{dt^2}(T) > 0,$

ii) $\dfrac{d}{dt} f'(u_+(t))|_{t=T} = 0.$

The proof is the same as that of Lemma 4.2. Using this lemma, we get

$$\frac{d\gamma_2}{dt}(t) > f'(u_+(t)), \qquad t > T.$$

Therefore we get the following

Theorem 5.1. *The shock S_2 starting from the point (t_2, X_2) always satisfies the entropy condition.*

It may happen that the shocks S_1 and S_2 collide in finite time. Then we see that a new, uniquely defined shock appears. The method of the proof is almost the same as that in Refs. 9 and 10.

Summing up these results, we can say that our program is true for the above Cauchy problem.

References

1. J. Guckenheimer, *Solving a single conservation law*, Lecture Notes in Math., Vol. 468, (Springer-Verlag, 1975), pp. 108-134.
2. S. Izumiya and G. T. Kossioris, *Bull. Sci. math.* **121**, 619 (1997).
3. G. Jennings, *Adv. in Math.* **33**, 192 (1979).
4. S. N. Kruzhkov, *Math. USSR Sb.* **1**, 93 (1967).
5. S. N. Kruzhkov, *Math. USSR Sb.* **10**, 217 (1970).
6. P. D. Lax, *Comm. Pure Appl. Math.* **10**, 537 (1957).
7. O. A. Oleinik, *Uspekhi Mat. Nauk* **12**, 3 (1957) (in Russian). English transl. in *Amer. Math. Soc. Transl.* **26**, 95 (1963).
8. D. G. Schaeffer, *Adv. in Math.* **11**, 358 (1973).
9. Tran D. V., M. Tsuji and Nguyen Duy T. S., *The characteristic method and its generalizations for first-order non-linear partial differential equations* (Chapman & Hall/CRC, USA, 1999).
10. M. Tsuji, *C. R. Acad. Sci. Paris* **289**, 397 (1979).
11. M. Tsuji, *J. Math. Kyoto Univ.* **26**, 299 (1986).
12. M. Tsuji, *Ann. Inst. H. Poincaré - Analyse nonlinéaire* **7**, 505(1990).
13. M. Tsuji, *Pitman Research Notes in Math.* **381**, 164 (1998)(Longman).
14. M. Tsuji and T. S. Nguyen Duy, *Acta Math. Vietnamica*, **27**, 97 (2002).
15. M. Tsuji, Some remarks on nonlinear hyperbolic equations and systems, in *Abstract and Applied Analysis* edited by N. M. Chuong, L. Nirenberg and W. Tutschke (World Scientific, 2004), pp. 355-364.

Advances in Deterministic and Stochastic Analysis
Eds. N. M. Chuong *et al.* (pp. 103–113)

§7. ITERATIVE METHOD FOR SOLVING A MIXED BOUNDARY VALUE PROBLEM FOR BIHARMONIC TYPE EQUATION*

DANG QUANG A[‡] and LE TUNG SON

Institute of Information Technology
Vietnam Academy of Science and Technology,
18 Hoang Quoc Viet Road, Cau Giay Dist., 10307 Hanoi, Vietnam
[‡]*E-mail: dangqa@ioit.ncst.ac.vn*

The solution of boundary value problems (BVP) for fourth order differential equations by their reduction to BVP for second order equations with the aim to use the achievements for the latter ones attracts attention from many researchers. In this paper, using the technique developed by ourselves in recent works, we construct iterative method for a mixed BVP for biharmonic type equation. The convergence rate of the method is proved and some numerical experiments are performed for testing it in dependence on the choice of an iterative parameter.

Keywords: Iterative method; Mixed boundary value problem; Biharmonic equation

1. Introduction

The solution of fourth order differential equations by their reduction to boundary value problems (BVP) for the second order equations, with the aim of using efficient algorithms for the latter ones, attracts attention from many researchers. Namely, for the biharmonic equation with the Dirichlet boundary condition, there is intensively developed the iterative method, which leads the problem to two problems for the Poisson equation at each iteration (see e.g. Refs. 2,8,9,11). Recently, Abramov and Ulijanova[1] proposed an iterative method for the Dirichlet problem for the biharmonic type equation, but the convergence of the method is not proved. In our previous works[5–7] with the help of boundary or mixed boundary-domain operators

*This work is supported in part by the National Basic Research Program in Natural Sciences, Vietnam

appropriately introduced, we constructed iterative methods for biharmonic and biharmonic type equations associated with the Dirichlet boundary condition. It is proved there that the methods are convergent with the rate of geometric progression. In this paper we develop our technique in Refs. 4-7 for a mixed BVP for the biharmonic type equation. Namely, we consider the following problem

$$\Delta^2 u + bu = f \quad \text{in } \Omega, \tag{1}$$

$$u = g \quad \text{on } \Gamma, \tag{2}$$

$$\frac{\partial u}{\partial \nu} = g_1 \text{ on } \Gamma_1, \tag{3}$$

$$\Delta u = g_2 \text{ on } \Gamma_2, \tag{4}$$

where Ω is a bounded domain in $\mathbb{R}^n (n \geqslant 2)$ with the Lipschits boundary Γ consisting of two smooth pieces Γ_1 and Γ_2, ν is the outward normal to Γ, b is a positive number and Δ is the Laplace operator. We propose an iterative method for reducing the problem to a sequence BVP for the Poisson equation, study its convergence and finally, present the results of numerical experiments for the illustrating the effectiveness of the method.

2. Reduction of the Problem to Boundary - Domain Operator Equation

As in Refs. 5 and 6 we set

$$\Delta u = v, \tag{5}$$

$$\varphi = -bu \tag{6}$$

and denote

$$v|_{\Gamma_1} = v_0. \tag{7}$$

Then the problem (1)-(4) is reduced to the problems:

$$\begin{aligned} \Delta v &= f + \varphi, \quad x \in \Omega, \\ v|_{\Gamma_1} &= v_0, \quad v|_{\Gamma_2} = g_2, \end{aligned} \tag{8}$$

$$\begin{aligned} \Delta u &= v, \quad x \in \Omega, \\ u|_{\Gamma} &= g. \end{aligned} \tag{9}$$

where v_0 and φ are temperorily unknown functions.
The solution u found from the above problems should satisfy the boundary condition (3) and the relation (6). To find v_0, φ, we introduce an operator

B as follows:

$$B : w \rightarrow Bw,$$

where

$$w = \begin{pmatrix} v_0 \\ \varphi \end{pmatrix}, \quad Bw = \begin{pmatrix} b\dfrac{\partial u}{\partial \nu}|_{\Gamma_1} \\ \varphi + bu \end{pmatrix}, \tag{10}$$

and u and v are the solutions to the problems

$$\begin{aligned} \Delta v = \varphi, \quad & x \in \Omega, \\ v|_{\Gamma_1} = v_0, \quad & v|_{\Gamma_2} = 0, \end{aligned} \tag{11}$$

$$\begin{aligned} \Delta u = v, \quad & x \in \Omega, \\ u|_{\Gamma} = 0. & \end{aligned} \tag{12}$$

The operator B primarily defined on couples of smooth functions is extended by continuity on whole space $L^2(\Gamma_1) \times L^2(\Omega)$. Its properties will be investigated later.

For the reduction of BVP (1)-(4) to an equation with the operator B we set $u = u_1 + u_2, v = v_1 + v_2$, where u_1, v_1 and u_2, v_2 are the solutions to the following problems

$$\begin{aligned} \Delta v_2 = f, \quad & x \in \Omega, \\ v_2|_{\Gamma_1} = 0, \quad & v_2|_{\Gamma_2} = g_2, \end{aligned} \tag{13}$$

$$\begin{aligned} \Delta u_2 = v_2, \quad & x \in \Omega, \\ u_2|_{\Gamma} = g, & \end{aligned} \tag{14}$$

$$\begin{aligned} \Delta v_1 = \varphi, \quad & x \in \Omega, \\ v_1|_{\Gamma_1} = v_0, \quad & v_1|_{\Gamma_2} = 0, \end{aligned} \tag{15}$$

$$\begin{aligned} \Delta u_1 = v_1, \quad & x \in \Omega, \\ u_1|_{\Gamma} = 0. & \end{aligned} \tag{16}$$

From (13) and (14) we can determine v_2 and u_2, and from (15), (16), by the definition of B, we have

$$Bw = \begin{pmatrix} b\dfrac{\partial u_1}{\partial \nu}|_{\Gamma_1} \\ \varphi + bu_1 \end{pmatrix}. \tag{17}$$

For u to be the solution of (1)-(4), as mentioned above, the relations (3) and (6) must be satisfied, i.e., there should be conditions

$$\frac{\partial u}{\partial \nu}|_{\Gamma_1} = g_1, \tag{18}$$

$$\varphi + bu = 0 \quad \text{in } \Omega. \tag{19}$$

From these relations we derive

$$\frac{\partial u_1}{\partial \nu}|_{\Gamma_1} = g_1 - \frac{\partial u_2}{\partial \nu}|_{\Gamma_1}, \tag{20}$$

$$\varphi + bu_1 = -bu_2 \quad \text{in } \Omega. \tag{21}$$

Setting

$$F = \begin{pmatrix} b\left(g_1 - \frac{\partial u_2}{\partial \nu}|_{\Gamma_1}\right) \\ -bu_2 \end{pmatrix} \tag{22}$$

from (17) we get the equation

$$Bw = F. \tag{23}$$

Now, we study properties of B. First, we introduce the space $H = L^2(\Gamma_1) \times L^2(\Omega)$ with the scalar product

$$(w, \overline{w})_H = (v_0, \overline{v_0})_{L^2(\Gamma_1)} + (\varphi, \overline{\varphi})_{L^2(\Omega)}$$

for the elements

$$w = \begin{pmatrix} v_0 \\ \varphi \end{pmatrix}, \quad \overline{w} = \begin{pmatrix} \overline{v_0} \\ \overline{\varphi} \end{pmatrix}.$$

Property 1. *B is symmetric in H*

Proof. We have

$$\begin{aligned}(Bw, \overline{w})_H &= \int_{\Gamma_1} b\overline{v_0}\frac{\partial u}{\partial \nu} d\Gamma_1 + \int_\Omega \overline{\varphi}(\varphi + bu)dx \\ &= b\int_\Gamma \overline{v}\frac{\partial u}{\partial \nu} d\Gamma + b\int_\Omega \overline{\varphi} u dx + \int_\Omega \overline{\varphi}\varphi dx\end{aligned}$$

since $\overline{v}|_{\Gamma 1} = \overline{v}_0$ and $\overline{v}|_{\Gamma 2} = 0$ in the problems (11), (12) for the definition of $B\overline{w}$. □

Now, we transform the boundary intergral, taking into account of (11) and (12). We have

$$I_1 = \int_\Gamma \overline{v}\frac{\partial u}{\partial \nu} d\Gamma = \int_\Gamma (\overline{v}\frac{\partial u}{\partial \nu} - u\frac{\partial \overline{v}}{\partial \nu})d\Gamma$$
$$= \int_\Omega (\overline{v}\Delta u - u\Delta\overline{v})dx = \int_\Omega v\overline{v}dx - \int_\Omega u\overline{\varphi}dx.$$

Therefore,

$$(Bw, \overline{w})_H = b\int_\Omega \overline{v}vdx + \int_\Omega \overline{\varphi}\varphi dx = (B\overline{w}, w). \tag{24}$$

Thus, the symmetry of B is proved.

Property 2. *B is positive in H.*

Indeed, from (24) we have

$$(Bw, w) = \int_\Omega (bv^2 + \varphi^2)dx \geqslant 0.$$

The equality appears if and only if $w = 0$.

Property 3. *B can be decomposed into the sum of a positive, symmetric and completely continuous operator and a projection operator, namely,*

$$B = B_0 + I_2, \tag{25}$$

where B_0 and I_2 are defined as follows

$$w = \begin{pmatrix} v_0 \\ \varphi \end{pmatrix}, \quad B_0 w = \begin{pmatrix} b\frac{\partial u}{\partial \nu}|_{\Gamma_1} \\ bu \end{pmatrix}, \quad I_2 w = \begin{pmatrix} 0 \\ \varphi \end{pmatrix}, \tag{26}$$

u being defined from (11), (12).

The complete continuity of B_0 easily follows from the theory of elliptic problems and embedding theorems of Sobolev spaces (see Ref. 10).

Property 4. *B is bounded in H .*

This fact is a direct corollary of Property 3.

Since $B = B* > 0$ but is not completely continuous in H the use of two-layer iterative scheme[13] to the equation (23) does not guarantee its convergence. Hence, in the next section we will disturb this equation and apply the parametric extrapolation technique (see Refs. 4 and 5) for constructing approximate solution for problem (1)-(4).

3. Construction of Approximate Solution of the Original Problem Via a Perturbed Problem

We associate with the original problem (1)-(4) the following perturbed problem

$$\Delta^2 u_\delta + bu_\delta = f(x), \quad x \in \Omega, \tag{27}$$

$$u_\delta|_\Gamma = g, \tag{28}$$

$$\left(\delta \Delta u_\delta + b\frac{\partial u_\delta}{\partial \nu}\right)|_{\Gamma_1} = bg_1, \tag{29}$$

$$\Delta u_\delta|_{\Gamma_2} = g_2, \tag{30}$$

where δ is a small positive parameter.

Theorem 3.1. *Suppose that $f \in H^{n-4}(\Omega)$, $g \in H^{n-1/2}(\Gamma)$, $g_1 \in H^{n-3/2}(\Gamma_1)$, $g_2 \in H^{n-5/2}(\Gamma_2); n \geqslant 4$. Then for the solution of the problem (27)-(30) there holds the following asymptotic expanssion:*

$$u_\delta = u + \sum_{i=1}^{N} \delta^i y_i + \delta^{N+1} z_\delta, \quad x \in \Omega, \quad 0 \leqslant N \leqslant n - \frac{5}{2}, \tag{31}$$

where $y_0 = u$ is the solution of (1)-(4), y_i $(i = 1, 2, ..., N)$ are functions independent of δ, $y_i \in H^{n-i}(\Omega)$, $z_\delta \in H^{n-N}(\Omega)$ and

$$||z_\delta||_{H^2(\Omega)} \leqslant C_1, \tag{32}$$

C_1 being independent of δ.

Proof. Under the assumption of the theorem there exists a unique solution $u \in H^n(\Omega)$ of the problem (27)-(30). After substituting (31) into (27)-(30) and balancing coefficients of like powers of δ, we see that y_i and z_δ satisfy the following problems:

$$\begin{aligned} &\Delta^2 y_i + by_i = 0, \quad x \in \Omega, \\ &y_i|_\Gamma = 0, \\ &b\frac{\partial y_i}{\partial \nu}|_{\Gamma_1} = -\Delta y_{i-1}|_{\Gamma_1} \quad , i = 1, 2, ..., N, \\ &\Delta y_i|_{\Gamma_2} = 0. \end{aligned} \tag{33}$$

$$\begin{aligned} &\Delta^2 z_\delta + bz_\delta = 0, \quad x \in \Omega, \\ &z_\delta|_\Gamma = 0, \\ &(\delta \Delta z_\delta + b\frac{\partial z_\delta}{\partial \nu}|_{\Gamma_1} = -\Delta y_N|_{\Gamma_1}, \\ &\Delta z_\delta|_{\Gamma_2} = 0. \end{aligned} \tag{34}$$

It is possible to establish successively that (33) has a unique solution $y_i \in H^{n-i}(\Omega)$ and (34) has a unique solution $z_\delta \in H^{n-N}(\Omega)$. Clearly, y_i ($i = 1, ..., N$) do not depend on δ. It remains to estimate z_δ. For this purpose, we reduce (34) to a boundary operator equation. Setting

$$\Delta z_\delta = v_\delta, \quad \varphi_\delta = -b z_\delta, \quad v_\delta|_{\Gamma_1} = v_{\delta 0} \tag{35}$$

we obtain

$$\begin{aligned} &\Delta v_\delta = \varphi_\delta, \quad x \in \Omega, \\ &v_\delta|_{\Gamma_1} = v_{\delta 0}, \quad v_\delta|_{\Gamma_2} = 0. \\ &\Delta z_\delta = v_\delta, \quad x \in \Omega, \\ &z_\delta|_\Gamma = 0. \end{aligned}$$

Now, denote

$$w_\delta = \begin{pmatrix} v_{\delta 0} \\ \varphi_\delta \end{pmatrix},$$

then by definition of B, we obtain

$$Bw_\delta = \begin{pmatrix} b\dfrac{\partial z_\delta}{\partial \nu}|_{\Gamma_1} \\ \varphi_\delta + b z_\delta \end{pmatrix}.$$

Using the second condition of (34), we obtain

$$(B + \delta I_1) w_\delta = h, \tag{36}$$

where

$$h = \begin{pmatrix} -\Delta y_N \\ 0 \end{pmatrix}, \quad I_1 w_\delta = \begin{pmatrix} w_\delta \\ 0 \end{pmatrix}.$$

It follows that (see Lemma 1 in Ref. 3)

$$(Bw_\delta, w_\delta) \leqslant (Bw, w), \tag{37}$$

where w is the solution of the equation $Bw = h$. This equation has a solution because it is the equation to which the problem (34) with $\delta = 0$ may be reduced.

In Sec. 2, when investigating the properties of B, we have established that

$$(Bw_\delta, w_\delta) = \int_\Omega (b{v_\delta}^2 + {\varphi_\delta}^2) dx. \tag{38}$$

In view of (35), we have

$$(Bw_\delta, w_\delta) = \int_\Omega (b|\Delta v_\delta|^2 + b|z_\delta|^2) dx. \tag{39}$$

Since the right-hand side of the above quality defines a norm in the class of functions vanishing on the boundary, which is equivalent to the norm $||.||_{H^2(\Omega)}$, we have $||z_\delta||_{H^2(\Omega)} \leqslant C_1$, where $C_1 = C_2\sqrt{(Bw,w)}$, C_2 being independent of δ.

Thus the theorem is proved. □

As usual, we construct an approximate solution U^E of the original problem (1)-(4) by the formula

$$U^E = \sum_{i=1}^{N+1} \gamma_i u_{\delta/i}, \tag{40}$$

where

$$\gamma_i = \frac{(-1)^{N+1-i} i^{N+1}}{i!(N+1-i)!}, \tag{41}$$

$u_{\delta/i}$ is the solution of (27)-(30) with the parameter δ/i $(i = 1, ..., N+1)$. Then it is easy to obtain the following estimate

$$||U^E - u||_{H^2(\Omega)} \leqslant C_2\delta^{N+1}, \tag{42}$$

where u is the solution of the problem (1)-(4) and C_2 is a constant independent of δ.

4. Iterative Method for Solving the Perturbed Problem

First we notice that in the same way as for the original problem (1)-(4), the problem (27)-(30) may be reduced to the operator equation

$$B_\delta w_\delta = F \tag{43}$$

for $w_\delta = (v_{\delta 0}, \varphi_\delta)^T$, where $v_{\delta 0} = \Delta u_\delta|_{\Gamma 1}$, $\varphi_\delta = -bu_\delta, B_\delta = B + \delta I_1, B$ and F are defined by (10) and (22), respectively. Clearly, B_δ is bounded and

$$B_\delta = B_\delta^* \geq \delta I, \tag{44}$$

where I is the identity operator.

For solving (43) we can apply the general theory of two-layer iterative scheme for equation with symmetric, positive definite operator.[13] Namely, we consider the iterative scheme

$$\frac{w_\delta^{(k+1)} - w_\delta^{(k)}}{\tau_{\delta,k+1}} + B_\delta w_\delta^{(k)} = F, \tag{45}$$

where $\{\tau_{\delta,k+1}\}$ is the Chebyshev collection of parameters according to bounds $\gamma_\delta^{(1)} = \delta, \gamma_\delta^{(2)} = \delta + ||B||$ (see Ref. 13 for details). In the case of simple iteration

$$\tau_{\delta,k} \equiv \tau_{\delta,0} = \frac{2}{\gamma_\delta^{(1)} + \gamma_\delta^{(2)}} \tag{46}$$

we obtain

$$||w_\delta{}^{(k)} - w_\delta||_H \leqslant (\rho_\delta)^k ||w_\delta{}^{(0)} - w_\delta||_H, \tag{47}$$

where

$$w_\delta^{(k)} = \begin{pmatrix} v_{\delta 0}^{(k)} \\ \varphi_\delta^{(k)} \end{pmatrix}, \quad w_\delta = \begin{pmatrix} v_{\delta 0} \\ \varphi_\delta \end{pmatrix},$$

$$\rho_\delta = \frac{1-\xi_\delta}{1+\xi_\delta}, \quad \xi_\delta = \frac{\gamma_\delta^{(1)}}{\gamma_\delta^{(2)}},$$

and as above $H = L^2(\Gamma_1) \times L^2(\Omega)$.

Using estimates for the solution of elliptic problems[10] and taking into account (47), we obtain the estimate

$$||u_\delta^{(k)} - u_\delta||_{H^{5/2}(\Omega)} \leqslant C(\rho_\delta)^k ||w_\delta^{(0)} - w_\delta||_H, \tag{48}$$

where C is a constant independent of δ.

The iterative scheme (45) can be realized by the following process:

(i) Given a couple $(v_{\delta 0}{}^{(0)}, \varphi_\delta{}^{(0)})$.

(ii) Knowing $v_{\delta 0}{}^{(k)}$ and $\varphi_\delta{}^{(k)}$, $k = 0, 1, \ldots$ solves successively two problems

$$\begin{aligned} &\Delta v_\delta{}^{(k)} = f + \varphi_\delta{}^{(k)}, \quad x \in \Omega, \\ &v_\delta^{(k)}|_{\Gamma_1} = v_{\delta 0}^{(k)}, \quad v_\delta^{(k)}|_{\Gamma_2} = g_2. \end{aligned} \tag{49}$$

$$\begin{aligned} &\Delta u_\delta^{(k)} = v_\delta^{(k)}, \quad x \in \Omega, \\ &u_\delta^{(k)}|_\Gamma = g. \end{aligned} \tag{50}$$

(iii) Calculate the new approximation of $v_{\delta 0}$ and φ_δ

$$v_{\delta 0}^{(k+1)} = v_{\delta 0}^{(k)} - \tau_{\delta,k+1}\left(b\frac{\partial u_\delta^{(k)}}{\partial \nu} + \delta v_{\delta 0}^{(k)} - bg_1\right), \quad x \in \Gamma_1, \tag{51}$$

$$\varphi_\delta^{(k+1)} = \varphi_\delta^{(k)} - \tau_{\delta,k+1}(\varphi_\delta^{(k)} + bu_\delta^{(k)}), \quad x \in \Omega. \tag{52}$$

5. Numerical Experiments

We perform some limited experiments in MATLAB for testing the convergenve of the approximate solution given by (40) with fixed $N = 2$ and δ will be chosen in agreement with the steps of the grid for discretizing differential problems. We consider the computational domain $\Omega = (0,1)^2$ with uniform grid and steps of the grid are $h_1 = h_2 = 1/M$. The part Γ_2 is the top side and Γ_1 is the remaining part of the boundary of the square Ω . For the iterative process we choose by experimental way the parameter $\tau = \frac{2}{2\delta+0.75\sqrt{2}}$. The Dirichlet problems (49), (50) are discretized by difference schemes of second order approximation and the normal derivative in (51) is approximated by difference formula of the same order of accuracy. The stopping criterion for the iterative process (49)-(52) is

$$\max(||v_{\delta 0}^{(k+1)} - v_{\delta 0}^{(k)}||_\infty, ||\varphi_\delta^{(k+1)} - \varphi_\delta^{(k)}||_\infty) < \varepsilon = h_1 h_2.$$

and in view of the estimate (42) we choose $\delta = \varepsilon^{1/3}$. For solving systems of grid equations the method of complete reduction, whose idea is the successive odd-even eliminations,[12] is used.

We take some exact solutions and for $b = 1$ construct the right-hand sides and the boundary data in respect with them. The results of computation on PC Pentium 4 with CPU 1.80 GHZ for these examples are presented in the following tables, where $Error = ||U^E - u||_\infty$ and in the column "Number of iterations " we report the number of iterations for finding the basic solution $u_{\delta/i}$ $(i = 1, 2, 3)$, the time is in seconds.

Table 1. Case $u = (x_1^2 - 1)(x_2^2 - 1)$

Grid	Number of iterations			Error	Time
16 × 16	13	19	24	0.0023	0.88
32 × 32	23	37	48	0.0007	4.95
64 × 64	42	70	93	0.0003	42.51

Table 2. Case $u = \sin(\pi x_1)\sin(\pi x_2)$

Grid	Number of iterations			Error	Time
16 × 16	10	16	21	0.0050	0.78
32 × 32	17	26	33	0.0013	3.52
64 × 64	27	39	47	0.0003	23.59

It is interesting to notice that for the above examples in the case if $\delta = 0$ our experiments show that the iterative process (49)-(52) does not

Table 3. Case $u = 0.25x_1^4 + 0.25x_2^4 + x_1^2 + x_2^2$

Grid	Number of iterations			Error	Time
16 × 16	16	25	32	0.0056	1.08
32 × 32	28	46	61	0.0022	6.17
64 × 64	50	84	115	0.0009	51.14

converge. It justifies the reason, why we have to consider the perturbed problem (27)-(30) and extrapolate its solution by the parameter δ.

References

1. A. A. Abramov and V. I. Ulijanova, *Journal of Comput. Math. and Math. Physics*, **32**, 567 (1992) (Russian).
2. Dang Quang A, *Math. Physics and Nonlinear Mechanics*, **44**, 54 (1988) (Russian).
3. Dang Quang A, *Vietnam Journal of Math.* **33**, 9 (2005).
4. Dang Quang A, *Journal of Comput. and Appl. Math.*, **51**, 193 (1994).
5. Dang Quang A, *Vietnam Journal of Math.* **26**, 243 (1998)
6. Dang Quang A, *Journal of Comp. Sci. and Cyber.* No. 4, 66 (1998).
7. Dang Quang A, *Journal of Comp. and Appl. Math.* **196**, 634 (2006).
8. A. Dorodnisyn, N. Meller, *Journal of Comp. Math. and Math. Physics* **8**, 393 (1968) (Russian).
9. R. Glowinski, J-L. Lions and R. Tremoliere, *Analyse Numerique des Inequations Variationelles*, (Dunod, Paris, 1976).
10. J-L. Lions and E. Magenes, *Problemes aux Limites non Homogenes et Applications*, Vol. 1, (Dunod, Paris, 1968).
11. B. V. Palsev, *Journal of Comput. Math. and Math. Physics* **6**, 43 (1966) (Russian).
12. A. Samarskii and E. Nikolaev, *Numerical Methods for Grid Equations*, Vol. 1: Direct Methods, (Birkhäuser, Basel, 1989).
13. A. Samarskii and E. Nikolaev, *Numerical Methods for Grid Equations*, Vol. 2: Iterative Methods, (Birkhäuser, Basel, 1989).

Advances in Deterministic and Stochastic Analysis
Eds. N. M. Chuong *et al.* (pp. 115–129)

§8. NUMERICAL SOLUTION TO A NON-LINEAR PARABOLIC BOUNDARY CONTROL PROBLEM

DINH NHO HAO*‡, NGUYEN TRUNG THANH and H. SAHLI

Vrije Universiteit Brussel,
Department of Electronics and Informatics,
Pleinlaan 2, 1050 Brussel, Belgium
‡*E-mail: hao@math.ac.vn*

The so-called difference of convex functions algorithm and the continuation technique are applied to solving the nonlinear parabolic boundary control problem

$$f(u) = \frac{1}{2}\int_0^1 |y(x,T;u) - z(x)|^2 dx + \frac{\alpha}{2}\int_0^T (u(t) - u^*(t))^2 dt,$$

where $\alpha > 0$ is given and $y(x,t) = y(x,t;u)$ is the solution to the non-linear parabolic problem

$$\begin{aligned}
&y_t(x,t) = y_{xx}(x,t), \quad 0 < x < 1, 0 < t < T,\\
&y(x,0) = y_0(x), \quad 0 < x < 1,\\
&y_x(0,t) = 0, \quad y_x(1,t) = g(y(1,t)) + u(t), \quad 0 < t < T
\end{aligned}$$

with $\{u \in L^\infty([0,T]) | u_{\min}(t) \le u(t) \le u_{\max}(t), \text{ for a.e. } t \in [0,T]\}$. Numerical examples are given to show the efficiency of the method.

1. Introduction

In Ref. 8 Kelley and Sachs proposed a trust region method for the problem of minimizing

$$f(u) = \frac{1}{2}\int_0^1 |y(x,T;u) - z(x)|^2 dx + \frac{\alpha}{2}\int_0^T u^2(t) dt, \tag{1}$$

where $\alpha > 0$ is given and $y(x,t) = y(x,t;u)$ is the solution to the non-linear parabolic problem

$$\begin{aligned}
&y_t(x,t) = y_{xx}(x,t), \quad 0 < x < 1, 0 < t < T,\\
&y(x,0) = y_0(x), \quad 0 < x < 1,\\
&y_x(0,t) = 0, \quad y_x(1,t) = g(y(1,t)) + u(t), \quad 0 < t < T.
\end{aligned} \tag{2}$$

*Also at Institute of Mathematics, 18 Hoang Quoc Viet Road, 10307 Hanoi, Vietnam

In (1)–(2) u is constrained to be in the set

$$\mathcal{U} = \{u \in L^\infty([0,T]) | u_{\min}(t) \leq u(t) \leq u_{\max}(t), \text{ for a.e. } t \in [0,T]\} \tag{3}$$

and the nonlinear function g is assumed to satisfy

$$g \in C^2(R), \quad g', g'' \in L^\infty(R). \tag{4}$$

In order to use some known results by Sachs[16] and Schmidt[17] of the problem (1)–(2) we will need some monotonicity conditions on g that we will describe in section 2.

The problem (1)–(2) can be understood as a variational formulation of the inverse problem where one wishes to find the heating regime $u(t)$ at a part of the boundary such that at the final time the temperature of the system reaches a given desired goal z. The particular case $g(y) = -y^4$ corresponding to the Stefan-Boltzmann radiation condition seems to have many applications in engineering.[6,7,16]

The method of Kelley and Sachs[8] is a projected form of the Steihaug trust-region-CG method with a smoothing step added at each iteration to improve performance in the global phase and provide mesh-independent sup-norm convergence in the terminal phase. However, whether their method delivers a global solution to (1)–(2) is not clear.

In this paper we approach the control problem (1)–(2) from a quite different point of view. Namely, from that of DC programming. We realize that f is a DC one, i.e. it can be represented by the *difference of two convex functions*. Therefore, we can consider (1)–(2) as a DC program and apply the DC programming techniques to solve the problem. In fact we apply the *difference of convex functions algorithm* (DCA).[12,13] As DCA is a local method, it cannot guarantee to deliver a global solution, in this work we combine it with a so called *continuation* or *cooling*[1] technique for the search of a global solution. We note that DCA is applicable to finite-dimensional problems only, but in order to solve (1)–(2) we need to disretize it somehow. The convergence of the solutions of the discretized control problems of type (1)–(2) to that of the continuous one has been proved for some cases in Refs. 15 and 10. The problem (1)–(2) has been first considered by Sachs[16] and then by Schmidt,[17] see also Refs. 3 and 14.

In the next section we prove that the functional f is twice Fréchet differentiable and deliver the formulas for the gradients as well as prove that f can be represented by the difference of two convex functionals. In section 3 the difference of convex functions algorithm (DCA) and the combination of DCA with the continuation technique for solving non-convex optimization problems will be described. Finally, in the last section numerical results based on our method will be given.

2. The Control Problem and Its dc Representation

In this section we prove that (1)–(2) is a DC program. Further, to make the algorithm more efficient we change the functional f by

$$F(u) = \frac{1}{2}\int_0^1 |y(x,T;u) - z(x)|^2 dx + \frac{\alpha}{2}\int_0^T (u(t) - u^*(t))^2 dt. \qquad (5)$$

Here u^* plays the role of a guess for the solutions or a selection criterion for the solutions: as (1)–(2) could have many solutions to, we choose that one which is nearest to u^*, by minimizing (5).

Now we review some results on the solution of (2). Let I be a bounded subinterval of the reals R, say, $I = [a,b], -\infty < a < b < \infty$. Let B be a norm space. By writing $y \in B_I$ we mean that $y \in B$ with ranges in I. We denote $\|\cdot\|_{L^2(0,1)}$ by $\|\cdot\|$.

By $H^1(0,1)$ we denote the Sobolev space of all elements $y \in L^2(0,1)$ having weak derivative u_x in $L^2(0,1)$. The space $H^1(Q)$ is similarly defined. Here, $Q := (0,1) \times (0,T]$. Further, let

$$V^{1,0}(Q) := C([0,T]; L^2(0,1)) \cap L^2((0,T); H^1(0,1))$$

with the norm

$$|y| := \text{vrai} \max_{0 \le t \le T} \|u(\cdot,t)\|_{L^2(0,1)} + \|u_x\|_{L^2(Q)}.$$

On the notion of Sobolev spaces we refer the reader to Ref. 9.

Let $u_{\min}, u_{\max}$ be given functions with ranges in I belonging to $L^\infty([0,T])$. To be consistent with the results of Schmidt[17] we additionally suppose that g is strictly decreasing in I.

Definition 2.1. Let $y_0 \in L^2_I(0,1)$. Then $y \in V_I^{1,0}(Q)$ is said to be a weak solution of (2) if for all $\phi \in H^1((0,1) \times (0,T))$ with $\phi(\cdot,T) = 0$, then the following integral identity holds

$$\int_0^T \int_0^1 (-y(x,t)\phi_t(x,t) + y_x(x,t)\phi_x(x,t))dxdt$$
$$= \int_0^1 y_0(x)\phi(x,0)dx + \int_0^T (g(y(1,t)) + u(t))\phi(1,t)dt.$$

It is proved in Ref. 17 that there is a unique weak solution to (2). And it is proved in Refs. 3 and 14 that $y \in L^\infty(Q) \cap C(\overline{Q_{\epsilon,T}})$ for every $\epsilon > 0$. Here $Q_{\epsilon,T} = (0,1) \times (\epsilon,T)$.

Now we prove that F is DC and propose a DC decomposition for it, in doing so we note that $F \in C^2(\mathcal{U})$. Denote by $G : L^\infty(0,T) \to V^{1,0}(Q)$ the mapping associating the solution of (2) to every function u. In what follows, we denote the first- and second-order Fréchet derivatives of G at a point u in the direction h or (h,k) by $G'(u)h$ and $G''(u)(h,k)$, respectively. Following Refs. 16,17 and 4 it can be proved that G is twice continuously Fréchet differentiable. If $u, h \in L^\infty(0,T), \varphi := G'(u)h$, then $\varphi = \varphi(x,t;h)$ is the solution of the problem

$$\begin{aligned}
&\varphi_t(x,t;h) = \varphi_{xx}(x,t;h), \quad 0 < x < 1, 0 < t < T,\\
&\varphi(x,0;h) = 0, \quad 0 < x < 1,\\
&\varphi_x(0,t;h) = 0, \quad \varphi_x(1,t;h) = g'(y(1,t;u))\varphi(1,t;h) + h, 0 < t < T.
\end{aligned} \tag{6}$$

If $h, k \in L^\infty(0,T)$ and $v = v(x,t;h,k) := G''(u)(h,k)$, then v is the solution of the problem

$$\begin{aligned}
&v_t = v_{xx}, \quad 0 < x < 1, 0 < t < T,\\
&v(x,0;h,k) = 0, \quad 0 < x < 1,\\
&v_x(0,t;h,k) = 0, \quad 0 < t < T,\\
&v_x(1,t;h,k) = g'(y(1,t;u))v(1,t;h,k)\\
&\qquad\qquad + g''(y(1,t;u))\varphi(1,t;h)\varphi(1,t;k), \quad 0 < t < T.
\end{aligned} \tag{7}$$

From these results we obtain that F is twice continuously Fréchet differentiable and, if $h \in L^\infty(0,T)$, then

$$F'(u)h = \int_0^1 \varphi(x,T;h)(y(x,T;u) - z(x))dx + \alpha \int_0^T (u - u^*)hdt, \tag{8}$$

and

$$\begin{aligned}
F''(u)(h,h) = &\int_0^1 |\varphi(x,T;h)|^2 dx\\
&+ \int_0^1 v(x,T;h,h)(y(x,T;u) - z(x))dx + \alpha \|h\|^2_{L^2(0,T)}.
\end{aligned} \tag{9}$$

It can be also proved that the gradient of F is

$$F'(u) = d(1,t) + \alpha(u - u^*), \tag{10}$$

where $d(x,t)$ is the solution of the adjoint problem

$$\begin{aligned}
&-d_t(x,t) = d_{xx}(x,t), \quad 0 < x < 1, 0 < t < T,\\
&d(x,T) = y(x,T;u) - z(x), \quad 0 < x < 1,\\
&d_x(0,t) = 0, \quad d_x(1,t) = g'(y(1,t;u))d(1,t), 0 < t < T.
\end{aligned} \tag{11}$$

Now we estimate $\|F''(u)\|$. We need the following result by Schmidt in Ref. 17, Corollary A.3.

Lemma 2.1. *Let w be the weak solution of the problem*

$$\begin{aligned} &w_t = w_{xx}, \quad 0 < x < 1, 0 < t \le T, \\ &-w_x(0,t) + \sigma_1(t)w(0,t) = f_1, \quad w_x(1,t) + \sigma_2(t)w(1,t) = f_2, \quad 0 < t \le T, \\ &w(\cdot, 0) = w_0, \quad 0 < x < 1, \end{aligned} \tag{12}$$

where σ_1, σ_2 are bounded, measurable, non-negative functions defined on $[0,T]$, $w_0 \in L^2(0,1) \cap L^\infty(0,1)$, $f_1, f_2 \in L^2(0,T) \cap L^\infty(0,T)$. Then there exists a positive constant C such that

$$\|w\|_{L^\infty(Q)} \le C \max\{\|w_0\|_{L^\infty(0,1)}, \|f_1\|_{L^\infty(0,T)}, \|f_2\|_{L^\infty(0,T)}\}.$$

Remark 2.1. By inspecting the proof of Schmidt we get

$$C = 2T + \frac{5}{4}.$$

Let

$$\max_{u \in I} |g''(u)| = \mathcal{G},$$

Applying Lemma 2.1 to (6), we get

$$\|\varphi(\cdot,\cdot;h)\|_{L^\infty(Q)} \le C\|h\|_{L^\infty(0,T)}. \tag{13}$$

Again applying Lemma 2.1 and the last inequality to (7) for $k = h$, we get

$$\|v(\cdot,\cdot;h,h)\|_{L^\infty(Q)} \le C^3\mathcal{G}^3\|h\|^2_{L^\infty(0,T)}. \tag{14}$$

From (9) and the two last inequalities we have

$$|F''(u)(h,h)| \le \left[TC^3(1+\mathcal{G})(\max\{|a|,|b|\} + \|z\|_{L^\infty(0,T)}) + \alpha\right] \|h\|^2_{L^\infty(0,T)}.$$

Let

$$\rho := TC^3(1+\mathcal{G})(\max\{|a|,|b|\} + \|z\|_{L^\infty(0,T)}) + \alpha. \tag{15}$$

Thus, the function F has now the following DC representation

$$F(u) = \left\{\frac{\rho}{2}\|u\|^2_{L^2(0,T)} + F(u)\right\} - \left\{\frac{\rho}{2}\|u\|^2_{L^2(0,T)}\right\}. \tag{16}$$

3. DCA and Continuation Techniques

3.1. *DCA*

3.1.1.

A function f defined in a convex set is called DC, if it can be written in the form $f(x) = g(x) - h(x)$. Here, DC means *Difference of Convex functions.* We suppose that f is defined in a Hilbert space X, although in the future we mainly work with $\mathbb{R}^n$. Note that the above representation of f is not unique, since $[g(x) + \alpha\|x\|^2] - [h(x) + \alpha\|x\|^2]$ for any positive α is also a DC representation for it.

3.1.2.

We say that a convex function g is proper, if $g > -\infty$ and there exists x such that $g(x) \neq +\infty$. In this case, we write $g \in \Gamma_0(X)$.

3.1.3.

When we work with a convex function g defined in a convex set C we can extend its domain into the whole X by setting its values to $+\infty$ outside C so that the new function remains convex. Hence in the future we will work with convex functions defined in the whole X.

3.1.4. *Let g be a convex function.*

Its *subdifferential* is defined by

$$\partial g(x) := \{p \in X | g(y) \geq g(x) + < y - x, p >, \forall y \in X\}.$$

Here, $< \cdot, \cdot >$ is the inner product in X.
It is known that g attains its minimum at x^* if and only if $0 \in \partial g(x^*)$.

3.1.5. *Let $g \in \Gamma_0(X)$ and lower semi-continuous (lsc).*

The *Fenchel transform* of g is defined by

$$g^*(y) := \sup_x \{< x, y > - g(x)\}.$$

It can be proved that g^* belongs to $\Gamma_0(X)$ and is lower semi-continuous. Furthermore, $g^{**} = g$.

Note that $y \in \partial g(x)$ if and only if $x \in \partial g^*(y)$, if and only if $< x, y >= g(x) + g^*(y)$.

3.1.6. *Consider the DC program*

$$(P) \qquad \alpha := \inf_x \{f(x) = g(x) - h(x)\}.$$

Here g and h are proper convex functions and lower semi-continuous. With the notation dom $g := \{x|g(x) < +\infty\}$ we see that if $|\alpha| < \infty$, then dom $g \subset$ dom h and dom $h^* \subset$ dom g^*.

The program (P) has its dual part

$$(D) \qquad \alpha = \inf_y \{h^*(y) - g^*(y)\}.$$

3.1.7. *DC algorithm:*

Since (P) and (D) are not convex programs, we approximate them by convex ones. Namely, the algorithm for solving (P) will use its dual part (D) and their convex approximations. We construct two sequences $\{x_k\}$ and $\{y_k\}$ as follows.

Let $x_0 \in$ dom g.

$$(D) \Rightarrow (D_k) \qquad \min_y \{h^*(y) - [g^*(y_{k-1}) + < x_k, y - y_{k-1} >]\}.$$

We see that $g^*(y_{k-1}) + < x_k, y - y_{k-1} >$ is an affine approximation to $g^*(y)$ at y_{k-1}. Hence $h^*(y) - [g^*(y_{k-1}) + < x_k, y - y_{k-1} >]$ is a convex function. It follows that the minimum of the last function attains at y_k which is characterized by

$$0 \in \partial h^*(y_k) - x_k$$

The last is equivalent to

$$x_k \in \partial h^*(y_k)$$

which is the condition that

$$y_k \in \partial h(x_k).$$

So from x_k and y_{k-1} we constructed y_k. Again to update x_k, with the same idea, we consider the k-th convex approximation to (P)

$$(P_k) \qquad \min_x \{g(x) - [h(y_k) + < x - x_k, y_k >]\}.$$

The solution of the last is x_{k+1} which is characterized by the chain of equivalencies

$$0 \in \partial g(x_{k+1}) - y_k \Leftrightarrow y_k \in \partial g(x_{k+1}) \Leftrightarrow x_{k+1} \in \partial g^*(y_k).$$

We summarize the algorithm as follows

$$x_0 \in \text{dom } g,$$

$$y_k = \text{Arg} \min_y \{h^*(y) - [g^*(y_{k-1}) + < x_k, y - y_{k-1} >\},$$

$$x_{k+1} = \operatorname{Arg}\min_x\{g(x) - [h(y_k) + < x - x_k, y_k >]\}.$$

The last equality is equivalent to

$$y_k \in \partial h(x_k), \quad x_{k+1} \in \partial g^*(y_k).$$

The convergence properties of DCA have been proved in Refs. 12 and 13.

3.2. *The Continuation Technique*

DCA provides however only critical points of f. In order to find a global minimizer of f we consider the regularized function $f_\alpha(x) := f(x)+\alpha/2\|x\|^2$ with $\alpha > 0$. When α is very large, in many cases it is obvious that f_α is convex. In these cases we can find a global minimizer of f_α. After finding this minimizer, which we denote by x_{α_0}, we reduce α to $\alpha_1 < \alpha_0$ and minimize f_{α_1} by an iterative method starting from x_{α_0} and so on. This is the continuation method.[1] As DCA is globally convergent, if we apply DCA to this procedure we can hope that we reach a global minimizer of f.

The combination of DCA and the continuation technique is summarized in the following algorithm

Algorithm 3.1. $\min_x f(x)$

(1) Initialization: $x_0 \in domf$, $\rho \geq \|f''\|$, *and a finite decreasing positive sequence* $\{\alpha_i\}_{i=0}^M$.
(2) Choose $h(x)$ *such that* $f(x) + h(x)$ *is convex (e.g.,* $h(x) = \frac{\rho}{2}\|x\|^2$*)*
(3) ***For*** $i = 0, 1, 2, \ldots, M$

3.1 $g_i(x) = f(x) + \frac{\alpha_i}{2}\|x\|^2 + h(x)$, $k = -1$
3.2 ***While*** *stop criterion has not been satisfied*

3.2.1 $k = k + 1$
3.2.2 $y_k \in \partial h(x_k)$
3.2.3 $x_{k+1} = argmin\{g_i(x) - < y_k, x >\}$

End (While)
3.3 $x_{\alpha_i} = x_{k+1}$ *(minimizer of* $f(x) + \frac{\alpha_i}{2}\|x\|^2$*)*
3.4 $x_0 = x_{\alpha_i}$ *(change the initial guess in the continuation technique)*

End (for).

The sub optimization problems in 3.2.3 of Algorithm 3.1 are convex. In this paper, we used the Fletcher-Reeves nonlinear conjugate gradient method with Amijo-type line search, which ensures global convergence of the algorithm, for solving these sub-problems (see Ref. 5 for more details).

4. Numerical Results

To solve (2) we use an iterative technique to deal with the non-linear boundary condition and the finite difference method for the linear boundary value problem. Namely, let

$$y^0(x,t) := y_0(x)$$

and $y^n(x,t)(n = 0, 1, 2, \ldots)$ be the n-th iteration. Solve

$$\begin{aligned}
&y_t^{n+1}(x,t) = y_{xx}^{n+1}(x,t), \quad 0 < x < 1, 0 < t < T,\\
&y^{n+1}(x,0) = y_0(x), \quad 0 < x < 1, \\
&y_x^{n+1}(0,t) = 0, \quad y_x^{n+1}(1,t) = g(y^n(1,t)) + u(t), \quad 0 < t < T.
\end{aligned} \tag{17}$$

We stop the iteration procedure when some tolerance between y^n and y^{n+1} is achieved. We note that for $n > 1$, the difference $w = y^{n+1} - y^n$ satisfies the simpler system

$$\begin{aligned}
&w_t(x,t) = w_{xx}(x,t), \quad 0 < x < 1, 0 < t < T,\\
&w(x,0) = 0, \quad 0 < x < 1, \\
&w_x(0,t) = 0, \quad w_x(1,t) = g(y^n(1,t)) - g(y^{n-1}(1,t)), \quad 0 < t < T.
\end{aligned} \tag{18}$$

than y^1 and $y^n, n > 1$. For the convergence of this method we refer the reader to Ref. 2.

To solve (17) we use the Crank-Nicholson finite difference scheme. We devide $[0,1]$ by $N-1$ equal subintervals: $x_0 = 0, x_1 = h, \ldots, x_i = ih, \ldots, x_N = 1$, where $h = 1/N$, and $[0,T]$ by $M-1$ equal subintervals: $t_0 = 0, t_1 = \tau, \ldots, t_j = j\tau, \ldots, t_M = T$. In order to approximate the boundary conditions by central finite difference quotients, we extend the interval $[0,1]$ one more layer to each direction and denote the extended grid points by x_{-1} and x_{N+1}. For a function y defined on $[0,1] \times [0,T]$, we set $y_i^j := y(x_i, t_j)$. For simplicity of notion, we denote the solution of (17) by Y and the inhomogeneous boundary condition by V. We discretize (17) as follows

$$\begin{aligned}
&\frac{Y_i^{j+1} - Y_i^j}{\tau} = \frac{Y_{i+1}^{j+1} - 2Y_i^{j+1} + Y_{i-1}^{j+1}}{2h^2} + \frac{Y_{i+1}^j - 2Y_i^j + Y_{i-1}^j}{2h^2},\\
&\qquad\qquad 0 \le i \le N, 0 \le j \le M-1,\\
&Y_i^0 = y_{0,i}, \quad 0 \le i \le N,\\
&\frac{Y_1^j - Y_{-1}^j}{2h} = 0, \quad \frac{Y_{N+1}^j - Y_{N-1}^j}{2h} = V^j, \quad 1 \le j \le M.
\end{aligned} \tag{19}$$

It is well-known that this scheme is absolutely stable and has second order approximation with respect to the both variables.[11] The Crank-Nicholson scheme is also used to solve the adjoint problem (11).

4.1. *Numerical Examples*

In this subsection we illustrate DCA and the continuation technique for the control problem (5) subject to (2) with some concrete examples.

Example 4.1. Let a be a number in $(0, \pi]$. Then

$$y(x,t) = e^{-a^2 t}\cos(ax)$$

is the solution of the problem

$$\begin{aligned} &y_t(x,t) = y_{xx}(x,t), \quad 0 < x < 1, 0 < t < T, \\ &y(x,0) = \cos(ax), \quad 0 < x < 1, \\ &y_x(0,t) = 0, \quad y_x(1,t) = -ae^{-a^2 t}\sin a, \quad 0 < t < T. \end{aligned} \tag{20}$$

Let

$$u(t) := e^{-4a^2 t}\cos^4 a - ae^{-a^2 t}\sin a.$$

We see that

$$y_x(1,t) = -y^4(1,t) + u(t).$$

Thus, in this example we set $g(y) = -y^4$. To test our method we set

$$z(x) = y(x,T) = e^{-a^2 T}\cos(ax).$$

In this example, T is set to be 1 and the space and time grid sizes are taken to be 0.01 and 0.025, respectively, resulting in a nonlinear optimization problem of 41 free variables. The control function u is then approximated by linear interpolation from the discrete values for solving the main problem (2) as well as the adjoint problem (11). We tested the algorithm with two different values of the parameter a.

In the first case, we set $a = \pi$, then $u(t) = e^{-4\pi^2 t}$ and $z(x) = e^{-\pi^2 T} cos(\pi x)$. The lower and upper bounds of the control function are taken to be $u_{min}(t) = -1$, and $u_{max}(t) = 2, \ t \in [0,1]$. In implementing the continuation technique, the regularization parameter α is reduced five times at each iteration from 5 at the first iteration to $5e-9$ at the last one.

To analyze the effects of the initial guess and the regularizing function $u^*(t)$ on the reconstruction of the control function, the algorithm was run with two different initial guesses

$$u_0^1(t) = 1 - t, \quad u_0^2(t) = 2, \ t \in [0,1],$$

and two different regularizing functions

$$u^{*,1}(t) = 0, \ t \in [0,1],$$

$$u^{*,2}(t) = 1 - 8t, \; t \in [0, 0.1], u^{*,2}(t) = \frac{2}{9}(1 - t), \; t \in (0.1, 1].$$

The results of the four cases are given in Table 1 and Figure 1.

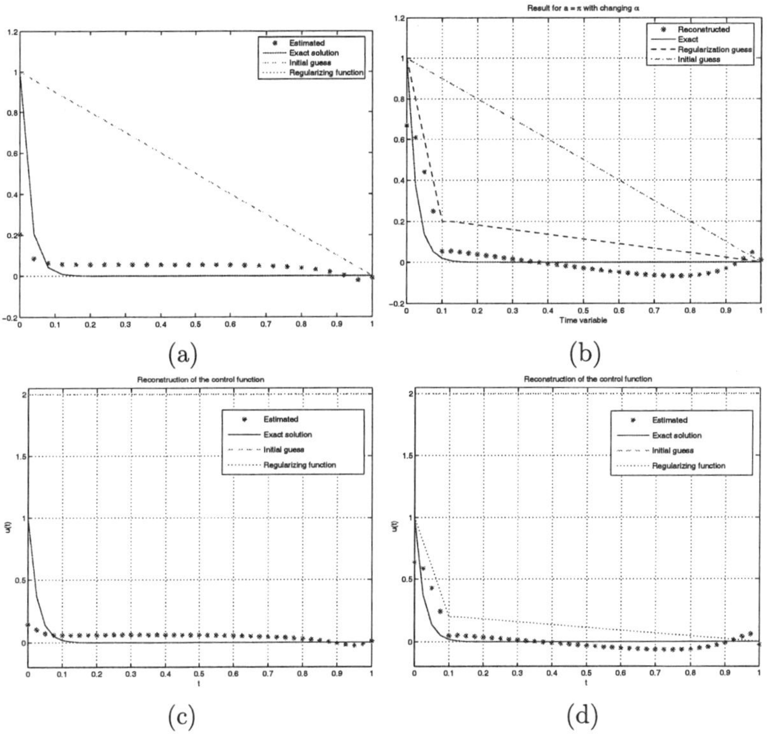

Fig. 1. Reconstruction of the control function of example 1, with a $= \pi$

As we can see from the figures that the regularizing function $u^*(t)$ has certain effects on the results. In the case that the regularizing function $u^*(t)$ is taken to be zero, the estimated control function does not approximate very well the exact solution at the points for which the exact solution is far from zero (Figures 1-(a) and 1-(c)). In the cases that the regularizing function $u^*(t)$ is taken to be $u^{*,2}(t)$, which is near to the exact solution, the results are better (Figures 1-(b) and 1-(d)). The figures also show that the algorithm possibly converges to the exact solution with different values of initial guess, so we hope that it can converge to the global minimum of the regularized objective function. However, we could not test the algorithm with so many different cases of the initial guess because of its time-consuming calculation.

Table 1. Result of example 1, with a $= \pi$

Test case	Objective function value	Obj. func. gradient's norm	Number of iterations
a	3.6840e-008	8.3236e-007	2102
b	2.1032e-007	2.2014e-006	1129
c	3.4826e-008	8.3246e-007	2317
d	3.1505e-007	3.0112e-006	971

It is noted from Table 1 that the problem is severely ill-posed. The values of the objective function is very small with different values of the control function $u(t)$, particularly, it is much smaller than the discretization error in solving the main and the adjoint problems. It is well-known that for severely ill-posed problems, as in this example, we cannot expect a very good estimation for the control function. However, for the optimal control problem, in which one only need to find a function $u(t)$ such that the temperature of the system at the final state is close enough to a given function, the algorithm works very well.

We also tested the algorithm on the previous example for $a = \pi/2$, then $u(t) = -(\pi/2)e^{-\pi^2 t/4}$, $z(x) = e^{-\pi^2 T/4}\cos(\pi x/2)$. The lower and upper bounds of the control function are $u_{min}(t) = -2$, and $u_{max}(t) = 0$, $t \in [0,1]$. All the setup of the test is similar to that of the previous case. The initial guesses and the regularizing functions in this case are taken to be

$$u_0^1(t) = 0, \quad u_0^1(t) = -1.58 + 1.4t, \; t \in [0,t],$$

$$u^{*,1}(t) = 0, \quad u^{*,2}(t) = -1.58 + 1.4t, \; t \in [0,t].$$

The results are given in Figure 2 and Table 2.

The figures show the estimated control function compared with the exact one. In cases (a) and (b) ((b) and (d) resp.), the algorithm was started at the same initial guess $u_0^1(t)$ ($u_0^1(t)$ resp.) while in cases (a) and (b) ((c) and (d) resp.) it used the same regularizing function $u^{*,1}(t)$ ($u^{*,2}(t)$ resp.). It is clear that the cases (c) and (d) gives better reconstructions than the other two cases as the effect of the regularizing function $u^*(t)$, while the initial guess does not effect so much the result. However, as we can see from Table 2, the values of the objective function are very close to one another and close to zero. This behavior again shows the severe ill-posedness of the problem.

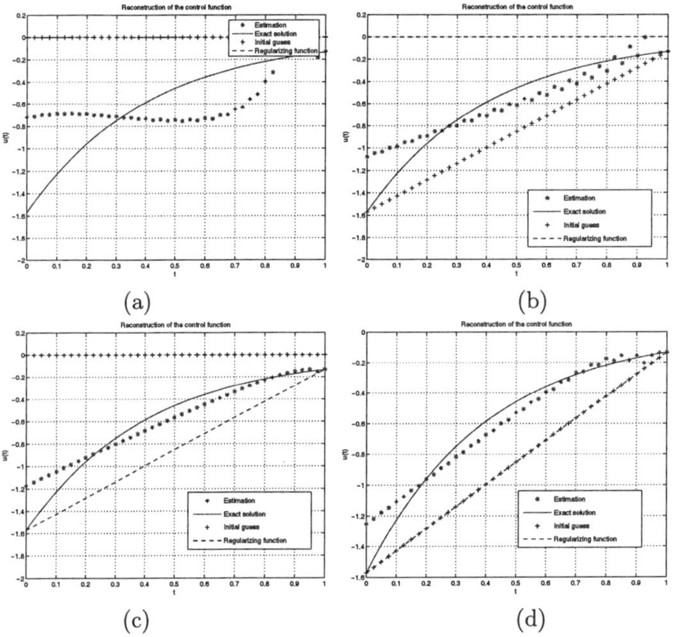

Fig. 2. Reconstruction of the control function of example 1, with a $= \pi/2$

Table 2. Result of example 1, with a $= \pi/2$

Test case	Objective function value	Obj. func. gradient's norm	Number of iterations
a	4.7510e-008	9.5507e-007	16848
b	1.1387e-009	3.7500e-007	2866
c	2.0380e-009	1.2505e-007	40230
d	7.5335e-010	1.2505e-007	32760

Example 4.2. In order to compare our results with those of Kelley and Sachs in Ref. 8, we consider here the same example as mentioned in Ref. 8, in which $g(y) = \frac{10y^4}{1000+y^4}$, $y_0(x) = 0$, $z(x) = 1$, and $T = 1$. The constraint of the control function is $0 \leq u(t) \leq 0.1 + 4t$.

All the discretization parameters in this example are as in Example 1. Unfortunately, the exact solution is not available in this example so we cannot have any guess about the solution of the control problem. We also tested the algorithm with two different initial guesses given by

$$u_0^1(t) = 0, \quad u_0^2(t) = 4t, \ t \in [0, 1],$$

and two different cases of the regularizing function $u^*(t)$

$$u^{*,1}(t) = 0, \quad u^{*,2}(t) = 4t, \; t \in [0, 1].$$

The results are given in Figure 3 and Table 3.

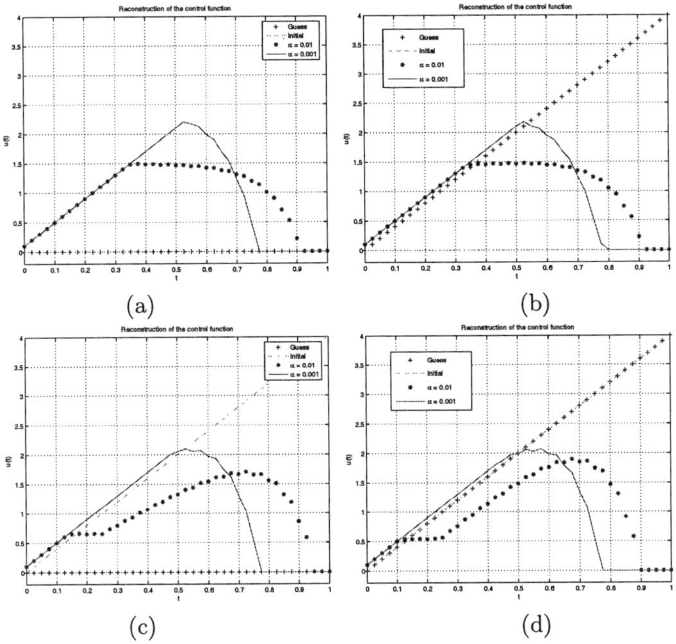

Fig. 3. Reconstruction of the control function of example 2

Table 3. Result of example 2

Test case	Objective function value	Obj. func. gradient's norm	Number of iterations
a	9.9149e-004	4.3683e-004	2816
b	7.2189e-004	4.3675e-004	3108
c	9.9563e-004	4.5075e-004	2740
d	7.2604e-004	4.5646e-004	2710

It follows from Figure 3 that, for the initial guess $u_0^1(t)$ and regularizing function $u^{*,1}(t)$, (Figure 3-(a)), and $\alpha = 0.01$, the result is the same as that of Kelley and Sachs. The result is also very similar in the case of the second initial guess $u_0^2(t)$ but with the same regularizing function (Figure 3-(b)). However, the results are different for the regularizing function $u^{*,2}(t)$ for

$\alpha = 0.01$ (see the difference between (a) and (b) with (c) and (d)). Those results show the effect of the regularizing function on the results. However, when α is reduced, the four tests converge to almost the same solution. This behavior ensures the efficiency of DCA and the continuation technique in searching for the global solution of the problem.

In conclusion, the combination of DCA and the continuation technique has been demonstrated to be possible for obtaining the global solution of the control problem (1)–(2). We also showed that the obtained result by this method is consistent with the result obtained by the trust region method proposed by Kelly and Sachs[8] in the case of the same regularization parameter and function, and the same initial guess. However, our technique is apparently proved to behave better because it reaches the same solution with different initial guesses as well as different regularizing functions.

References

1. J. C. Alexander and J. A. Yorke, *Trans. Amer. Math. Soc.* **242**, 271 (1978).
2. W. F. Ames, *Numerical Methods for Partial Differential Equations*, Third edition, Computer Science and Scientific Computing, (Academic Press, Inc., Boston, MA, 1992).
3. E. Casas, *SIAM J. Control Optim.* **35**, 1297 (1997).
4. E. Casas and F. Tröltzsch, *Appl. Math. Optim.* **39**, 211 (1999).
5. Yu-hong Dai, *Journal of Conputational Mathematics*, **9**, 539 (2001).
6. Dinh Nho Hào: *Methods for Inverse Heat Conduction Problems*, (Peter Lang, Frankfurt am Main, 1998).
7. H. Jiang, T. H. Nguyen and M. Prud'homme, *Control of the boundary heat flux during the heating process of a solid material*, Manuscript.
8. C. T. Kelley and E. W. Sachs, *SIAM J. Optimization* **9**, 1064 (1999).
9. O. A. Ladyzhenskaya, *The Boundary Value Problems of Mathematical Physics*, (Springer-Verlag, New York, Berlin, Heidelberg, Tokyo, 1985).
10. P. Neittaanmki and D. Tiba: *Optimal control of nonlinear parabolic systems. Theory, algorithms, and applications*, (Marcel Dekker, Inc., New York, 1994).
11. M. Necati Özisik: *Boundary value problems of heat conduction*, (Dover Publications, Inc., New York, 1968).
12. T. Pham Dinh and L. T. H. An, *SIAM J. Optim.* **8**, 476 (1998).
13. T. Pham Dinh and El Bernoussi Souad, *Trends in mathematical optimization* (Irsee, 1986), pp. 277–293, *Internat. Schriftenreihe Numer. Math.*, (Birkhäuser, Basel, 1988).
14. J. P. Raymond and H. Zidani, *Appl. Math. Optim.* **39**, 143 (1999).
15. F. Tröltzsch, *Appl. Math. Optim.* **29**, 309 (1994).
16. E. Sachs, *ZAMM* **58**, 443 (1978).
17. E. J. P. G. Schmidt, *J. Diff. Eq.* **78**, 89 (1989).

Advances in Deterministic and Stochastic Analysis
Eds. N. M. Chuong *et al.* (pp. 131–175)

§9. A CLASS OF SOLUTIONS TO MAXWELL'S EQUATIONS IN MATTER AND ASSOCIATED SPECIAL FUNCTIONS

PETER MASSOPUST

GSF - Institute for Biomathematics and Biometry
Neuherberg, Germany
and
Centre of Mathematics, M6
Technical University of Munich
Garching, Germany
massopust@ma.tum.de

The solutions for a class of Maxwell's Equations in matter are presented. These solutions describe the magnetic fields as generated by a hard ferromagnet of finite length with missing mass and are important in the area of nondestructive evaluation.

Keywords: Maxwell's equations, ferromagnetism, Poisson equation, Bessel functions, spherical harmonics, Mathieu functions.

1. Introduction and Preliminaries

In this section the basic model for a hard ferromagnet is presented. The basic model is subject to modifications that describe the observed physical phenomena more closely but these modifications result in sets of solutions that are no longer expressible in simple closed analytical forms. For the purposes of this paper, however, the basic model contains all the important aspects of ferromagnetic theory.

1.1. *Basic Magnetostatics*

Let $\Omega \subset \mathbb{R}^3$ denote a hard ferromagnet* and $\partial\Omega$ its piecewise smooth oriented surface. We assume that Ω is a compact subset of $\mathbb{R}^3$ and has

*A ferromagnet is called *hard* if its magnetization $\mathbf{M}$ is essentially independent of any exterior applied field for moderate field strengths. One may treat such materials as having a fixed specific magnetization $\mathbf{M}$.

nonempty simply-connected interior. Applying an exterior field **H** to Ω induces a magnetization vector field $\mathbf{M} : \Omega \to \mathbb{R}^3$ in the interior $\mathring{\Omega}$ of Ω. It is advantageous to *assume* that **M** vanishes on the boundary $\partial\Omega$: $\mathbf{M}|_{\partial\Omega} \equiv 0$. Note that this introduces a discontinuity of **M** on the boundary, which is not observed in the physical reality but provides a convenient mathematical idealization. The vector fields **H** and **M** produce the magnetic field **B** in $\mathbb{R}^3$ according to the constitutive relationship†

$$\mathbf{B} = \mathbf{H} + 4\pi\,\mathbf{M}. \tag{1}$$

In addition, there exists also a constitutive relation between the vector fields **H** and **B** of the form

$$\mathbf{B} = \boldsymbol{f}(\mathbf{H}),$$

where $\boldsymbol{f}$ is a multiple-valued function referred to as the hysteresis function. Figure 1 shows an example of a hysteresis function. In the neighborhood of

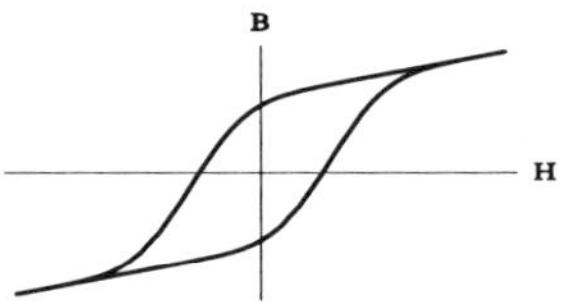

Fig. 1. An hysteresis function.

any point $\mathbf{H}_0$ where $\boldsymbol{f}$ is single-valued, one may define the derivative

$$\mu(\mathbf{H}) := \frac{\partial \mathbf{B}}{\partial \mathbf{H}},$$

called the differential magnetic permeability. For an isotropic ferromagnet and for small values of **H** and **B**, there exists a *linear* relationship between the fields of the form

$$\mathbf{B} = \mu\,\mathbf{H}. \tag{2}$$

Maxwell's equations, which are based on theoretical and experimental results, describe the behavior of electric and magnetic fields in matter. Their differential and integral forms as follows.‡(Here $\langle\,,\,\rangle$ denotes the canonical

†In what follows, cgs units will be used.

‡In the cgs system, the integration of the electric field over the surface of a sphere produces the factor 4π (times the electric charge enclosed). This explains the ubiquitous occurrence of 4π.

inner product on $L^2(\mathbf{R}^3)$.)

$$\langle \nabla, \epsilon \mathbf{E} \rangle = 4\pi \varrho, \qquad \oint_S \langle \epsilon \mathbf{E}, d\boldsymbol{\sigma} \rangle = Q, \tag{3}$$

$$\langle \nabla, \mathbf{B} \rangle = 0, \qquad \oint_S \langle \mathbf{B}, d\boldsymbol{\sigma} \rangle = 0, \tag{4}$$

$$\nabla \times \mathbf{H} = \frac{4\pi}{c} \mathbf{J} + \frac{1}{c} \frac{\partial(\epsilon \mathbf{E})}{\partial t}, \qquad \oint_\gamma \langle \mathbf{H}, d\mathbf{s} \rangle = \frac{4\pi}{c} I + \frac{1}{c} \frac{\partial \Phi^e}{\partial t}, \tag{5}$$

$$\nabla \times \mathbf{E} = -\frac{\partial \mathbf{B}}{\partial t}, \qquad \oint_\gamma \langle \mathbf{E}, d\mathbf{s} \rangle = -\frac{1}{c} \frac{\partial \Phi^m}{\partial t}. \tag{6}$$

Here the vector function $\mathbf{E} : \mathbb{R}^3 \to \mathbb{R}^3$ is called the electric field and ϵ the electric permittivity. The constant c is the speed of light, $\varrho : \mathbb{R}^3 \to \mathbb{R}$ the electric charge density and Q the total electric charge inside the closed surface S. The vector function $\mathbf{J} : \mathbb{R}^3 \to \mathbb{R}^3$ denotes the macroscopic current density and $I := \int_S \mathbf{J} \cdot d\boldsymbol{\sigma}$ the (macroscopic) current encircled by the closed curve γ bounding the surface S. Given any closed orientable piecewise smooth curve γ bounding a surface S, the quantities $\Phi^e, \Phi^m : \mathbb{R}^3 \to \mathbb{R}$ are called the electric and magnetic flux, respectively:

$$\Phi^e := \int_S \langle \mathbf{E}, d\boldsymbol{\sigma} \rangle, \qquad \Phi^m := \int_S \langle \mathbf{B}, d\boldsymbol{\sigma} \rangle. \tag{7}$$

Here the orientation on S is induced by that of γ. Eqs. (7) allows the interpretation of $\mathbf{B}$ as a magnetic flux density, i.e., as the "number" of magnetic field lines enclosed by the curve γ, whereas Φ^e can be interpreted as the "number" of electric field lines piercing the surface S . For similar interpretations of electrodynamic quantities and their modern differential-geometric interpretations in terms of differential forms, the reader is referred to Refs. 13,17.

Eqs. (3) and Eqs. (4) are referred to as Gauss's law. In particular, Eqs. (4) expresses the fact that there are no magnetic monopoles, i.e. that $\mathbf{B}$ is an (axial) vector field or, more precisely, a 2-form: $\mathbf{B} = B_x(\boldsymbol{x})\, dy \wedge dz + B_y(\boldsymbol{x})\, dz \wedge dx + B_z(\boldsymbol{x})\, dx \wedge dy$. The functions $B_x, B_y, B_z : \mathbb{R}^3 \to \mathbb{R}$ are called the x, y, and z-component of $\mathbf{B}$, respectively. Eqs. (5) is commonly called Ampère's law with Maxwell's extension and Eqs. (6) Faraday's Law of Induction.

For the purposes of this paper, it is assumed that there are no electric fields present and that the magnetic field $\mathbf{B}$ is static, i.e., $\partial \mathbf{B}/\partial t \equiv 0$. Hence the set of Maxwell's equations reduces to

$$\langle \nabla, \mathbf{B} \rangle = 0, \qquad \oint_S \langle \mathbf{B}, d\boldsymbol{\sigma} \rangle = 0,$$

$$\langle \nabla, \mathbf{H} \rangle = \frac{4\pi}{c}\, \mathbf{J}, \qquad \oint_\gamma \langle \mathbf{H}, d\mathbf{s} \rangle = \frac{4\pi}{c}\, I.$$

In the case when $\mathbf{J}$ vanishes, $\nabla \times \mathbf{H} = \mathbf{0}$ and thus there exists locally a continuously differentiable scalar function $\Psi : \mathbb{R}^3 \to \mathbb{R}$, called the magnetic scalar potential, so that

$$\mathbf{H} = -\nabla\Psi. \tag{8}$$

Using Eq. (8) together with Eq. (11) transforms $\langle \nabla, \mathbf{B} \rangle = 0$ into the Poisson equation

$$\Delta\Psi = 4\pi \langle \nabla, \mathbf{M} \rangle, \tag{9}$$

where Δ denotes the Laplacian. The right-hand side of Eq. (12) can be thought of as a magnetic charge density $\rho : \mathbb{R}^3 \to \mathbb{R}$, $\boldsymbol{x} \mapsto -\langle \nabla, \mathbf{M} \rangle(\boldsymbol{x})$: Choose a point $p \in \partial\Omega$ and apply the Divergence Theorem to a cylinder of radius $\varepsilon > 0$ and height 2ε centered at p making use of the assumption that $\mathbf{M}$ vanishes on $\partial\Omega$ to obtain that the quantity $\langle \mathbf{M}, \mathbf{n} \rangle$ equals

$$\langle \mathbf{M}, \mathbf{n} \rangle = \frac{1}{4\pi}\, (\nabla\Psi)\Big|_{\partial\Omega} \cdot \mathbf{n}$$

and can therefore be thought of as a magnetic surface charge density. (Here $\mathbf{n}$ denotes the unit (outward) normal to $\partial\Omega$.) A solution to (12) is then given by (see for instance in Ref. 5)[§]

$$\Psi(\boldsymbol{x}) = -\int_\Omega \frac{\langle \nabla_{\boldsymbol{x}'}, \mathbf{M}' \rangle}{|\boldsymbol{x} - \boldsymbol{x}'|}\, dv + \oint_{\partial\Omega} \frac{\langle \mathbf{M}', d\boldsymbol{\sigma}' \rangle}{|\boldsymbol{x} - \boldsymbol{x}'|} \tag{10}$$

(Here $\nabla_{\boldsymbol{x}'}$ refers to the gradient with respect to the variable $\boldsymbol{x}' \in \Omega$ and, for the remainder of this paper, all primed quantities are evaluated at $\boldsymbol{x}' \in \Omega$.) If we assume a linear relationship of the form (2), then one can compute the magnetic field $\mathbf{B}$ from the above equation via

$$\mathbf{B} = -\nabla(\mu\,\Psi) = -\mu\nabla\Psi,$$

Assuming that the interchange of ∇ and $\int$ is allowed, one obtains

$$\mathbf{B}(\boldsymbol{x}) = -\mu \int_\Omega \frac{(\boldsymbol{x} - \boldsymbol{x}')\,\langle \nabla_{\boldsymbol{x}'}, \mathbf{M}' \rangle}{|\boldsymbol{x} - \boldsymbol{x}'|^3}\, dv + \mu \oint_{\partial\Omega} \frac{(\boldsymbol{x} - \boldsymbol{x}')\,\langle \mathbf{M}', d\boldsymbol{\sigma}' \rangle}{|\boldsymbol{x} - \boldsymbol{x}'|^3} \tag{11}$$

[§]Here one needs to assume that $\Psi \in C^2(\mathring{\Omega}) \cap C(\Omega)$.

For the above-described idealized situation of magnetizing Ω, the magnetization field $\mathbf{M}$ may, in first approximation, be considered to be *uniform* and *constant*: $\mathbf{M} = M_0\, \boldsymbol{e}_z$, $M_0 > 0$, $\boldsymbol{e}_z$ unit vector in the z-direction. Thus, (11) simplifies to

$$\mathbf{B}(\boldsymbol{x}) = \mu\, M_0 \oint_{\partial\Omega} \frac{(\boldsymbol{x} - \boldsymbol{x}')\,\langle\, \boldsymbol{e}_z,\, d\boldsymbol{\sigma}'\rangle}{|\boldsymbol{x} - \boldsymbol{x}'|^3}. \tag{12}$$

For $0 < a < b < L$, $L \gg 1$, and $A(r_1, r_2) := \{(x, y) : 0 \le r_1 \le \sqrt{x^2 + y^2} \le r_2\}$ assume now that $\Omega := A(r_1, r_2) \times [a, b]$. (For $r_1 > 0$, such a hard ferromagnet is encountered, for instance, in nondestructive evaluations of pipes.) The surface $\partial\Omega$ of Ω consists of the annuli $A(r_1, r_2) \times \{a\}$, $A(r_1, r_2) \times \{b\}$, and the two cylindrical surfaces $\partial\Omega_1 := \{(x, y) : x^2 + y^2 = r_1^2\} \times [a, b]$, $r_1 > 0$, and $\partial\Omega_2 := \{(x, y) : x^2 + y^2 = r_2^2\} \times [a, b]$. Now suppose mass, represented by a volume element V, is removed from Ω in such a way that part of ∂V belongs to either $\partial\Omega_1$ or $\partial\Omega_2$, but not both. Denote by Σ that part of the boundary of V which does not belong to $\partial\Omega$ and assume Σ is piecewise smooth and orientable. Note that Σ is also part of the boundary of $\Omega \setminus V$.

Now if the dimensions of Σ are big enough and the applied field $\mathbf{H}$ is sufficiently large, then the magnetic field lines of $\mathbf{B}$ will leave the interior of Ω, penetrate its exterior, and re-enter Ω. This process is called *magnetic flux leakage.* As removing mass from Ω is a local phenomenon in regards to the overall geometry, focus is entirely given to the magnetic field $\mathbf{B}$ in the vicinity of Σ. This field, $\mathbf{B}_\Sigma$, is called the *magnetic flux leakage field* and its shape can be explained using *Ampère's Dipole Model.* One assumes that $\mathbf{B}_\Sigma$ is generated by magnetic dipoles situated on opposite sides of the surface $\Sigma \subset \Omega \setminus V$. In other words, the distribution of magnetic dipoles on the surface Σ, regarded as a Radon measure on Σ, generates $\mathbf{B}_\Sigma$. An explicit expression for $\mathbf{B}_\Sigma$ is given by Eq. (11) with only the slight modification of replacing $\partial\Omega$ by Σ.

$$\mathbf{B}_\Sigma(\boldsymbol{x}) = -\int_\Omega \frac{(\boldsymbol{x} - \boldsymbol{x}')\,\langle\, \nabla_{\boldsymbol{x}'},\, \mathbf{M}'\rangle}{|\boldsymbol{x} - \boldsymbol{x}'|^3}\, dv + \oint_\Sigma \frac{(\boldsymbol{x} - \boldsymbol{x}')\,\langle\, \mathbf{M}',\, d\boldsymbol{\sigma}'\rangle}{|\boldsymbol{x} - \boldsymbol{x}'|^3} \tag{13}$$

Here we also used the fact that the magnetic permeability μ is approximately equal to one outside the ferromagnet Ω. The magnetic dipole distribution is given by the Radon measure $\mu_\Sigma := \langle\, \mathbf{M},\, d\boldsymbol{\sigma}_\Sigma\rangle$ whose support is Σ. The distribution of the magnetic dipoles, i.e., μ_Σ, is determined by the magnetization $\mathbf{M}$ in a neighborhood of Σ in Ω and the specific material properties of the ferromagnet Ω. Precise knowledge of these two properties together with exact information about the geometry of Σ would uniquely determine the Radon measure μ_Σ. On the other hand, μ_Σ does determine the geometry of Σ once **M** and the ferromagnet's material properties are known. In this case, one has the one-to-one correspondence $\Sigma \longleftrightarrow \mu_\Sigma$. Unfortunately, **M** and the material properties of Ω are only approximately, if at all, known.

1.2. *Coordinate Systems*

Here a summary of the five types of coordinate systems employed in this paper is given and the relevant quantities necessary for the evaluation of the integral in Eq. (13) are presented. Since the integrals in Eq. 13 are additive with respect to volumes and surfaces, the volume elements V associated with these coordinate systems make up the building blocks for more complex removed masses.

1.2.1. *Cartesian Coordinates*

The unit vectors in the cartesian coordinate system (x, y, z) are denoted by $\{\widehat{\imath}, \widehat{\jmath}, \widehat{\boldsymbol{k}}\}$. Throughout the remainder of this paper, it is assumed that the y-axis is the symmetry axis for Ω, that the magnetization **M** is along the y-axis, and that $\widehat{\boldsymbol{k}}$ is perpendicular to $\partial\Omega_2$.

Definition 1.1. If V has the form

$$V_{car} = [-w, w] \times [-\ell, \ell] \times [-d, d],$$

where $2w > 0$, $2\ell > 0$, and $2d > 0$ denotes the width, length, and depth of the missing mass element V, then it is called *cartesian.*

1.2.2. *Circular Cylindrical Coordinates*

Definition 1.2. V is called *cylindrical* if it is of the form

$$V_{cyl} = \{(r, \psi, z) \,:\, r = r_0,\ 0 \le \phi \le 2\pi,\ 0 \le z \le d\},$$

where $r_0 > 0$ and $d > 0$ denote the radius and depth.

1.2.3. *Spherical Coordinates*

Definition 1.3. V is called *(hemi)spherical* if it is of the form

$$V_{sph} = \{(\rho, \phi, \theta) \,:\, \rho = \rho_0,\ 0 \le \phi \le 2\pi,\ \pi/2 \le \theta \le \pi\},$$

with $\rho_0 > 0$ denoting the radius/depth.

1.2.4. *Parabolic Coordinates*

The parabolic coordinate system (ξ, η, φ) consists of

(1) Confocal paraboloids about the positive z-axis: $\xi = \text{constant}$, $0 \leq \xi < \infty$.
(2) Confocal paraboloids about the negative z-axis: $\eta = \text{constant}$, $0 \leq \eta < \infty$.
(3) Half planes through the z-axis: $\varphi = \text{constant}$, $0 \leq \varphi \leq 2\pi$.

As usual, the azimuth angle φ is measured counter-clockwise from the positive x-axis when viewed from the positive z-axis. The relationship between cartesian coordinate (x, y, z) and parabolic coordinates (ξ, η, φ) is given by

$$\begin{cases} x = \xi\eta\cos\varphi, & \xi = \sqrt{\sqrt{x^2+y^2+z^2} - z} \\ y = \xi\eta\sin\varphi, & \text{and } \eta = \sqrt{\sqrt{x^2+y^2+z^2} + z} \\ z = \frac{1}{2}(\eta^2 - \xi^2) & \varphi = \arctan y/x. \end{cases}$$

It is noted, that the parabolic coordinate system introduced above is a *left-handed* system, i.e., $\hat{\boldsymbol{\xi}} \times \hat{\boldsymbol{\eta}} = -\hat{\boldsymbol{\varphi}}$. The unit vectors are denoted by $\hat{\boldsymbol{\xi}}$, $\hat{\boldsymbol{\eta}}$, and $\hat{\boldsymbol{\varphi}}$, and the nabla operator ∇ and the Laplacian Δ are

$$\nabla = \frac{1}{\sqrt{\xi^2+\eta^2}}\frac{\partial}{\partial\xi}\hat{\boldsymbol{\xi}} + \frac{1}{\sqrt{\xi^2+\eta^2}}\frac{\partial}{\partial\eta}\hat{\boldsymbol{\eta}} + \frac{1}{\xi\eta}\frac{\partial}{\partial\varphi}\hat{\boldsymbol{\varphi}}$$

and

$$\Delta = \frac{1}{\xi^2\eta^2(\xi^2+\eta^2)}\left[\frac{\partial}{\partial\xi}\left(\xi\eta\frac{\partial}{\partial\xi}\right) + \frac{\partial}{\partial\eta}\left(\xi\eta\frac{\partial}{\partial\eta}\right)\right] + \frac{1}{\xi^2\eta^2}\frac{\partial^2}{\partial\varphi^2},$$

respectively. The oriented surface element $d\boldsymbol{\sigma}$ for a paraboloid $\xi = \xi_0 =$ const. is given by

$$d\boldsymbol{\sigma} = \xi_0\eta\begin{pmatrix}\eta\cos\varphi \\ \eta\sin\varphi \\ -\xi_0\end{pmatrix} d\eta d\varphi, \tag{14}$$

where, due to the left-handedness of the parabolic coordinate system, $\boldsymbol{n}$ is the *inward* normal

$$\boldsymbol{n} = \frac{1}{\sqrt{\xi_0^2+\eta^2}}\begin{pmatrix}\eta\cos\varphi \\ \eta\sin\varphi \\ -\xi_0\end{pmatrix}.$$

Definition 1.4. V is called *circular parabolic of radius* r_0 *and depth* d if it is of the form

$$V_{par} = \{(\xi, \eta, \varphi) : \xi = \xi_0,\ 0 \leq \eta \leq \eta_0,\ 0 \leq \varphi \leq 2\pi\},$$

where $\xi_0 = r_0/\sqrt{2d}$ and $\eta_0 = \sqrt{2d}$.

Remark 1.1.

(1) The values for ξ and η in the above definition are obtained as follows. When $\xi_0^2\eta_0^2 = x^2 + y^2 = r_0^2$ the depth d should equal the difference between $z_1 = z|_{x^2+y^2=r_0^2} = (1/2)(r_0^2/\xi_0^2 - \xi_0^2)$ and $z_0 = z|_{x=y=0} = -\xi_0^2/2$. This requirement then gives the values for ξ_0 and η_0 in terms of r_0 and d.
(2) A circular parabolic V is expressed then in cylindrical coordinates (η, φ, z) by

$$V_{par}: \quad z = \frac{1}{2}\left(\eta^2 - \frac{r_0^2}{2d}\right), \; 0 \le \eta \le \sqrt{2d}, \; 0 \le \varphi \le 2\pi.$$

Note that $-r_0^2/4d \le z \le d - r_0^2/4d$.

1.2.5. *Elliptic Cylindrical Coordinates*

The elliptic cylindrical coordinate system (u, v, z) consists of

(1) Elliptic cylinders $u =$ constant, $0 \le u < \infty$.
(2) Hyperbolic cylinders $v =$constant, $0 \le v \le 2\pi$.
(3) Planes parallel to the xy-plane, $z =$ constant, $-\infty < z < +\infty$.

The transformation equations between cartesian coordinates (x, y, z) and elliptic cylindrical coordinates (u, v, z) are given by

$$\begin{cases} x = a\cosh u \cos v, & u = \mathfrak{Re}\left[\operatorname{arccosh}\left(\frac{x+iy}{a}\right)\right] \\ y = a\sinh u \sin v, & \text{and } v = \mathfrak{Im}\left[\operatorname{arccosh}\left(\frac{x+iy}{a}\right)\right] \\ z = z & \end{cases} \tag{15}$$

Here $\mathfrak{Re}$ and $\mathfrak{Im}$ denotes the real and imaginary part, respectively. The azimuth angle v is again measured in a counterclockwise fashion from the positive x-axis when viewed from the positive z-axis. The parameter a is the *focal length* of the ellipse given by $a^2 =$ (major semi-axis)^2 − (minor semi-axis)^2.

The nabla operator ∇ and the Laplacian Δ in this coordinate system are

$$\nabla = \frac{\sqrt{2}}{a\sqrt{\cosh 2u - \cos 2v}}\left(\frac{\partial}{\partial u}\,\hat{\boldsymbol{u}} + \frac{\partial}{\partial v}\,\hat{\boldsymbol{v}}\right) + \frac{\partial}{\partial z}\,\hat{\boldsymbol{z}}$$

and

$$\Delta = \frac{2}{a^2(\cosh 2u - \cos 2v)}\left[\frac{\partial^2}{\partial u^2} + \frac{\partial^2}{\partial v^2}\right] + \frac{\partial^2}{\partial z^2}$$

respectively, with $\hat{\boldsymbol{u}}$, $\hat{\boldsymbol{v}}$, and $\hat{\boldsymbol{z}}$ denoting the unit vectors.

Definition 1.5. V is called *elliptic parabolic of width w, length ℓ, and depth d* if it is of the form

$$V_{ecyl} = \{(u, v, z) : 0 \leq u \leq \operatorname{arctanh}(\ell/w),\ 0 \leq v \leq 2\pi,\ 0 \leq z \leq d\},$$

or, more precisely,

$$V_{ecyl}: \quad z = a^2(\cosh^2 u \cos^2 v + \sinh^2 u \sin^2 v), \quad 0 \leq u \leq u_0, \quad 0 \leq v \leq 2\pi,$$

where

$$a = \sqrt{d(w^2 - \ell^2)}, w > \ell, \quad \text{and} \quad u_0 = \operatorname{arctanh}(\ell/w).$$

Remark 1.2. The values for a and u_0 are obtained by requiring that for $z = d$ the boundary of V in this plane is an ellipse with major semi-axis equal to w and minor semi-axis equal to ℓ.

The surface element $d\boldsymbol{\sigma}$ for an elliptic parabolic V is given as

$$d\boldsymbol{\sigma} = -\frac{a^3}{2}\begin{pmatrix} \cos 3v \cosh u - \cos v \cosh 3u \\ \sin 3v \sinh u - \sin v \sinh 3u \\ \cos 2v - \cosh 2v \end{pmatrix} du dv,$$

with the unit normal $\boldsymbol{n}$ being

$$\boldsymbol{n} = \frac{1}{\sqrt{1 + 2a^2(\cosh 2u + \cos 2v)}}\begin{pmatrix} -2a \cosh u \cos v \\ -2a \sinh u \sin v \\ 1 \end{pmatrix}.$$

Remark 1.3. Setting $r = (a/2)e^v$, Eqs. (15) can be expressed in the form

$$\begin{cases} x = \left(r + \frac{a^2}{4r}\right) \cos u \\ y = \left(r - \frac{a^2}{4r}\right) \sin u \end{cases}$$

If the focal length $a \to 0$, $v \to \infty$ in such a way that ae^v remains finite, and the usual cylindrical coordinates are recovered.

2. Cartesian V

In this section, the magnetic field $\mathbf{B}_\Sigma$ generated by a cartesian V is explicitly computed. For this purpose, consider a cartesian V centered at the origin with length 2ℓ, width $2w$, and depth $2d$. To compute the magnetic field $\mathbf{B}_\Sigma$

generated by this type of V, (13) is employed. For a function $f : \mathbb{R}^3 \to \mathbb{R}$ and a triple $\boldsymbol{a} := (a_1, a_2, a_3) \in \mathbb{R}^3$, denote by $\{f, \boldsymbol{a}\}$ the function

$$\begin{aligned}\{f,\boldsymbol{a}\}(x,y,z) :=& [f(x-a_1, y+a_2, z-a_3) - f(x-a_1, y+a_2, z+a_3)]\\ &- [f(x+a_1, y+a_2, z-a_3) - f(x+a_1, y+a_2, z+a_3)]\\ &- [f(x-a_1, y-a_2, z-a_3) - f(x-a_1, y-a_2, z+a_3)]\\ &+ [f(x+a_1, y-a_2, z-a_3) - f(x+a_1, y-a_2, z+a_3)]\end{aligned}$$

Theorem 2.1. *Suppose that V_{car} is cartesian of width $2w$, length 2ℓ, and depth $2d$ and the magnetization is of the form $\mathbf{M} = M\widehat{\jmath}$. Then the magnetic flux leakage field $\mathbf{B}_\Sigma = (B_x, B_y, B_z)$ generated by V_{car} can be explicitly and exactly calculated. The exact expressions for the components are given by $B_x(x,y,z) = M\{\varphi, (w,\ell,d)\}(x,y,z)$, $B_y(x,y,z) = M\{\psi, (w,\ell,d)\}(x,y,z)$ and $B_z(x,y,z) = M\{\varphi, (w,\ell,d)\}(z,y,x)$, where $\varphi(x,y,z) = \log(-z + \sqrt{x^2+y^2+z^2})$ and $\psi(x,y,z) = \operatorname{sgn}(y)\arcsin(\frac{x}{\sqrt{x^2+y^2}}\,\frac{z}{\sqrt{y^2+z^2}})$.*

Proof. Under the assumption on $\mathbf{M}$, $\langle \mathbf{M}', d\boldsymbol{\sigma}'\rangle = M(d\sigma'|_{y'=-\ell} - d\sigma'|_{y'=\ell})$, and thus

$$\begin{aligned}\Psi(\boldsymbol{x}) &= \oint_B \frac{\langle \mathbf{M}', d\boldsymbol{\sigma}\rangle'}{|\boldsymbol{x}-\boldsymbol{x}'|}\\ &= M\left(\int_{-w}^{+w}\int_{-d}^{d} \frac{dz'dx'}{\sqrt{(x-x')^2+(y+\ell)^2+(z-z')^2}}\right.\\ &\qquad \left. - \int_{-w}^{+w}\int_{-d}^{d} \frac{dz'dx'}{\sqrt{(x-x')^2+(y-\ell)^2+(z-z')^2}}\right).\end{aligned}$$

Since the functions $r_-(\boldsymbol{x},\boldsymbol{x}') = 1/\sqrt{(x-x')^2+(y+\ell)^2+(z-z')^2}$ and $r_+(\boldsymbol{x},\boldsymbol{x}') = 1/\sqrt{(x-x')^2+(y-\ell)^2+(z-z')^2}$ satisfy all the hypotheses for the application of the theorem (Interchangeability of Integral and Partial Derivative) in Ref. 15, p. 59, the x-, y, and z-components of $\mathbf{B}_\Sigma$ are obtained as

$$B_x(\boldsymbol{x}) = -\frac{\partial \Psi}{\partial x} = -M\int_{-w}^{+w}\int_{-d}^{d}\left(\frac{\partial r_-}{\partial x} - \frac{\partial r_+}{\partial x}\right) dz'dx',$$

$$B_y(\boldsymbol{x}) = -\frac{\partial \Psi}{\partial y} = -M\int_{-w}^{+w}\int_{-d}^{d}\left(\frac{\partial r_-}{\partial y} - \frac{\partial r_+}{\partial y}\right) dz'dx',$$

and

$$B_z(\boldsymbol{x}) = -\frac{\partial \Psi}{\partial z} = -M\int_{-w}^{+w}\int_{-d}^{d}\left(\frac{\partial r_-}{\partial z} - \frac{\partial r_+}{\partial z}\right) dz'dx',$$

respectively.

Performing the partial differentiation in the expression for B_x yields

$$B_x = M\left[\int_{-w}^{+w}\int_{-d}^{d}\frac{(x-x')\,dz'dx'}{[(x-x')^2+(y+\ell)^2+(z-z')]^{3/2}}\right.$$
$$\left.-\int_{-w}^{+w}\int_{-d}^{d}\frac{(x-x')\,dz'dx'}{[(x-x')^2+(y-\ell)^2+(z-z')]^{3/2}}\right].$$

Setting $\xi = x - x'$ and integrating over ξ between $\xi = x + w$ and $\xi = x - w$ gives

$$B_x = M\left[\int_{-d}^{d}\frac{dz'}{\sqrt{(x-w)^2+(y+\ell)^2+(z-z')^2}}\right.$$
$$-\int_{-d}^{d}\frac{dz'}{\sqrt{(x+w)^2+(y+\ell)^2+(z-z')^2}}$$
$$-\int_{-d}^{d}\frac{dz'}{\sqrt{(x-w)^2+(y-\ell)^2+(z-z')^2}}$$
$$\left.+\int_{-d}^{d}\frac{dz'}{\sqrt{(x+w)^2+(y-\ell)^2+(z-z')^2}}\right].$$

Substituting $\zeta = z - z'$ in each of the four integrals reduces each one to an integral of the form

$$\int\frac{dX}{\sqrt{a^2+X^2}} = \log(X+\sqrt{a^2+X^2}) + \text{constant}$$

with a different value for a^2. Defining a function $\varphi : \mathbf{R}^3\setminus\{0,0,0\}\to\mathbf{R}^+$ by

$$\varphi(x,y,z) = \log(-z+\sqrt{x^2+y^2+z^2}),$$

yields the stated expression for B_x.

To obtain the y-component B_y of $\mathbf{B}_\Sigma$, one proceeds similarly.

$$B_y = -M\left[\int_{-w}^{+w}\int_{-d}^{d}\frac{(y+\ell)\,dz'dx'}{\sqrt{(x-x')^2+(y+\ell)^2+(z-z')^2}^3}\right.$$
$$\left.-\int_{-w}^{+w}\int_{-d}^{d}\frac{(y-\ell)\,dz'dx'}{\sqrt{(x-x')^2+(y-\ell)^2+(z-z')^2}^3}\right].$$

Setting $\xi = x - x'$ and integrating over ξ between $\xi = x + w$ and $\xi = x - w$ and then replacing z' by $\zeta = z - z'$ and integrating over ζ from $z + d$ to

$z-d$ yields

$$B_y = -M\left[(x-w)(y+\ell)\int_{z+d}^{z-d}\frac{d\zeta}{[(y+\ell)^2+\zeta^2]\sqrt{(x-w)^2+(y+\ell)^2+\zeta^2}}\right.$$
$$-(x+w)(y+\ell)\int_{z+d}^{z-d}\frac{d\zeta}{[(y+\ell)^2+\zeta^2]\sqrt{(x+w)^2+(y+\ell)^2+\zeta^2}}$$
$$-(x-w)(y-\ell)\int_{z+d}^{z-d}\frac{d\zeta}{[(y-\ell)^2+\zeta^2]\sqrt{(x-w)^2+(y-\ell)^2+\zeta^2}}$$
$$\left.+(x+w)(y-\ell)\int_{z+d}^{z-d}\frac{d\zeta}{[(y-\ell)^2+\zeta^2]\sqrt{(x+w)^2+(y-\ell)^2+\zeta^2}}\right].$$

Each of the four integrals is of the form

$$\int\frac{dX}{(a^2+X^2)\sqrt{X^2+b^2}}$$

with different values for a^2 and b^2. Such integrals may be solved employing the substitution $X = a\tan u$. This substitution yields

$$\int\frac{a\sec^2 u du}{a^2\sec^2 u\sqrt{b^2+a^2\tan^2 u}} = \frac{1}{a}\int\frac{du}{\sqrt{b^2+a^2\tan^2 u}}.$$

The latter integral evaluates to

$$\frac{1}{a\sqrt{b^2-a^2}}\arcsin\left(\sqrt{\frac{b^2-a^2}{b^2}}\sin u\right), \qquad b^2 > a^2.$$

Introducing the function $\psi : \mathbf{R}^3\setminus\{0,0,0\}\to\mathbf{R}$

$$\psi(x,y,z) = \operatorname{sgn}(y)\arcsin\left(\frac{x}{\sqrt{x^2+y^2}}\,\frac{z}{\sqrt{y^2+z^2}}\right),$$

where $\operatorname{sgn}(x)$ denotes the signum function, gives the exact solution for B_y.

The z-component B_z is given by

$$B_z(x,y,z) = M\left[\int_{-w}^{+w}\int_{-d}^{d}\frac{(z-z')\,dz'dx'}{[(x-x')^2+(y+\ell)^2+(z-z')^2]^{3/2}}\right.$$
$$\left.-\int_{-w}^{+w}\int_{-d}^{d}\frac{(z-z')\,dz'dx'}{[(x-x')^2+(y-\ell)^2+(z-z')^2]^{3/2}}\right].$$

Notice that B_z is just B_x with x replaced by z. Hence, the conclusion follows. □

Figure 2 below shows an example of magnetic field components for cartesian V. The shapes of the depicted surfaces are characteristic of general magnetic flux leakage fields: The field component B_z is bipolar and the component B_x of quadrupole nature. The field component B_y is symmetric about a plane perpendicular to the axis of Ω (for a symmetric defect.)

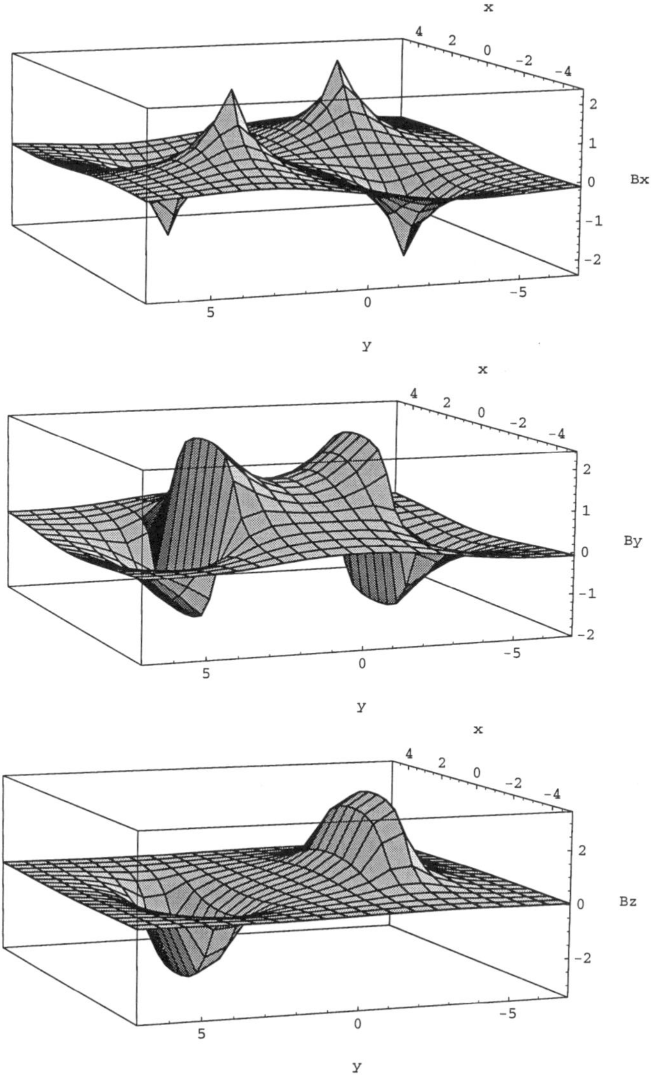

Fig. 2. The x, y, z-component of $\mathbf{B}_\Sigma$ for a cartesian V.

3. Cylindrical V

Suppose that V is cylindrical with radius r_0 and depth d, i.e., $V_{cyl} = \{(r, \phi, z) : r = r_0, \; 0 \leq \phi \leq 2\pi, \; 0 \leq z \leq d\}$. Assume again that $\mathbf{M}$ is again of the form $\mathbf{M} = M\widehat{\jmath}$, and thus $\langle \mathbf{M}', d\boldsymbol{\sigma}' \rangle = -Mr_0 \sin\phi' d\phi' dz'$. Note that since the magnetic field leaks into the interior of V_{cyl}, $\boldsymbol{n}$ is the *inward normal* $\boldsymbol{n} = (-\cos\phi', -\sin\phi', 0)$. Thus,

$$\Psi(r, \phi, z) = \oint_{\partial V_{cyl}} \frac{\langle \mathbf{M}', d\boldsymbol{\sigma}' \rangle}{|\boldsymbol{x} - \boldsymbol{x}'|} = -Mr_0 \int_0^d \int_0^{2\pi} \frac{\sin\phi' d\phi' dz'}{|\boldsymbol{x} - \boldsymbol{x}'|}.$$

In order to evaluate the latter integral, the expansion of the Green's function $|\boldsymbol{x} - \boldsymbol{x}'|^{-1}$ in terms of *Bessel* or *cylinder functions* is employed. (See, for instance, Ref. 8.)

Lemma 3.1. *The Green's function in cylindrical coordinates for the equation $\Delta\Psi = 0$ is given by*

$$|\boldsymbol{x} - \boldsymbol{x}'|^{-1} = \frac{4}{\pi} \int_0^\infty \cos[k(z - z')] \left\{ \frac{1}{2} I_0(kr_<) K_0(kr_>) + \sum_{\nu=1}^{\infty} \cos\nu(\phi - \phi') I_\nu(kr_<) K_\nu(kr_>) \right\} dk, \quad (16)$$

where $r_< = \min\{r, r_0\}$ and $r_> = \max\{r, r_0\}$.

Proof. See Ref. 8 □

Here I_ν denotes the *modified Bessel function of the first kind of order ν* and K_ν the *modified Bessel function of the third kind of order ν or Macdonald's function.* Exact definitions and properties of these functions may be found in Refs.1,3,5,9.

With this expansion of $|\boldsymbol{x} - \boldsymbol{x}'|^{-1}$, the magnetic scalar potential becomes

$$\Psi(r, \phi, z) = -\frac{4Mr_0}{\pi} \int_0^\infty \int_0^d \int_0^{2\pi} \cos[k(z - z')] \sin\phi' \left[\frac{1}{2} I_0(kr_<) K_0(kr_>) + \sum_{\nu=1}^{\infty} \cos[\nu(\phi - \phi')] I_\nu(kr_<) K_\nu(kr_>) \right] d\phi' dz' dk.$$

Note that the above – and all subsequent – interchanges of the integrals and the infinite sum is justified. The above integral can be further rewritten as

$$\Psi(r,\phi,z) = -\frac{4Mr_0}{\pi}\int_0^\infty\int_0^d\int_0^{2\pi}\cos[k(z-z')]\sin\phi' \times\left[\frac{1}{2}I_0(kr_<)K_0(kr_>)\right]d\phi'dz'dk$$
$$-\frac{4Mr_0}{\pi}\sum_{\nu=1}^\infty\int_0^\infty\int_0^d\int_0^{2\pi}\cos[k(z-z')]\sin\phi'\cos[\nu(\phi-\phi')] \times I_\nu(kr_<)K_\nu(kr_>)d\phi'dz'dk. \tag{17}$$

As

$$\int_0^{2\pi}\sin\phi'd\phi' = 0$$

and

$$\int_0^{2\pi}\cos[\nu(\phi-\phi')]\sin\phi'd\phi' = \frac{-2\sin[\nu\pi]\sin[\nu(\pi-\phi)]}{\nu^2-1} = 0, \text{ for } \nu\neq\pm1,$$

only the $\nu = 1$ term in (17) contributes. For this value of ν

$$\int_0^{2\pi}\cos(\phi-\phi')\sin\phi'd\phi' = \pi\sin\phi$$

and therefore,

$$\Psi(r,\phi,z) = -4Mr_0\sin\phi\int_0^\infty\int_0^d\cos[k(z-z')]I_1(kr_<)K_1(kr_>)dz'dk$$
$$= -4Mr_0\sin\phi\int_0^\infty\left(\int_0^d\cos[k(z-z')]dz'\right)I_1(kr_<)K_1(kr_>)dk.$$

Hence, the next result is established.

Theorem 3.1. *Suppose that V is cylindrical with radius r_0 and depth d. Then the magnetic scalar potential generated by V_{cyl} is given by*

$$\Psi(r,\phi,z) = -4Mr_0\sin\phi\int_0^\infty\frac{\sin kz-\sin[k(z-d)]}{k}I_1(kr_<)K_1(kr_>)dk \tag{18}$$

In the case that $r_< \neq r_0 \neq r_>$, this above representation may be expressed in terms of an infinite series of the form

$$\Psi(r,\phi,z) = -\frac{\pi Mr_0r_<\sin\phi}{r_>^2}\sum_{n=0}^\infty(-1)^n\frac{z^{2n+1}-(z-d)^{2n+1}}{r_>^{2n}}\frac{(2n+1)!!}{(2n)!!} \times {}_2F_1\left(n+1/2,n+3/2;2;\frac{r_<^2}{r_>^2}\right).$$

Proof. The first part is obvious from the above calculations; the second part is established as follows. The Taylor series for the function $(\sin kz - \sin[k(z-d)])/k$ is given by

$$\frac{\sin kz - \sin[k(z-d)]}{k} = \sum_{n=0}^{\infty} (-1)^n \frac{z^{2n+1} - (z-d)^{2n+1}}{(2n+1)!} k^{2n}$$

Substitution of this series into (18) and interchange of infinite sum and integral yields

$$\begin{aligned}\Psi(r,\phi,z) &= -4Mr_0 \sin\phi \sum_{n=0}^{\infty} \frac{(-1)^n[z^{2n+1} - (z-d)^{2n+1}]}{(2n+1)!} \\ &\quad \times \int_0^{\infty} k^{2n}\, I_1(kr_<)K_1(kr_>)dk \\ &= -4Mr_0 \sin\phi \sum_{n=0}^{\infty} \frac{(-1)^n[z^{2n+1} - (z-d)^{2n+1}]}{(2n+1)!} 2^{2n-1} r_< r_>^{-2n-2} \\ &\quad \times \Gamma(n+1/2)\,\Gamma(n+3/2)\,{}_2F_1(n+1/2, n+3/2; 2; (r_</r_>)^2),\end{aligned}$$

for $r_< < r_>$. Here ${}_2F_1(a,b;c;x)$ denotes *Gauß's hypergeometric function:*

$${}_2F_1(a,b;c;x) = \frac{\Gamma(c)}{\Gamma(a)\,\Gamma(b)} \sum_{m=0}^{\infty} \frac{\Gamma(a+m)\,\Gamma(b+m)}{\Gamma(c+m)} \frac{x^m}{m!}.$$

Note that

$$\begin{aligned}\frac{\Gamma(n+1/2)\,\Gamma(n+3/2)}{(2n+1)!} &= \frac{\Gamma(n+1/2)\,\Gamma(n+3/2)}{\Gamma(2n+2)} \\ &= \frac{\Gamma(n+1/2)\,\Gamma(n+3/2)}{(2\pi)^{-1/2}2^{2n+3/2}\Gamma(n+1)\,\Gamma(n+3/2)} \\ &= \frac{\sqrt{\pi}\,\Gamma(n+1/2)}{2^{2n+1}\Gamma(n+1)} = \frac{\pi\,(2n-1)!!}{2^{3n+1}\Gamma(n+1)} \\ &= \frac{\pi}{2^{2n+1}} \frac{(2n-1)!!}{(2n)!!},\end{aligned}$$

where the duplication formula for Gamma functions was used (Ref. 1, 6.1.18). Thus, after simplification,

$$\begin{aligned}\Psi(r,\phi,z) = &-\frac{\pi M r_0 r_< \sin\phi}{r_>^2} \sum_{n=0}^{\infty} (-1)^n \frac{z^{2n+1} - (z-d)^{2n+1}}{r_>^{2n}} \frac{(2n+1)!!}{(2n)!!} \\ &\times {}_2F_1(n+1/2, n+3/2; 2; (r_</r_>)^2),\end{aligned}$$

which proves the statement. □

In order to obtain formulae for the magnetic field $\mathbf{B}_\Sigma$, two cases need to be considered.

CASE I: $r > r_0$

In this case, $r_< = \min\{r, r_0\} = r_0$ and $r_> = \max\{r, r_0\} = r$. Hence,

$$\Psi^{(I)}(r,\phi,z) = -4Mr_0 \sin\phi \int_0^\infty \frac{\sin kz - \sin k(z-d)}{k} I_1(kr_0)K_1(kr)dk.$$

The radial component B_r of $\mathbf{B}_\Sigma$ is the given by

$$B_r^{(I)}(r,\phi,z) = 4Mr_0 \sin\phi \int_0^\infty \frac{\sin kz - \sin[k(z-d)]}{k} I_1(kr_0)k\frac{dK_1}{dr}(kr)dk.$$

Using the fact that $dK_1/dr = (-1/2)[K_0+K_2]$ (Ref. 3, p. 401, 11.4.6) gives

$$B_r^{(I)}(r,\phi,z) = -2Mr_0\sin\phi\int_0^\infty\{\sin kz - \sin[k(z-d)]\}\, I_1(kr_0)[K_0(kr)+K_2(kr)]dk$$

The azimuthal component B_ϕ of $\mathbf{B}_\Sigma$ is obtained in a similar fashion. As $B_\phi = (-1/r)(\partial\Psi/\partial\phi)$, the result is

$$B_\phi^{(I)}(r,\phi,z) = 4M\left(\frac{r_0}{r}\right)\cos\phi \int_0^\infty \frac{\sin kz - \sin[k(z-d)]}{k} I_1(kr_0)K_1(kr)dk$$

The z-component B_z of $\mathbf{B}_\Sigma$ is $-\partial\Psi/\partial z$ yielding

$$B_z^{(I)}(r,\phi,z) = 4Mr_0 \sin\phi \int_0^\infty \{\cos kz - \cos[k(z-d)]\}\, I_1(kr_0)K_1(kr)dk$$

CASE II: $r < r_0$

In this case, $r_< = \min\{r, r_0\} = r$ and $r_> = \max\{r, r_0\} = r_0$. Hence,

$$\Psi^{(II)}(r,\phi,z) = -4Mr_0 \sin\phi \int_0^\infty \frac{\sin kz - \sin[k(z-d)]}{k} I_1(kr)K_1(kr_0)dk.$$

The radial component B_r of $\mathbf{B}_\Sigma$ is the given by

$$B_r^{(II)}(r,\phi,z) = 4Mr_0 \sin\phi \int_0^\infty \frac{\sin kz - \sin[k(z-d)]}{k} K_1(kr_0)k\frac{dI_1}{dr}(kr)dk.$$

Using the fact that $dI_1/dr = (1/2)[I_0 + I_2]$ (Ref. 3, p. 397, 11.116) gives

$$B_r^{(II)}(r,\phi,z) = 2Mr_0 \sin\phi \int_0^\infty \{\sin kz - \sin[k(z-d)]\}\, K_1(kr_0)[I_0(kr)+I_2(kr)]dk$$

As above in Case I, the azimuthal component B_ϕ is

$$B_\phi^{(II)}(r,\phi,z) = 4M\left(\frac{r_0}{r}\right)\cos\phi \int_0^\infty \frac{\sin kz - \sin[k(z-d)]}{k} I_1(kr)K_1(kr_0)dk$$

and the z-component

$$B_z^{(II)}(r,\phi,z) = 4Mr_0 \sin\phi \int_0^\infty \{\cos kz - \cos[k(z-d)]\}\, I_1(kr)K_1(kr_0)dk$$

The cartesian components B_x and B_y of $\mathbf{B}_\Sigma$ are obtained via

$$\begin{pmatrix} B_x(x,y,z) \\ B_y(x,y,z) \\ B_z(x,y,z) \end{pmatrix} = \begin{pmatrix} \cos\phi & -\sin\phi & 0 \\ \sin\phi & \cos\phi & 0 \\ 0 & 0 & 1 \end{pmatrix} \begin{pmatrix} B_r(r,\phi,z) \\ B_\phi(r,\phi,z) \\ B_z(r,\phi,z) \end{pmatrix}$$

4. Spherical *V*

In this section, an exact solution for the magnetic flux leakage field due to a (hemi)spherical V is derived. To this end, assume that the radius/depth of said V is ρ_0. Then $V_{sph} = \{(\rho,\phi,\theta) : \rho = \rho_0,\ 0 \le \phi \le 2\pi,\ \pi/2 \le \theta \le \pi\}$. The magnetization $\mathbf{M}$ is again assumed to be along the y-axis: $\mathbf{M} = M\widehat{\jmath}$. The normal to Σ is given by $\langle \mathbf{M}', d\boldsymbol{\sigma}' \rangle = -M\rho_0^2 \sin\phi' \sin^2\theta' d\phi' d\theta'$ (inward normal!). Thus,

$$\Psi(\rho,\phi,\theta) = \oint_{\partial S} \frac{\langle \mathbf{M}', d\boldsymbol{\sigma}' \rangle}{|\boldsymbol{x} - \boldsymbol{x}'|} = -M\rho_0^2 \int_{\pi/2}^{\pi} \int_0^{2\pi} \frac{\sin^2\theta' \sin\phi' d\phi' d\theta'}{|\boldsymbol{x} - \boldsymbol{x}'|}.$$

To proceed the next lemma is needed.

Lemma 4.1. *In spherical coordinates, the Green's function $|\boldsymbol{x} - \boldsymbol{x}'|^{-1}$ for the potential equation $\Delta\Psi = 0$ has an expansion in terms of spherical harmonics $Y_{\ell m}$ of the form*

$$\frac{1}{|\boldsymbol{x} - \boldsymbol{x}'|} = 4\pi \sum_{\ell=0}^{\infty} \sum_{m=-\ell}^{\ell} \frac{1}{2\ell+1} \left(\frac{\rho_<^\ell}{\rho_>^{\ell+1}} \right) Y_{\ell m}^*(\theta',\phi') Y_{\ell m}(\theta,\phi),$$

with $\rho_< = \min\{\rho,\rho_0\}$ and $\rho_> = \max\{\rho,\rho_0\}$ (Refs. 3,5,6,8,15). Here * denotes complex conjugation. The spherical harmonics $Y_{\ell m}(\theta,\phi)$ are expressible as

$$Y_{\ell m}(\theta,\phi) = \sqrt{\frac{2\ell+1}{4\pi} \frac{(\ell-m)!}{(\ell+m)!}}\, P_{\ell m}(\cos\theta)\, e^{im\phi} \tag{19}$$

where $P_{\ell m}(x)$ are the so-called associated Legendre functions of degree ℓ and order m:

$$P_{\ell m}(x) = \frac{1}{2^\ell \ell!} (1-x^2)^{m/2} \frac{d^{m+\ell}}{dx^{m+\ell}} (x^2-1)^\ell, \quad -\ell \le m \le \ell \tag{20}$$

Proof. See Ref. 5 or Ref. 8, for instance. For definitions and results see Ref. 1 (Ch. 8) or Ref. 3 (Ch. 12). □

The right-hand side of (19) may be expressed completely in terms of real functions by splitting off the $\ell = 0$ and $(\ell, m) = (\ell, 0)$ term, combining those terms in the sum over m that correspond to positive and negative indexes, and using the fact that $e^{-im\phi} + e^{im\phi} = 2\cos\phi$. This yields

$$\frac{1}{|\boldsymbol{x}-\boldsymbol{x}'|} = \frac{1}{\rho_>} + \sum_{\ell=1}^{\infty} \frac{\rho_<^\ell}{\rho_>^{\ell+1}} P_\ell(\cos\theta')P_\ell(\cos\theta)$$
$$+ \sum_{\ell=1}^{\infty}\sum_{m=1}^{\ell} 2\frac{(\ell-m)!}{(\ell+m)!}\left(\frac{\rho_<^\ell}{\rho_>^{\ell+1}}\right) P_{\ell m}(\cos\theta')P_{\ell m}(\cos\theta)\cos[m(\phi-\phi')]$$

The functions $P_\ell(x) = P_{\ell,0}(x)$ are the *Legendre polynomials of degree* ℓ. With the above expression for $|\boldsymbol{x}-\boldsymbol{x}'|^{-1}$, the magnetic scalar potential now reads

$$\Psi = -M\rho_0^2\Bigg[\int_{\pi/2}^{\pi}\int_0^{2\pi} \frac{\sin^2\theta' \sin\phi' d\phi' d\theta'}{\rho_>}$$
$$+\sum_{\ell=1}^{\infty}\frac{\rho_<^\ell}{\rho_>^{\ell+1}}\int_{\pi/2}^{\pi}\int_0^{2\pi} \sin^2\theta'\sin\phi' P_\ell(\cos\theta')d\phi'd\theta'\, P_\ell(\cos\theta)$$
$$+2\sum_{\ell=1}^{\infty}\sum_{m=1}^{\ell}\frac{(\ell-m)!}{(\ell+m)!}\frac{\rho_<^\ell}{\rho_>^{\ell+1}}$$
$$\times\int_{\pi/2}^{\pi}\int_0^{2\pi} \sin^2\theta'\sin\phi'\cos[m(\phi-\phi')]P_{\ell m}(\cos\theta')d\phi'd\theta'\, P_{\ell m}(\cos\theta)\Bigg]. \tag{21}$$

The first and second term inside the bracket vanishes since the integral over ϕ' is equal to zero. Moreover,

$$\int_0^{2\pi} \sin\phi'\cos[m(\phi-\phi')]d\phi' = \begin{cases} 0 & \text{if } m > 1 \\ \pi\sin\phi & \text{if } m = 1. \end{cases}$$

Therefore, in the third term inside the bracket only the $m = 1$ summand contributes to the sum over m. Thus,

$$\Psi(\rho,\phi,\theta) = -2\pi M\rho_0^2 \sum_{\ell=1}^{\infty}\frac{1}{\ell(\ell+1)}\frac{\rho_<^\ell}{\rho_>^{\ell+1}}\int_{\pi/2}^{\pi} \sin^2\theta' P_{\ell,1}(\cos\theta')d\theta'\, P_{\ell,1}(\cos\theta)\sin\phi.$$

The integral over θ' can be exactly evaluated using the following approach. Let $u = \cos\theta'$. Then

$$\int_{\pi/2}^{\pi} \sin^2\theta' P_{\ell,1}(\cos\theta')d\theta' = -\int_0^{-1} \sqrt{1-u^2}P_{\ell,1}(u)du$$
$$= \int_0^1 \sqrt{1-u^2}P_{\ell,1}(-u)du$$
$$= (-1)^{\ell+1}\int_0^1 \sqrt{1-u^2}P_{\ell,1}(u)du.$$

Here the fact that $P_{\ell m}(-x) = (-1)^{\ell+m}P_{\ell m}(x)$ was employed (Ref. 3, p. 437, 12.93). Using the definition of the associated Legendre functions $P_{\ell m}$ (Ref. 3, p. 435, 12.84),

$$P_{\ell m}(x) = (1-x^2)^{m/2}\frac{d^m}{dx^m}P_\ell(x),$$

the function $P_{\ell,1}$ can be written as $P_{\ell,1}(u) = \sqrt{1-u^2}\,dP_\ell/du$. Hence,

$$(-1)^{\ell+1}\int_0^1 \sqrt{1-u^2}P_{\ell,1}(u)du = (-1)^{\ell+1}\int_0^1 (1-u^2)\frac{dP_\ell}{du}du$$
$$= (-1)^{\ell+1}\left[(1-u^2)P_\ell(u)|_0^1 + 2\int_0^1 uP_\ell(u)du\right]$$
$$= (-1)^{\ell+1}\left[-P_\ell(0) + 2\int_0^1 uP_\ell(u)du\right].$$

Now,

$$P_\ell(0) = \begin{cases} (-1)^\lambda \frac{(2\lambda-1)!!}{(2\lambda)!!} & \text{if } \ell = 2\lambda \\ 0 & \text{if } \ell = 2\lambda+1 \end{cases} \tag{22}$$

(Ref. 3, p. 424, 12.34 and 12.35) and the moment integral is equal to

$$\int_0^1 uP_\ell(u)du = \begin{cases} 1/3 & \text{if } \ell = 1 \\ \frac{\sqrt{\pi}}{4\Gamma(\frac{3}{2}-\lambda)\Gamma(2+\lambda)} & \text{if } \ell = 2\lambda \\ 0 & \text{if } \ell = 2\lambda+1, \end{cases} \tag{23}$$

(Ref. 1, 8.14.15). In the equation above, Γ denotes the real *Gamma function*, which is the extension of the factorial to real numbers: $\Gamma(x) = (x+1)!$. Using the fact that (Ref. 1, 6.1.17 and 6.1.12)

$$\Gamma\left(\frac{3}{2}-\lambda\right) = \frac{\sqrt{\pi}\,2^{\lambda-1}}{(-1)^{\lambda+1}(2\lambda-3)!!},$$

combining (22) and (23), and simplifying yields

$$\int_{\pi/2}^{\pi} \sin^2\theta' P_{\ell,1}(\cos\theta')d\theta' = \begin{cases} \frac{(-1)^{\lambda+1}\,\lambda(2\lambda+1)(2\lambda-3)!!}{(\lambda+1)(2\lambda)!!} & \text{if } \ell = 2\lambda \\ \frac{2}{3} & \text{if } \ell = 1 \\ 0 & \text{if } \ell = 2\lambda+1. \end{cases} \quad (24)$$

Therefore one arrives at the following result.

Theorem 4.1. *The magnetic scalar potential due to a (hemi)spherical defect of radius ρ_0 is explicitly given by*

$$\Psi(\rho,\phi,\theta) = -2\pi M\rho_0^2 \left[\frac{1}{3}\frac{\rho_<}{\rho_>^2}\sin\theta\sin\phi + \sum_{\lambda=1}^{\infty}\frac{(-1)^{\lambda+1}(2\lambda-3)!!}{(2\lambda+2)!!}\frac{\rho_<^{2\lambda}}{\rho_>^{2\lambda+1}}P_{2\lambda,1}(\cos\theta)\sin\phi\right]$$

or, more specifially,

$$\Psi(\rho,\phi,\theta) = \begin{cases} -\frac{2\pi M\rho}{3}\sin\theta\sin\phi \\ \quad +2\pi M\rho_0\sum_{\lambda=1}^{\infty} N_\lambda \left(\frac{\rho}{\rho_0}\right)^{2\lambda} P_{2\lambda,1}(\cos\theta)\sin\phi & (\rho < \rho_0) \\ -\frac{2\pi M}{3}\frac{\rho_0^3}{\rho^2}\sin\theta\sin\phi \\ \quad +2\pi M\rho_0\sum_{\lambda=1}^{\infty} N_\lambda \left(\frac{\rho_0}{\rho}\right)^{2\lambda+1} P_{2\lambda,1}(\cos\theta)\sin\phi & (\rho > \rho_0) \end{cases}$$

where

$$N_\lambda = \frac{(-1)^{\lambda+1}(2\lambda-3)!!}{(2\lambda+2)!!}.$$

Proof. Only $P_{1,1}(\cos\theta) = \sin\theta$ remains to be shown. This, however, follows immediately from (20). □

The ρ-, ϕ-, and θ-component of the magnetic field $\mathbf{B}_\Sigma$ are thus

$$B_\rho(\rho,\phi,\theta) = \begin{cases} \frac{2\pi M}{3}\sin\theta\sin\phi \\ \quad +2\pi M\sum_{\lambda=1}^{\infty} 2\lambda N_\lambda \left(\frac{\rho}{\rho_0}\right)^{2\lambda-1} P_{2\lambda,1}(\cos\theta)\sin\phi & (\rho < \rho_0) \\ \frac{-4\pi M}{3}\frac{\rho_0^3}{\rho^3}\sin\theta\sin\phi \\ \quad -2\pi M\sum_{\lambda=1}^{\infty}(2\lambda+1) N_\lambda \left(\frac{\rho_0}{\rho}\right)^{2\lambda+2} P_{2\lambda,1}(\cos\theta)\sin\phi & (\rho > \rho_0) \end{cases}$$

$$B_\phi(\rho,\phi,\theta) = \begin{cases} \frac{2\pi M}{3}\cos\phi + 2\pi M \sum_{\lambda=1}^{\infty} N_\lambda \left(\frac{\rho}{\rho_0}\right)^{2\lambda-1} \\ \quad \times P_{2\lambda,1}(\cos\theta)\frac{\cos\phi}{\sin\theta} & (\rho < \rho_0) \\ \\ \frac{2\pi M}{3}\frac{\rho_0^3}{\rho^3}\cos\phi + 2\pi M \sum_{\lambda=1}^{\infty} N_\lambda \left(\frac{\rho_0}{\rho}\right)^{2\lambda+2} \\ \quad \times P_{2\lambda,1}(\cos\theta)\frac{\cos\phi}{\sin\theta} & (\rho > \rho_0) \end{cases}$$

$$B_\theta(\rho,\phi,\theta) = \begin{cases} \frac{2\pi M}{3}\cos\theta\sin\phi - 2\pi M \sum_{\lambda=1}^{\infty} N_\lambda \left(\frac{\rho}{\rho_0}\right)^{2\lambda-1} \\ \qquad \times[\frac{1}{2}P_{2\lambda,2}(\cos\theta) - \lambda(2\lambda+1)P_{2\lambda}(\cos\theta)]\sin\phi & (\rho < \rho_0) \\ \\ \frac{2\pi M}{3}\frac{\rho_0^3}{\rho^3}\cos\theta\sin\phi - 2\pi M \sum_{\lambda=1}^{\infty} N_\lambda \left(\frac{\rho_0}{\rho}\right)^{2\lambda+2} \\ \qquad \times[\frac{1}{2}P_{2\lambda,2}(\cos\theta) - \lambda(2\lambda+1)P_{2\lambda}(\cos\theta)]\sin\phi & (\rho > \rho_0) \end{cases}$$

In the derivation of B_θ the following recursion relation of associated Legendre functions was used to replace the derivative of $P_{1,1}$, respectively, $P_{2\lambda,1}$ with respect to θ (Ref. 3, p. 437, 12.90).

$$\frac{dP_{\ell m}(\cos\theta)}{d\theta} = -\frac{1}{2}[P_{\ell,m+1}(\cos\theta) - (\ell+m)(\ell-m+1)P_{\ell,m-1}(\cos\theta)]. \quad (25)$$

The cartesian components B_x, B_y, and B_z of $\mathbf{B}_\Sigma$ are obtained via

$$\begin{aligned} B_x &= \sin\theta\cos\phi\, B_\rho + \cos\theta\cos\phi\, B_\theta - \sin\phi\, B_\phi \\ B_y &= \sin\theta\sin\phi\, B_\rho + \cos\theta\sin\phi\, B_\theta + \cos\phi\, B_\phi \\ B_z &= \cos\theta\, B_\rho - \sin\theta\, B_\theta. \end{aligned} \quad (26)$$

4.1. *Semi-spherical V*

To complete the discussion of spherical V's, an exact expression for the magnetic flux leakage field $\mathbf{B}_\Sigma$ generated by a semi-spherical V is derived.

Definition 4.1. V is called *semi-spherical* if it is either of the form $V_l = \{(\rho,\phi,\theta) : \rho = \rho_0,\ \pi \le \phi \le 2\pi,\ \pi/2 \le \theta \le \pi\}$ or $V_r = \{(\rho,\phi,\theta) : \rho = \rho_0,\ 0 \le \phi \le \pi,\ \pi/2 \le \theta \le \pi\}$.

In the derivation of the exact solution for magnetic flux leakage, the first form V_l is used and the expansion in (21) is employed with the appropriate modifications. (The magnetic flux leakage field for V_r is obtained from that of V_l via the reflection $y \mapsto -y$.)

The scalar magnetic potential Ψ is comprised of three terms $T_1 = T_1(\rho,\phi,\theta)$, $T_2 = T_2(\rho,\phi,\theta)$, and $T_3 = T_3(\rho,\phi,\theta)$. More precisely,

$$\Psi = -M\rho_0^2\left[T_1(\rho,\phi,\theta) + T_2(\rho,\phi,\theta) + T_3(\rho,\phi,\theta)\right].$$

with

$$T_1 = \int_{\pi/2}^{\pi}\int_{\pi}^{2\pi} \frac{\sin^2\theta' \sin\phi' d\phi' d\theta'}{\rho_>} = -\frac{\pi}{2\rho_>},$$

$$T_2 = \sum_{\ell=1}^{\infty} \frac{\rho_<^\ell}{\rho_>^{\ell+1}} \int_{\pi/2}^{\pi}\int_{\pi}^{2\pi} \sin^2\theta' \sin\phi' P_\ell(\cos\theta') d\phi' d\theta'\, P_\ell(\cos\theta)$$

$$= -2\sum_{\ell=1}^{\infty} \frac{\rho_<^\ell}{\rho_>^{\ell+1}} \cdot \mathcal{K}_\ell \cdot P_\ell(\cos\theta)$$

where

$$\mathcal{K}_\ell = \int_{\pi/2}^{\pi} \sin^2\theta' P_\ell(\cos\theta') d\theta',$$

and

$$T_3 = 2\sum_{\ell=1}^{\infty}\sum_{m=1}^{\ell} \frac{(\ell-m)!}{(\ell+m)!} \frac{\rho_<^\ell}{\rho_>^{\ell+1}}$$

$$\times \int_{\pi/2}^{\pi}\int_{\pi}^{2\pi} \sin^2\theta' \sin\phi' \cos[m(\phi-\phi')] P_{\ell m}(\cos\theta') d\phi' d\theta'\, P_{\ell m}(\cos\theta)$$

$$= 2\left[\frac{1}{2}\frac{\rho_<}{\rho_>^2} \cdot \frac{\pi}{2}\sin\phi \cdot \frac{2}{3} \cdot P_{1,1}(\cos\theta)\right.$$

$$+ \sum_{\ell=2}^{\infty} \frac{(\ell-1)!}{(\ell+1)!} \frac{\rho_<^\ell}{\rho_>^{\ell+1}} \cdot \frac{\pi}{2}\sin\phi \cdot \int_{\pi/2}^{\pi} \sin^2\theta' P_{\ell,1}(\cos\theta) d\theta' \cdot P_{\ell,1}(\cos\theta)$$

$$\left. + \sum_{\ell=2}^{\infty}\sum_{\mu=1}^{[\![\frac{\ell}{2}]\!]} \frac{\rho_<^\ell}{\rho_>^{\ell+1}} \frac{(\ell-2\mu)!}{(\ell+2\mu)!} \frac{2\cos 2\mu\phi}{4\mu^2-1} \cdot \int_{\pi/2}^{\pi} \sin^2\theta' P_{\ell,2\mu}(\cos\theta') d\theta' \cdot P_{\ell,2\mu}(\cos\theta)\right]$$

$$= \frac{\pi}{3}\frac{\rho_<}{\rho_>^2} \cdot \sin\phi \cdot \sin\theta$$

$$+ \pi\sin\phi \sum_{\lambda=1}^{\infty} \frac{(-1)^{\lambda+1}(2\lambda-3)!!}{(2\lambda+2)!!} \frac{\rho_<^{2\lambda}}{\rho_>^{2\lambda+1}} \cdot P_{2\lambda,1}(\cos\theta)$$

$$+ 4\sum_{\ell=2}^{\infty}\sum_{\mu=1}^{[\![\frac{\ell}{2}]\!]} \frac{\rho_<^\ell}{\rho_>^{\ell+1}} \frac{(\ell-2\mu)!}{(\ell+2\mu)!} \frac{\cos 2\mu\phi}{4\mu^2-1} \cdot \mathcal{K}_{\ell,2\mu} \cdot P_{\ell,2\mu}(\cos\theta),$$

where

$$\mathcal{K}_{\ell,2\mu} = \int_{\pi/2}^{\pi} \sin^2\theta' \, P_{\ell,2\mu}(\cos\theta') d\theta'$$

and the fact that

$$\int_{\pi}^{2\pi} \sin\phi' \cos m(\phi - \phi') d\phi' = \frac{2\cos\frac{m\pi}{2}\cos m\left(\frac{3\pi}{3} - \phi\right)}{m^2 - 1}$$
$$= \begin{cases} \frac{\pi}{2}\sin\phi & \text{for } m = 1 \\ \frac{2\cos 2\mu\phi}{4\mu^2 - 1} & \text{for } m = 2\mu \\ 0 & \text{otherwise} \end{cases}$$

was used. Here $[\![\cdot]\!] : \mathbb{R} \to \mathbb{Z}$ denotes the greatest integer function. In the above derivations previous results, in particular (24), was used.

To obtain the magnetic field components, the partials of T_1, T_2, and T_3 need to be calculated. To this end, note that

$$\frac{\partial T_1}{\partial \rho} = \begin{cases} 0 & (\rho < \rho_0), \\ \frac{\pi}{2\rho^2} & (\rho > \rho_0). \end{cases}$$

$$\frac{\partial T_1}{\partial \phi} = \frac{\partial T_1}{\partial \theta} = 0, \quad \rho < \rho_0 \text{ and } \rho > \rho_0.$$

$$\frac{\partial T_2}{\partial \rho} = \begin{cases} -2\sum_{\ell=1}^{\infty} \frac{\ell\rho^{\ell-1}}{\rho_0^{\ell+1}} \cdot \mathcal{K}_\ell \cdot P_\ell(\cos\theta) & (\rho < \rho_0), \\ 2\sum_{\ell=1}^{\infty} \frac{(\ell+1)\rho_0^\ell}{\rho^{\ell+2}} \cdot \mathcal{K}_\ell \cdot P_\ell(\cos\theta) & (\rho > \rho_0). \end{cases}$$

$$\frac{\partial T_2}{\partial \phi} = 0, \quad \rho < \rho_0 \text{ and } \rho > \rho_0.$$

$$\frac{\partial T_2}{\partial \theta} = \begin{cases} 2\sum_{\ell=1}^{\infty} \frac{\rho^\ell}{\rho_0^{\ell+1}} \cdot \mathcal{K}_\ell \cdot P_{\ell,1}(\cos\theta) & (\rho < \rho_0), \\ 2\sum_{\ell=1}^{\infty} \frac{\rho_0^\ell}{\rho^{\ell+1}} \cdot \mathcal{K}_\ell \cdot P_{\ell,1}(\cos\theta) & (\rho > \rho_0), \end{cases}$$

where

$$\frac{\partial P_\ell(\cos\theta)}{\partial \theta} = -P_{\ell,1}(\cos\theta).$$

$$\frac{\partial T_3}{\partial \rho} = \begin{cases} \frac{\pi}{3}\frac{1}{\rho_0^2}\cdot\sin\phi\cdot\sin\theta \\ +\pi\sin\phi\sum_{\lambda=1}^{\infty}\frac{(-1)^{\lambda+1}(2\lambda)(2\lambda-3)!!}{(2\lambda+2)!!}\frac{\rho^{2\lambda-1}}{\rho_0^{2\lambda+1}}\cdot P_{2\lambda,1}(\cos\theta) \\ +4\sum_{\ell=2}^{\infty}\sum_{\mu=1}^{\left[\!\left[\frac{\ell}{2}\right]\!\right]}\frac{\rho^{\ell-1}}{\rho_0^{\ell+1}}\frac{\ell(\ell-2\mu)!}{(\ell+2\mu)!}\frac{\cos 2\mu\phi}{4\mu^2-1}\cdot\mathcal{K}_{\ell,2\mu}\cdot P_{\ell,2\mu}(\cos\theta) \\ & (\rho<\rho_0) \\ \frac{-2\pi}{3}\frac{\rho_0}{\rho^3}\cdot\sin\phi\cdot\sin\theta \\ +\pi\sin\phi\sum_{\lambda=1}^{\infty}\frac{(-1)^{\lambda}(2\lambda+1)(2\lambda-3)!!}{(2\lambda+2)!!}\frac{\rho_0^{2\lambda}}{\rho^{2\lambda+2}}\cdot P_{2\lambda,1}(\cos\theta) \\ +4\sum_{\ell=2}^{\infty}\sum_{\mu=1}^{\left[\!\left[\frac{\ell}{2}\right]\!\right]}\frac{\rho_0^{\ell}}{\rho^{\ell+2}}\frac{-(\ell+1)(\ell-2\mu)!}{(\ell+2\mu)!}\frac{\cos 2\mu\phi}{4\mu^2-1}\cdot\mathcal{K}_{\ell,2\mu}\cdot P_{\ell,2\mu}(\cos\theta) \\ & (\rho>\rho_0) \end{cases}$$

$$\frac{\partial T_3}{\partial \phi} = \begin{cases} \frac{\pi}{3}\frac{\rho}{\rho_0^2}\cdot\cos\phi\cdot\sin\theta \\ +\pi\cos\phi\sum_{\lambda=1}^{\infty}\frac{(-1)^{\lambda+1}(2\lambda-3)!!}{(2\lambda+2)!!}\frac{\rho^{2\lambda}}{\rho_0^{2\lambda+1}}\cdot P_{2\lambda,1}(\cos\theta) \\ -8\sum_{\ell=2}^{\infty}\sum_{\mu=1}^{\left[\!\left[\frac{\ell}{2}\right]\!\right]}\frac{\rho^{\ell}}{\rho_0^{\ell+1}}\frac{(\ell-2\mu)!}{(\ell+2\mu)!}\frac{\mu\sin 2\mu\phi}{4\mu^2-1}\cdot\mathcal{K}_{\ell,2\mu}\cdot P_{\ell,2\mu}(\cos\theta) \\ & (\rho<\rho_0) \\ \frac{\pi}{3}\frac{\rho_0}{\rho^2}\cdot\cos\phi\cdot\sin\theta \\ +\pi\cos\phi\sum_{\lambda=1}^{\infty}\frac{(-1)^{\lambda+1}(2\lambda-3)!!}{(2\lambda+2)!!}\frac{\rho_0^{2\lambda}}{\rho^{2\lambda+1}}\cdot P_{2\lambda,1}(\cos\theta) \\ -8\sum_{\ell=2}^{\infty}\sum_{\mu=1}^{\left[\!\left[\frac{\ell}{2}\right]\!\right]}\frac{\rho_0^{\ell}}{\rho^{\ell+1}}\frac{(\ell-2\mu)!}{(\ell+2\mu)!}\frac{\mu\sin 2\mu\phi}{4\mu^2-1}\cdot\mathcal{K}_{\ell,2\mu}\cdot P_{\ell,2\mu}(\cos\theta) \\ & (\rho>\rho_0) \end{cases}$$

and

$$\frac{\partial T_3}{\partial \theta} = \begin{cases} \frac{\pi}{3}\frac{\rho}{\rho_0^2}\cdot\sin\phi\cdot\cos\theta+\pi\sin\phi\sum_{\lambda=1}^{\infty}\frac{(-1)^{\lambda}(2\lambda-3)!!}{2(2\lambda+2)!!}\frac{\rho^{2\lambda}}{\rho_0^{2\lambda+1}} \\ \quad\times\left[P_{2\lambda,2}(\cos\theta)-2\lambda(2\lambda+1)P_{2\lambda}(\cos\theta)\right] \\ -2\sum_{\ell=2}^{\infty}\sum_{\mu=1}^{\left[\!\left[\frac{\ell}{2}\right]\!\right]}\frac{\rho^{\ell}}{\rho_0^{\ell+1}}\frac{(\ell-2\mu)!}{(\ell+2\mu)!}\frac{\cos 2\mu\phi}{4\mu^2-1}\cdot\mathcal{K}_{\ell,2\mu} \\ \quad\times\left[P_{\ell,2\mu+1}(\cos\theta)-(\ell+2\mu)(\ell-2\mu+1)P_{\ell,2\mu-1}(\cos\theta)\right] \\ & (\rho<\rho_0). \\ \frac{\pi}{3}\frac{\rho_0}{\rho^2}\cdot\sin\phi\cdot\cos\theta+\pi\sin\phi\sum_{\lambda=1}^{\infty}\frac{(-1)^{\lambda}(2\lambda-3)!!}{2(2\lambda+2)!!}\frac{\rho_0^{2\lambda}}{\rho^{2\lambda+1}} \\ \quad\times\left[P_{2\lambda,2}(\cos\theta)-2\lambda(2\lambda+1)P_{2\lambda}(\cos\theta)\right] \\ -2\sum_{\ell=2}^{\infty}\sum_{\mu=1}^{\left[\!\left[\frac{\ell}{2}\right]\!\right]}\frac{\rho_0^{\ell}}{\rho^{\ell+1}}\frac{(\ell-2\mu)!}{(\ell+2\mu)!}\frac{\cos 2\mu\phi}{4\mu^2-1}\cdot\mathcal{K}_{\ell,2\mu} \\ \quad\times\left[P_{\ell,2\mu+1}(\cos\theta)-(\ell+2\mu)(\ell-2\mu+1)P_{\ell,2\mu-1}(\cos\theta)\right] \\ & (\rho>\rho_0). \end{cases}$$

The derivatives $dP_{\ell,m}/d\cos\theta$ were replaced according to (25).

Theorem 4.2. *The magnetic flux leakage field* $\mathbf{B}_\Sigma = (B_\rho, B_\phi, B_\theta)$ *generated by a semi-spherical* V *of radius* ρ_0 *is explicitly obtained from the expressions below.*

$$B_\rho(\rho,\phi,\theta) = M\rho_0^2\left[\frac{\partial T_1}{\partial\rho} + \frac{\partial T_2}{\partial\rho} + \frac{\partial T_3}{\partial\rho}\right],$$

$$B_\phi(\rho,\phi,\theta) = \frac{M\rho_0^2}{\rho\sin\theta}\left[\frac{\partial T_1}{\partial\phi} + \frac{\partial T_2}{\partial\phi} + \frac{\partial T_3}{\partial\phi}\right],$$

$$B_\theta(\rho,\phi,\theta) = \frac{M\rho_0^2}{\rho}\left[\frac{\partial T_1}{\partial\theta} + \frac{\partial T_2}{\partial\theta} + \frac{\partial T_3}{\partial\theta}\right],$$

where the derivatives of T_i, $i = 1, 2, 3$, *are given above.*

The cartesian components of $\mathbf{B}_\Sigma$ can be obtained via (26).

5. Parabolic V

In this section the magnetic flux leakage field generated by the two types of parabolic V's introduced in this paper is calculated.

5.1. *Circular Parabolic V*

Assume that V_{par} is a circular parabolic with radius r_0 and depth d:

$$V_{par} = \{(\xi,\eta,\varphi) : \xi = \xi_0,\ 0 \le \eta \le \eta_0,\ 0 \le \varphi \le 2\pi\},$$

with $\xi_0 = \frac{r_0}{\sqrt{2d}}$ and $\eta_0 = \sqrt{2d}$. Using the form of the oriented surface element as derived in (14) and $\mathbf{M} = M\widehat{\jmath}$, one finds for the magnetic scalar potential

$$\Psi(\xi,\eta,\varphi) = \int_0^{\eta_0}\int_0^{2\pi}\frac{\langle\mathbf{M}', d\boldsymbol{\sigma}'\rangle}{|\boldsymbol{x}-\boldsymbol{x}'|} = \xi_0\int_0^{\eta_0}\int_0^{2\pi}\frac{\eta'^2\sin\phi'\,d\eta'\,d\phi'}{|\boldsymbol{x}-\boldsymbol{x}'|}.$$

Next an expression for the Green's function $1/|\boldsymbol{x}-\boldsymbol{x}'|$ in parabolic coordinates is derived. To this end, recall that the *Dirac delta distribution* δ in parabolic coordinates is given by

$$\delta(\boldsymbol{x}-\boldsymbol{x}') = \frac{1}{\xi\eta(\xi^2+\eta^2)}\,\delta(\xi-\xi')\delta(\eta-\eta')\delta(\varphi-\varphi')$$

and that a solution to the (distributional) differential equation

$$\Delta G(\xi,\xi',\eta,\eta',\varphi,\varphi') = \frac{-4\pi}{\xi\eta(\xi^2+\eta^2)}\,\delta(\xi-\xi')\delta(\eta-\eta')\delta(\varphi-\varphi') \qquad (27)$$

is sought. For $(\xi, \eta, \varphi) \neq (\xi', \eta', \varphi')$ the above equation can be solved by the ansatz $G = \Xi(\xi)\mathrm{H}(\eta)\Phi(\varphi)$. Thus, upon substitution, simplification, and division by $\Xi\mathrm{H}\Phi$ one obtains

$$\frac{\xi^2\eta^2}{\xi^2+\eta^2}\left[\frac{1}{\Xi}\left(\frac{\partial^2\Xi}{\partial\xi^2}+\frac{1}{\xi}\frac{\partial\Xi}{\partial\xi}\right)+\frac{1}{\mathrm{H}}\left(\frac{\partial^2\mathrm{H}}{\partial\eta^2}+\frac{1}{\eta}\frac{\partial\mathrm{H}}{\partial\eta}\right)\right]+\frac{1}{\Phi}\frac{\partial^2\Phi}{\partial\varphi^2}=0$$

As the variables ξ, η, and φ are independent of each other, there exists a constant m^2 so that

$$\frac{1}{\Phi}\frac{\partial^2\Phi}{\partial\varphi^2}=-m^2$$

and

$$\frac{\xi^2\eta^2}{\xi^2+\eta^2}\left[\frac{1}{\Xi}\left(\frac{\partial^2\Xi}{\partial\xi^2}+\frac{1}{\xi}\frac{\partial\Xi}{\partial\xi}\right)+\frac{1}{\mathrm{H}}\left(\frac{\partial^2\mathrm{H}}{\partial\eta^2}+\frac{1}{\eta}\frac{\partial\mathrm{H}}{\partial\eta}\right)\right]=m^2.$$

The former equation has solutions of the form

$$\Phi(\varphi)=e^{\pm im\varphi}. \tag{28}$$

The latter equation can be rewritten as

$$\left[\frac{1}{\Xi}\left(\frac{\partial^2\Xi}{\partial\xi^2}+\frac{1}{\xi}\frac{\partial\Xi}{\partial\xi}\right)-\frac{m^2}{\xi^2}\right]+\left[\frac{1}{\mathrm{H}}\left(\frac{\partial^2\mathrm{H}}{\partial\eta^2}+\frac{1}{\eta}\frac{\partial\mathrm{H}}{\partial\eta}\right)-\frac{m^2}{\eta^2}\right]=0.$$

Separation of variables yields

$$\xi^2\frac{\partial^2\Xi}{\partial\xi^2}+\xi\frac{\partial\Xi}{\partial\xi}-(c\xi^2+m^2)\Xi=0 \tag{29}$$

and

$$\eta^2\frac{\partial^2\mathrm{H}}{\partial\eta^2}+\eta\frac{\partial\mathrm{H}}{\partial\eta}+(c\eta^2-m^2)\mathrm{H}=0 \tag{30}$$

for a constant c.

(29) is a *modified Bessel Equation* whose solutions are Bessel functions I_m and K_m of the third kind, and (30) is just Bessel's equation with solutions J_m and Y_m. The Green's function for a magnetic scalar potential concentrated on the surface $\xi = \xi_0 =$ const. must involve only the function J_m as Y_m is singular at the origin.

To solve (27), recall that

$$\delta(\varphi-\varphi')=\frac{1}{2\pi}\sum_{m=-\infty}^{+\infty}e^{\pm im(\varphi-\varphi')} \tag{31}$$

(this is just the *Fourier series* for the Dirac delta distribution) and

$$\frac{\delta(\eta-\eta')}{\eta}=\int_0^\infty J_m(k\eta)J_m(k\eta')k\,dk, \tag{32}$$

which is the *Hankel transform* or *Fourier Bessel Integral* of the Dirac delta distribution (Refs. 8,14). Therefore, a solution of the inhomogenous equation (27) in the form

$$G(\xi,\xi';\eta,\eta';\varphi,\varphi') = \frac{1}{2\pi}\sum_{m=-\infty}^{+\infty} e^{\pm im(\varphi-\varphi')}\int_0^\infty J_m(k\eta)J_m(k\eta')g_m(\xi,\xi')kdk$$

is sought. Substituting this ansatz into (27) and simplifying gives

$$\frac{1}{2\pi}\sum_{m=-\infty}^{+\infty}\int_0^\infty\left(\frac{1}{\xi^2+\eta^2}\left[\ddot{g}_m+\frac{1}{\xi}\dot{g}_m-\frac{m^2}{\xi^2}g_m\right]e^{\pm im(\varphi-\varphi')}J_m(k\eta)J_m(k\eta')\right.$$
$$\left.+\frac{1}{\xi^2+\eta^2}\left[k^2\ddot{J}_m+\frac{k}{\eta}\dot{J}_m-\frac{m^2}{\eta^2}J_m\right]e^{\pm im(\varphi-\varphi')}J_m(k\eta')\right)kdk$$
$$=\frac{-4\pi}{\xi\eta(\xi^2+\eta^2)}\delta(\xi-\xi')\delta(\eta-\eta')\delta(\varphi-\varphi')$$

(Here the dots indicate differentiation with respect to the variable upon which the function depends.) Using the fact that

$$k^2\ddot{J}_m+\frac{k}{\eta}\dot{J}_m-\frac{m^2}{\eta^2}J_m=-k^2J_m$$

and (31) and (32), one finally obtains

$$\frac{1}{2\pi}\sum_{m=-\infty}^{+\infty}\int_0^\infty\frac{e^{\pm im(\varphi-\varphi')}J_m(k\eta)J_m(k\eta')}{\xi^2+\eta^2}\left[\ddot{g}_m+\frac{1}{\xi}\dot{g}_m-\left(k^2+\frac{m^2}{\xi^2}\right)g_m\right.$$
$$\left.+\frac{4\pi\delta(\xi-\xi')}{\xi}\right]kdk=0.$$

Hence $g_m(\xi,\xi')$ is a Green's function for the ordinary distributional differential equation

$$\ddot{g}_m+\frac{1}{\xi}\dot{g}_m-\left(k^2+\frac{m^2}{\xi^2}\right)g_m=-\frac{4\pi\delta(\xi-\xi')}{\xi} \tag{33}$$

whose solution is known to be (cf. for instance,[8])

$$g_m(\xi,\xi')=4\pi I_m(k\xi_<)K_m(k\xi_>)$$

with $\xi_< = \min\{\xi,\xi'\}$ and $\xi_> = \max\{\xi,\xi'\}$. Finally, the following result is obtained.

Theorem 5.1. *The expansion of Green's functions in parabolic coordinates is given by*

$$G(\xi,\xi';\eta,\eta';\varphi,\varphi) = 2\sum_{m=0}^{\infty} \epsilon_m \cos[m(\varphi-\varphi')] \times \int_0^{\infty} J_m(k\eta)J_m(k\eta')I_m(k\xi_<)K_m(k\xi_>)kdk \tag{34}$$

where

$$\epsilon_m = \begin{cases} 1 & \text{if } m=0 \\ 2 & \text{if } m>0 \end{cases}$$

is the Neumann factor, $\xi_< = \min\{\xi,\xi'\}$, *and* $\xi_> = \max\{\xi,\xi'\}$.

With this expression for the Green's function, the magnetic scalar potential becomes

$$\Psi(\xi,\eta,\varphi) = 2\xi_0 \sum_{m=0}^{\infty} \epsilon_m \int_0^{\infty}\int_0^{\eta_0}\int_0^{2\pi} \cos[m(\varphi-\varphi')]\sin\varphi' d\varphi' \times \eta'^2 J_m(k\eta)J_m(k\eta')I_m(k\xi_<)K_m(k\xi_>)d\eta\, kdk$$

where $\xi_< = \min\{\xi,\xi_0\}$ and $\xi_> = \max\{\xi,\xi_0\}$. As

$$\int_0^{2\pi} \cos[m(\varphi-\varphi')]\sin\varphi' d\varphi' = \begin{cases} \pi\sin\phi & \text{if } m=1 \\ 0 & \text{otherwise} \end{cases}$$

only the term $m=1$ contributes to the sum. Moreover,

$$\int_0^{\eta_0} \eta'^2 J_1(k\eta')d\eta' = \eta_0^2 J_2(k\eta_0)/k,$$

and thus the next theorem holds.

Theorem 5.2. *The magnetic scalar potential due to a circular parabolic* V *of radius* r_0 *and depth* d *is given by*

$$\Psi(\xi,\eta,\varphi) = 4\pi r_0\sqrt{2d}\sin\varphi \int_0^{\infty} J_1(k\eta)J_2(k\eta_0)I_1(k\xi_<)K_1(k\xi_>)dk,$$

where $\xi_< = \min\{\xi,\xi_0\}$ *and* $\xi_> = \max\{\xi,\xi_0\}$. *In the case* $\xi_< < \xi_>$, *the above improper integral may be expressed as an infinite series of the form*

$$\Psi(\xi,\eta,\varphi) = 2\pi r_0\,\eta\,\eta_0\,\sqrt{2d}\left(\frac{\xi_<}{\xi_>^5}\right)\sin\varphi\sum_{n=0}^{\infty}(-1)^n(n+1)(n+2)\left(\frac{\eta}{\xi_>}\right)^{2n} \times {}_2F_1\left(2+n,3+n;2;\frac{\xi_<^2}{\xi_>^2}\right){}_2F_1\left(-n,-n-1;3;\frac{\eta_0^2}{\eta^2}\right)$$

Proof. Only the second statement needs to be shown. To this end, employ the following series representation of the product of two Bessel functions (Ref. 16, p. 148, (2)):

$$\begin{aligned} J_\mu(az)J_\nu(bz) &= \frac{(\frac{1}{2}az)^\mu(\frac{1}{2}bz)^\nu}{\Gamma(\nu+1)} \\ &\quad \times \sum_{n=0}^{\infty} \frac{(-1)^n(\frac{1}{2}az)^{2n}{}_2F_1(-n,-\mu-n;\nu+1;b^2/a^2)}{m!\Gamma(\mu+m+1)}. \end{aligned}$$

Setting $\mu = 1$, $\nu = 2$, $a = \eta$, and $b = \eta_0$ in the above equation, using the fact that

$$\begin{aligned} \int_0^\infty k^{2n+3} I_1(k\xi_<)K_1(k\xi_>)dk &= 4^{1+n} r_< r_>^{-5-2n}\,\Gamma(2+n)\,\Gamma(3+n) \\ &\quad \times {}_2F_1(2+n,3+n;2;r_<^2/r_>^2) \end{aligned}$$

yields

$$\begin{aligned} \int_0^\infty J_1(k\eta)J_2(k\eta_0)I_1(k\xi_<)K_1(k\xi_>)dk &= \frac{1}{8}\eta\,\eta_0^2\,\xi_< \\ &\quad \times \sum_{n=0}^{\infty} \frac{(-1)^n 2^{-2n}\eta^{2n}4^{1+n}(n+1)!(n+2)!}{n!(n+1)!\xi_>^{5+2n}} \\ &\quad \times {}_2F_1(2+n,3+n;2;\xi_<^2/\xi_>^2)\,{}_2F_1(-n,-n-1;3;\eta_0^2/\eta^2). \end{aligned}$$

Simplification now gives the result. □

Theorem 5.3. *The components of the magnetic field* $\mathbf{B}_\Sigma = (B_\xi, B_\eta, B_\varphi)$ *generated by a circular parabolic* V *of radius* r_0 *and depth* d *are then given by*

$$B_\xi(\xi,\eta,\varphi) = -\frac{4\pi r_0\sqrt{2d}\sin\varphi}{\sqrt{\xi^2+\eta^2}} \int_0^\infty J_1(k\eta)J_2(k\eta_0)\frac{\partial}{\partial\xi}I_1(k\xi_<)K_1(k\xi_>)dk$$

$$B_\eta(\xi,\eta,\varphi) = -\frac{2\pi r_0\sqrt{2d}\sin\varphi}{\sqrt{\xi^2+\eta^2}} \int_0^\infty [J_0(k\eta)-J_2(k\eta)]J_2(k\eta_0)I_1(k\xi_<)K_1(k\xi_>)k\,dk,$$

as $2J_1' = J_0 - J_2$, *and*

$$B_\varphi(\xi,\eta,\varphi) = -\frac{4\pi r_0\sqrt{2d}\cos\varphi}{\xi\eta} \int_0^\infty J_1(k\eta)J_2(k\eta_0)I_1(k\xi_<)K_1(k\xi_>)dk.$$

The cartesian components (B_x, B_y, B_z) of the magnetic field $\mathbf{B}_\Sigma$ are obtained from the parabolic components $(B_\xi, B_\eta, B_\varphi)$ via the transformation

$$\begin{pmatrix} B_x \\ B_y \\ B_z \end{pmatrix} = \frac{1}{\xi^2+\eta^2} \begin{pmatrix} \eta\cos\varphi & \xi\cos\varphi & -\left(\frac{\xi}{\eta}+\frac{\eta}{\xi}\right)\sin\varphi \\ \eta\sin\varphi & \xi\sin\varphi & \left(\frac{\xi}{\eta}+\frac{\eta}{\xi}\right)\cos\varphi \\ -\xi & \eta & 0 \end{pmatrix} \begin{pmatrix} B_\xi \\ B_\eta \\ B_\varphi \end{pmatrix}$$

5.2. *Elliptic Parabolic V*

Now assume that V is elliptic parabolic:

$$V_{epar}: \quad z' = \frac{a^2}{2}(\cosh 2u' + \cos 2v'), \quad 0 \le u' \le u_0, \quad 0 \le v' \le 2\pi,$$

with

$$a = \sqrt{d}\,\sqrt{w^2 - \ell^2} \quad \text{and} \quad u_0 = \operatorname{arctanh}(\ell/w).$$

Assuming that $\mathbf{M} = M_0\hat{\jmath}$, the magnetic scalar potential Ψ in elliptic cylindrical coordinates is given by

$$\begin{aligned} \Psi(u,v,z) &= \int_0^{u_0}\int_0^{2\pi} \frac{\langle \mathbf{M}', d\boldsymbol{\sigma}'\rangle}{|\boldsymbol{x}-\boldsymbol{x}'|} \\ &= \frac{a^3 M_0}{2}\int_0^{u_0}\int_0^{2\pi} \frac{(\sin 3v' \sinh u' - \sin v' \sinh 3u')du'dv'}{|\boldsymbol{x}-\boldsymbol{x}'|}. \end{aligned}$$

At this point, the Green's function $G(\boldsymbol{x},\boldsymbol{x}') = 1/|\boldsymbol{x}-\boldsymbol{x}'|$ must be expressed in elliptic cylindrical coordinates. For this purpose, note that this Green's function satisfies the (distributional) differential equation

$$\Delta G(\boldsymbol{x},\boldsymbol{x}') = \frac{\partial^2 G}{\partial x^2} + \frac{\partial^2 G}{\partial y^2} + \frac{\partial^2 G}{\partial z^2} = -4\pi\delta(\boldsymbol{x}-\boldsymbol{x}'),$$

where $\delta(\boldsymbol{x}-\boldsymbol{x}')$ denotes the Dirac delta distribution. In elliptic cylindrical coordinates the above equation is written as

$$\begin{aligned} \Delta G(\boldsymbol{u},\boldsymbol{u}') &= \frac{2}{a^2(\cosh 2u - \cos 2v)}\left[\frac{\partial^2 G}{\partial u^2} + \frac{\partial^2 G}{\partial v^2}\right] + \frac{\partial^2 G}{\partial z^2} \\ &= \frac{-8\pi}{a^2(\cosh 2u - \cos 2v)}\delta(u-u')\delta(v-v')\delta(z-z'). \end{aligned} \quad (35)$$

For $\boldsymbol{u} \neq \boldsymbol{u}'$ a solution of the above equation may be found by the ansatz $G(\boldsymbol{u},\boldsymbol{u}') = U(u)V(v)Z(z)$. Substitution into (35) and division through UVZ yields

$$\frac{1}{Z}\frac{d^2 Z}{dz^2} + G = 0,$$

where

$$G = \frac{2}{a^2(\cosh 2u - \cos 2v)} \left[\frac{1}{U} \frac{\partial^2 U}{\partial u^2} + \frac{1}{V} \frac{\partial^2 V}{\partial v^2} \right].$$

Since u, v, and z are independent variables, it follows that there exists a constant k^2 so that

$$\frac{d^2 Z}{dz^2} + k^2 Z = 0.$$

This equation has solutions of the form $Z(z) = e^{\pm ikz}$. Again, from the fact that $G = k^2$ and the independence of u and v there exists a constant p such that

$$\frac{1}{U} \frac{\partial^2 U}{\partial u^2} - \frac{a^2 k^2}{2} \cosh 2u = p$$

and

$$-\frac{1}{V} \frac{\partial^2 V}{\partial v^2} - \frac{a^2 k^2}{2} \cos 2v = p.$$

Simplified, these two equations reduce to a *Mathieu equation*

$$\frac{\partial^2 V}{\partial v^2} + (p - 2q \cos 2v) V = 0 \tag{36}$$

and an *associated Mathieu equation*

$$\frac{\partial^2 U}{\partial u^2} - (p - 2q \cosh 2u) U = 0, \tag{37}$$

with $q = -a^2 k^2 / 4 < 0$. (Note that $k = 0$ does not yield bounded solutions.)

It is known that in (36), for each given value of q, there exists a set of eigenvalues $p_n = p_n(q)$ and an associated set of complete (even or odd) periodic eigenfunctions of period π or 2π. (The second integral is non-periodic and for physical reasons ruled out.) The set of even eigenfunctions of period 2π is denoted by $\{\mathrm{ce}_{2n+1}(v) \ : \ n = 1, 2, \ldots\}$ and the set of odd eigenfunctions of period 2π by $\{\mathrm{se}_{2n+1}(v) \ : \ n = 1, 2, \ldots\}$.[¶] Moreover, ce_{2n+1} and se_{2n+1} satisfy the orthogonality conditions

$$\int_0^{2\pi} \begin{Bmatrix} \mathrm{se}_n(v)\, \mathrm{se}_m(v) \\ \mathrm{ce}_n(v)\, \mathrm{ce}_m(v) \end{Bmatrix} dv = \pi \delta_{nm}$$

and

$$\int_0^{2\pi} \mathrm{se}_n(v) \mathrm{ce}_n(v) du = 0,$$

[¶]The functions ce_{2n+1} and se_{2n+1} do also depend on q. To simplify notation, however, this dependence is suppressed when not explicitly needed.

and may be expressed as infinite series of cosines and sines

$$\mathrm{ce}_{2n+1}(v) = \sum_{\nu=0}^{\infty} (-1)^{n+\nu} B_{2\nu+1}^{(2n+1)} \cos(2\nu+1)v, \quad q < 0$$

$$\mathrm{se}_{2n+1}(v) = \sum_{\nu=0}^{\infty} (-1)^{n+\nu} A_{2\nu+1}^{(2n+1)} \sin(2\nu+1)v, \quad q < 0.$$

where the coefficients $A_{2\nu+1}^{(2n+1)} = A_{2\nu+1}^{(2n+1)}(q)$ and $B_{2\nu+1}^{(2n+1)} = B_{2\nu+1}^{(2n+1)}(q)$ depend on q and its sign. Note that

$$\mathrm{ce}_{2n+1}(0) = (-1)^n \sum_{\nu=0}^{\infty} (-1)^{\nu} B_{2\nu+1}^{(2n+1)}$$

and

$$\mathrm{se}_{2n+1}(\pi/2) = (-1)^n \sum_{\nu=0}^{\infty} A_{2\nu+1}^{(2n+1)}$$

Solutions for the modified Mathieu equation (37) for the *same values of p and q* are of the form $\mathrm{ceh}_{2n+1}(u)$ associated with $\mathrm{ce}_{2n+1}(v)$ and $\mathrm{seh}_{2n+1}(u)$ associated with $\mathrm{se}_{2n+1}(v)$. Non-periodic second integrals of Eq. (37) are given by $\mathrm{fekh}_{2n+1}(u)$ and $\mathrm{gekh}_{2n+1}(u)$ associated with $\mathrm{ceh}_{2n+1}(u)$ and $\mathrm{seh}_{2n+1}(u)$, respectively. It is noted that these *modified Mathieu functions* also depend on q and its sign.

The following behavior of unmodified and modified Mathieu functions applies when $q \to 0-$.

$$\begin{aligned}
\mathrm{ce}_n(v) &\to \cos nv, \text{ for } n \geq 1, \\
\mathrm{se}_n(v) &\to \sin nv, \text{ for } n \geq 1, \\
\mathrm{ceh}_n(u) &\to c_n I_n(kr), \ c_n \text{ constant}, \\
\mathrm{seh}_n(u) &\to c_n' I_n(kr), \ c_n' \text{ constant}, \\
\mathrm{fekh}_n(u) &\to c_n'' K_n(kr), \ c_n'' \text{ constant}, \\
\mathrm{gekh}_n(u) &\to c_n''' K_n(kr), \ c_n''' \text{ constant}.
\end{aligned}$$

For proofs see,[10] and for the verification of the argument in the modidied Bessel functions I_n and K_n recall Remark (1.3) and the fact that $k^2a^2 = -4q$.

More details and additional information about Mathieu functions can be found in,[1],[4] and.[10]

Set

$$V(v) = \mathrm{se}_{2n+1}(v, q), \text{ with } q = -a^2k^2/4 < 0.$$

Since the system $\{\mathrm{se}_{2n+1}\}$ forms an orthogonal complete set of eigenfunctions, the Dirac delta distribution $\delta(v-v')$ can be written in the form

$$\delta(v-v') = \frac{1}{\pi}\sum_{n=0}^{\infty} \mathrm{se}_{2n+1}(v)\mathrm{se}_{2n+1}(v').$$

(Refs. 3,8,14) In order to solve (35), the Dirac delta distribution $\delta(z-z')$ is represented by its Fourier transform as

$$\delta(z-z') = \frac{1}{2\pi}\int_{-\infty}^{\infty} e^{\pm ik(z-z')}dk = \frac{1}{\pi}\int_{0}^{\infty} \cos[k(z-z')]dk.$$

The following ansatz for the Green's function is then made.

$$G(u,u';v,v';z,z') = \frac{1}{\pi^2}\sum_{n=0}^{\infty}\int_0^{\infty} \cos[k(z-z')]\mathrm{se}_{2n+1}(v)\mathrm{se}_{2n+1}(v')g_n(k,u,u')dk,$$

where the function g_n is determined so that (35) is satisfied.

Substituting the above expression for G into (35) yields (As above, the dots indicate differentiation with respect to the variable upon which the function depends.)

$$\begin{aligned}\frac{1}{\pi^2}\sum_{n=0}^{\infty}\int_0^{\infty}\Bigg(&\frac{2}{a^2(\cosh 2u-\cos 2v)}\,[\cos[k(z-z')]\mathrm{se}_{2n+1}(v)\mathrm{se}_{2n+1}(v')\ddot{g}_n\\ &+\cos[k(z-z')]\ddot{\mathrm{se}}_{2n+1}(v)\mathrm{se}_{2n+1}(v')g_n]\\ &-k^2\cos[k(z-z')]\mathrm{se}_{2n+1}(v)\mathrm{se}_{2n+1}(v')g_n\Bigg)dk\end{aligned}$$

Employing (36), and simplifying gives

$$\begin{aligned}\frac{1}{\pi^2}\sum_{n=0}^{\infty}\int_0^{\infty}&\left(\frac{2\cos[k(z-z')]\mathrm{se}_{2n+1}(v)\mathrm{se}_{2n+1}(v')}{a^2(\cosh 2u-\cos 2v)}\right)\\ &\times\left[\ddot{g}_n-(p-q\cosh 2u)g_n+4\pi\delta(u-u')\right]dk=0.\end{aligned}$$

Hence, g_n satisfies the (distributional) differential equation

$$\ddot{g}_n-(p-q\cosh 2u)g_n = -4\pi\delta(u-u'). \tag{38}$$

As se_{2n+1} was chosen as a first integral of (36), the functions seh_{2n+1} and gekh_{2n+1} are the only acceptable solutions of (38). A solution $g_n(u,u')$ satisfying the symmetry requirement $g_n(u,u') = g_n(u',u)$ and the finiteness condition at $u=0$ and $u\to\infty$ must therefore be of the form $g_n(u,u') = c\,\mathrm{seh}_{2n+1}(u_<)\times\ \mathrm{gekh}_{2n+1}(u_>)$, where $u_< = \min\{u,u'\}$, $u_> = \max\{u,u'\}$, and c is a constant to be determined so that the Wronskian W of seh_{2n+1} and gekh_{2n+1} satisfies

$$\begin{aligned} &W(\mathrm{seh}_{2n+1}, \mathrm{gekh}_{2n+1}) \\ &= \mathrm{seh}_{2n+1}(u')\frac{d\mathrm{gekh}_{2n+1}}{du}(u') - \frac{d\mathrm{seh}_{2n+1}}{du}(u')\mathrm{gekh}_{2n+1}(u') \\ &= -4\pi/c. \end{aligned}$$

To determine c, results concerning the expansion of seh_{2n+1} and gekh_{2n+1} in terms of modified Bessel functions and the asymptotic behavior of seh_{2n+1} and gekh_{2n+1} and their derivatives are employed.

Lemma 5.1. *The functions* seh_{2n+1} *and* gekh_{2n+1} *have the following series representations in terms of modified Bessel functions (for* $q < 0$*). (*$\dot{} = d/du$*)*

$$\mathrm{seh}_{2n+1}(u) = \frac{(-1)^{n+1}\dot{\mathrm{ce}}_{2n+1}(\pi/2)}{\sqrt{|q|}A_1^{(2n+1)}} \sum_{\nu=0}^{\infty} A_{2\nu+1}^{(2n+1)} I_{2\nu+1}(2\sqrt{|q|}\sinh u)$$

and

$$\mathrm{gekh}_{2n+1}(u) = \frac{(-1)^{n+1}\dot{\mathrm{ce}}_{2n+1}(\pi/2)}{2\pi\sqrt{|q|}A_1^{(2n+1)}} \sum_{\nu=0}^{\infty} A_{2\nu+1}^{(2n+1)} K_{2\nu+1}(2\sqrt{|q|}\sinh u)$$

The functions seh_{2n+1} *and* gekh_{2n+1} *and their derivatives satisfy the following asymptotic estimates as* $u \to \infty$.

$$\mathrm{seh}_{2n+1}(u) \sim \frac{(-1)^n \dot{\mathrm{se}}_{2n+1}(0)\,\mathrm{se}_{2n+1}(\pi/2)}{\sqrt{2\pi}|q|^{3/4}B_1^{(2n+1)}}\, e^{-u/2}\, e^{\sqrt{|q|}e^u} \tag{39}$$

and

$$\mathrm{gekh}_{2n+1}(u) \sim \frac{(-1)^n \dot{\mathrm{se}}_{2n+1}(0)\,\mathrm{se}_{2n+1}(\pi/2)}{\sqrt{2\pi}|q|^{3/4}B_1^{(2n+1)}}\, e^{-u/2}\, e^{-\sqrt{|q|}e^u} \tag{40}$$

Moreover,

$$\frac{d\mathrm{seh}_{2n+1}}{du} \sim -\frac{\dot{\mathrm{ce}}_{2n+1}(\pi/2)\,\mathrm{se}_{2n+1}(\pi/2)}{\sqrt{2\pi}|q|^{1/4}A_1^{(2n+1)}}\, e^{u/2}\, e^{\sqrt{|q|}e^u} \tag{41}$$

and

$$\frac{d\mathrm{gekh}_{2n+1}}{du} \sim \frac{\dot{\mathrm{ce}}_{2n+1}(\pi/2)\,\mathrm{se}_{2n+1}(\pi/2)}{2\sqrt{2\pi}|q|^{1/4}A_1^{(2n+1)}}\, e^{u/2}\, e^{-\sqrt{|q|}e^u} \tag{42}$$

Proof. See Ref. 4 for the Bessel series expansions and the asymptotic estimates for seh_{2n+1} and gekh_{2n+1}. To obtain the estimates on the derivatives, consider (39) and (40) together with the following asymptotic estimates for the derivatives of the modified Bessel functions I_ν and K_ν (Refs. 1,3,9).

$$I'_\nu(x) \sim \frac{e^x}{\sqrt{2\pi x}}, \qquad x \to \infty$$

and

$$K'_\nu(x) \sim -\sqrt{\frac{\pi}{2x}}e^{-x}, \qquad x \to \infty. \qquad \square$$

Using the asymptotic estimates provided in the above lemma, the Wronskian of seh_{2n+1} and gekh_{2n+1} is computed as

$$\begin{aligned} W(\text{seh}_{2n+1}, \text{gekh}_{2n+1}) &= \left[\frac{(-1)^n \dot{\text{se}}_{2n+1}(0)\,\text{se}_{2n+1}(\pi/2)}{\sqrt{2\pi}|q|^{3/4}B_1^{(2n+1)}}\, e^{-u/2}\, e^{\sqrt{|q|e^u}}\right. \\ &\qquad \left.\times \frac{\dot{\text{ce}}_{2n+1}(\pi/2)\,\text{se}_{2n+1}(\pi/2)}{2\sqrt{2\pi}|q|^{1/4}A_1^{(2n+1)}}\, e^{u/2}\, e^{-\sqrt{|q|e^u}}\right] \\ &\quad + \left[\frac{\dot{\text{ce}}_{2n+1}(\pi/2)\,\text{se}_{2n+1}(\pi/2)}{\sqrt{2\pi}|q|^{1/4}A_1^{(2n+1)}}\, e^{u/2}\, e^{\sqrt{|q|e^u}}\right. \\ &\qquad \left.\times \frac{(-1)^n \dot{\text{se}}_{2n+1}(0)\,\text{se}_{2n+1}(\pi/2)}{\sqrt{2\pi}|q|^{3/4}B_1^{(2n+1)}}\, e^{-u/2}\, e^{-\sqrt{|q|e^u}}\right] \\ &= \frac{3(-1)^n \text{se}^2_{2n+1}(\pi/2)\, \dot{\text{se}}_{2n+1}(0)\, \dot{\text{ce}}_{2n+1}(\pi/2)}{4\pi|q|A_1^{(2n+1)}B_1^{(2n+1)}} \\ &= -\frac{4\pi}{c} \end{aligned}$$

Hence,

$$\begin{aligned} c &= \frac{16(-1)^{n+1}\pi^2|q|A_1^{(2n+1)}B_1^{(2n+1)}}{3\text{se}^2_{2n+1}(\pi/2)\,\dot{\text{se}}_{2n+1}(0)}\,\dot{\text{ce}}_{2n+1}(\pi/2) \\ &= \frac{4(-1)^n\pi^2a^2k^2A_1^{(2n+1)}B_1^{(2n+1)}}{3\text{se}^2_{2n+1}(\pi/2)\,\dot{\text{se}}_{2n+1}(0)}\,\dot{\text{ce}}_{2n+1}(\pi/2) \end{aligned}$$

and therefore

$$g_n(u,u') = c\,\text{seh}_{2n+1}(u_<)\,\text{gekh}_{2n+1}(u_>),$$

implying the next result.

Theorem 5.4. *The Green's function of the equation $\Delta\Psi = 0$ is elliptic cylindrical coordinates is given by*

$$G(u,u';v,v';z,z') = \frac{4a^2}{3}\sum_{n=0}^{\infty}\int_0^{\infty} c_n(k)\cos[k(z-z')]\,\mathrm{se}_{2n+1}(v,k)$$
$$\times\ \mathrm{se}_{2n+1}(v',k)\,\mathrm{seh}_{2n+1}(u_<,k)\,\mathrm{gekh}_{2n+1}(u_>,k)\,dk$$

with

$$c_n(k) = \frac{(-1)^n k^2 A_1^{(2n+1)} B_1^{(2n+1)}}{\mathrm{se}^2_{2n+1}(\pi/2)\,\dot{\mathrm{se}}_{2n+1}(0)\,\dot{\mathrm{ce}}_{2n+1}(\pi/2)}$$

and $u_< = \min\{u,u'\}$, $u_> = \max\{u,u'\}$.

Next the magnetic scalar potential for an elliptic parabolic V is computed. To this end, note that

$$\Psi(u,v,z) = \frac{2a^5 M_0}{3}\sum_{n=0}^{\infty}\int_0^{\infty}\int_0^{u_0}\int_0^{2\pi} c_n(k)\cos\left[z-\frac{a^2}{2}(\cosh 2u'-\cos 2v')\right]$$

$$\times\ [\sinh u'\sin 3v' - \sinh 3u'\sin v']\,\mathrm{se}_{2n+1}(v,k)\,\mathrm{se}_{2n+1}(v',k)$$

$$\times\ \mathrm{seh}_{2n+1}(u_<,k)\,\mathrm{gekh}_{2n+1}(u_>,k)\,du'dv'dk$$

$$= \frac{2a^5 M_0}{3}\sum_{n=0}^{\infty}\int_0^{\infty} c_n(k)\left(\int_0^{u}\int_0^{2\pi}\cos\left[z-\frac{a^2}{2}(\cosh 2u'-\cos 2v')\right]\right.$$

$$\times\ [\sinh u'\sin 3v' - \sinh 3u'\sin v']\,\mathrm{se}_{2n+1}(v,k)\,\mathrm{se}_{2n+1}(v',k)$$

$$\times\ \mathrm{seh}_{2n+1}(u',k)\,\mathrm{gekh}_{2n+1}(u,k)\,du'dv'$$

$$+\int_u^{u_0}\int_0^{2\pi}\cos\left[z-\frac{a^2}{2}(\cosh 2u'-\cos 2v')\right]$$

$$\times\ [\sinh u'\sin 3v' - \sinh 3u'\sin v']\,\mathrm{se}_{2n+1}(v,k)\,\mathrm{se}_{2n+1}(v',k)$$

$$\left.\times\ \mathrm{seh}_{2n+1}(u,k)\,\mathrm{gekh}_{2n+1}(u',k)\,du'dv'\right)dk$$

To simplify notation, the following definitions are made.

Definition 5.1. Let

$$\mathcal{S}_{2n+1}(u,z,k) = \frac{(-1)^n}{\pi}\int_0^u\int_0^{2\pi}\cos k\left[z - \frac{a^2}{2}(\cosh 2u' - \cos 2v')\right]$$
$$\times\,[\sinh u'\sin 3v' - \sinh 3u'\sin v']\,\mathrm{seh}_{2n+1}(u',k)\,\mathrm{se}_{2n+1}(v',k)\,du'dv'.$$

and

$$\mathcal{G}_{2n+1}(u,z,k;u_0) = \frac{(-1)^n}{\pi}\int_u^{u_0}\int_0^{2\pi}\cos k\left[z - \frac{a^2}{2}(\cosh 2u' - \cos 2v')\right]$$
$$\times\,[\sinh u'\sin 3v' - \sinh 3u'\sin v']\,\mathrm{gekh}_{2n+1}(u',k)\,\mathrm{se}_{2n+1}(v',k)\,du'dv'.$$

To further simplify $\mathcal{S}_{2n+1}$ and $\mathcal{G}_{2n+1}$, note that

$$\int_0^{2\pi}\cos k\left[z - \frac{a^2}{2}(\cosh 2u' - \cos 2v')\right]\sin 3v'\,\mathrm{se}_{2n+1}(v',k)du'$$
$$= \cos k\left[z - \frac{a^2}{2}\cosh 2u'\right]\int_0^{2\pi}\cos k\left[\frac{a^2}{2}\cos 2v'\right]\sin 3v'\,\mathrm{se}_{2n+1}(v',k)\,du'$$
$$- \sin k\left[z - \frac{a^2}{2}\cosh 2u'\right]\int_0^{2\pi}\sin k\left[\frac{a^2}{2}\cos 2v'\right]\sin 3v'\,\mathrm{se}_{2n+1}(v',k)\,du'$$

and, in particular,

$$\int_0^{2\pi}\cos k\left[\frac{a^2}{2}\cos 2v'\right]\sin 3v'\,\mathrm{se}_{2n+1}(v',k)\,du' = \sum_{m=0}^{\infty}\frac{(-1)^m\left(\frac{ka^2}{2}\right)^{2m}}{(2m)!}$$
$$\times\sum_{\nu=0}^{\infty}(-1)^{n+\nu}A_{2\nu+1}^{(2n+1)}\int_0^{2\pi}\cos^{2m}(2v')\,\sin 3v'\,\sin(2\nu+1)v'\,dv'$$

and

$$\int_0^{2\pi}\sin k\left[\frac{a^2}{2}\cos 2v'\right]\sin 3v'\,\mathrm{se}_{2n+1}(v',k)\,du' = \sum_{m=0}^{\infty}\frac{(-1)^m\left(\frac{ka^2}{2}\right)^{2m+1}}{(2m+1)!}$$
$$\times\sum_{\nu=0}^{\infty}(-1)^{n+\nu}A_{2\nu+1}^{(2n+1)}\int_0^{2\pi}\cos^{2m+1}(2v')\,\sin 3v'\,\sin(2\nu+1)v'\,dv'.$$

Using the trigonometric identity

$$\sin 3v'\sin(2\nu+1)v' = [\cos 2(\nu+2)v' - \cos 2(\nu-1)v']/2,$$

one obtains

$$
\begin{aligned}
\int_0^{2\pi} & \cos^{2m}(2v')\,\sin 3v'\,\sin(2\nu+1)v'\,dv' \\
&= \frac{1}{2}\int_0^{2\pi} \cos^{2m}(2v')\,[\cos 2(\nu+2)v' - \cos 2(\nu-1)v']\,dv' \\
&= \frac{1}{4}\int_0^{4\pi} \cos^{2m} x\,[\cos(\nu+2)x - \cos(\nu-1)x]\,dx
\end{aligned}
\tag{43}
$$

and

$$
\begin{aligned}
\int_0^{2\pi} & \cos^{2m+1}(2v')\sin 3v'\,\sin(2\nu+1)v'\,dv' \\
&= \frac{1}{2}\int_0^{2\pi} \cos^{2m+1}(2v')\,[\cos 2(\nu+2)v' - \cos 2(\nu-1)v']\,dv' \\
&= \frac{1}{4}\int_0^{4\pi} \cos^{2m+1} x\,[\cos(\nu+2)x - \cos(\nu-1)x]\,dx
\end{aligned}
\tag{44}
$$

To proceed, the following lemma is needed.

Lemma 5.2. *Let a and m be integers. Then, for $m > 0$,*

$$
\int_0^{4\pi} \cos^{2m} x\cos(ax)dx = \begin{cases} \frac{4\pi}{2^{2m}}\binom{2m}{m-b}, & \text{for } a = 2b \text{ and } b = 0,\ldots,m. \\ 0 & \text{otherwise.} \end{cases}
$$

and, for $m \geq 0$,

$$
\int_0^{4\pi} \cos^{2m+1} x\cos(ax)dx = \begin{cases} \frac{4\pi}{2^{2m+2}}\left[\binom{2m+1}{m-b} + \binom{2m+1}{m+1+b}\right], & \text{for } a = 2b+1 \text{ and } \\ & b = 0,\ldots,m. \\ 0 & \text{otherwise.} \end{cases}
$$

Thus, applying the above lemma to (43) and (44) and simplifying yields

$$
\begin{aligned}
\int_0^{2\pi} & \cos k\left[\frac{a^2}{2}\cos 2v'\right]\,\sin 3v'\,\mathrm{se}_{2n+1}(v',k)\,du' \\
&= \frac{(-1)^n\pi}{2^{2m}}\sum_{\nu=0}^{m-1}\left[A_{4\nu+1}^{(2n+1)}\binom{2m}{m-1-\nu} + A_{4\nu+3}^{(2n+1)}\binom{2m}{m-\nu}\right]
\end{aligned}
$$

and

$$\int_0^{2\pi} \sin k\left[\frac{a^2}{2}\cos 2v'\right]\sin 3v'\,\mathrm{se}_{2n+1}(v',k)\,du'$$
$$= -\frac{(-1)^n\pi}{2^{2m+2}}\sum_{\nu=0}^{m}\left(A_{4\nu+3}^{(2n+1)}\left[\binom{2m+1}{m-\nu-1}+\binom{2m+1}{m+2+\nu}\right]\right.$$
$$\left.-A_{4\nu+5}^{(2n+1)}\left[\binom{2m+1}{m-\nu}+\binom{2m+1}{m+1+\nu}\right]\right).$$

Hence,

$$\int_0^{2\pi}\cos k\left[z-\frac{a^2}{2}(\cosh 2u'-\cos 2v')\right]\sin 3v'\,\mathrm{se}_{2n+1}(v',k)du'$$
$$=(-1)^n\pi\left(\cos k\left[z-\frac{a^2}{2}\cosh 2u'\right]\sum_{m=1}^{\infty}\frac{(-1)^m k^{2m}a^{4m}}{2^{4m}(2m)!}\right.$$
$$\times\sum_{\nu=0}^{m-1}\left[A_{4\nu+1}^{(2n+1)}\binom{2m}{m-1-\nu}+A_{4\nu+3}^{(2n+1)}\binom{2m}{m-\nu}\right]$$
$$+\sin k\left[z-\frac{a^2}{2}\cosh 2u'\right]\sum_{m=0}^{\infty}\frac{(-1)^m k^{2m+1}a^{4m+2}}{2^{4m+3}(2m+1)!}$$
$$\times\sum_{\nu=0}^{m}\left(A_{4\nu+3}^{(2n+1)}\left[\binom{2m+1}{m-\nu-1}+\binom{2m+1}{m+2+\nu}\right]\right.$$
$$\left.\left.-A_{4\nu+5}^{(2n+1)}\left[\binom{2m+1}{m-\nu}+\binom{2m+1}{m+1+\nu}\right]\right)\right).$$

The integral

$$\int_0^{2\pi}\cos k\left[z-\frac{a^2}{2}(\cosh 2u'-\cos 2v')\right]\sin v'\,\mathrm{se}_{2n+1}(v',k)du'$$
$$=\cos k\left[z-\frac{a^2}{2}\cosh 2u'\right]\int_0^{2\pi}\cos k\left[\frac{a^2}{2}\cos 2v'\right]\sin v'\,\mathrm{se}_{2n+1}(v',k)\,du'$$
$$-\sin k\left[z-\frac{a^2}{2}\cosh 2u'\right]\int_0^{2\pi}\sin k\left[\frac{a^2}{2}\cos 2v'\right]\sin v'\,\mathrm{se}_{2n+1}(v',k)\,du'$$

is computed in a similar fashion, employing the trigonometric identity

$$\sin v'\sin(2\nu+1)v' = [\cos 2(\nu+1)v' - \cos 2\nu v']/2.$$

The result is

$$\int_0^{2\pi} \cos k\left[z - \frac{a^2}{2}(\cosh 2u' - \cos 2v')\right] \sin v' \,\mathrm{se}_{2n+1}(v', k) du'$$

$$= (-1)^n \pi \left(\cos k\left[z - \frac{a^2}{2}\cosh 2u'\right] \sum_{m=1}^{\infty} \frac{(-1)^{m+1} k^{2m} a^{4m}}{2^{4m}(2m)!} \right.$$

$$\times \left[A_1^{(2n+1)} \binom{2m}{m} + \sum_{\nu=1}^{m} \left(A_{4\nu-1}^{(2n+1)} + A_{4\nu+1}^{(2n+1)} \right) \binom{2m}{m-\nu} \right]$$

$$- \sin k\left[z - \frac{a^2}{2}\cosh 2u'\right] \sum_{m=0}^{\infty} \frac{(-1)^m k^{2m+1} a^{4m+2}}{2^{4m+3}(2m+1)!}$$

$$\left. \times \sum_{\nu=0}^{m} \left(A_{4\nu+1}^{(2n+1)} + A_{4\nu+3}^{(2n+1)} \right) \left[\binom{2m+1}{m-\nu} + \binom{2m+1}{m+1+\nu} \right] \right).$$

For the sake of notational simplicity, define

$$\Sigma_n^{(1)}(k) = \sum_{m=1}^{\infty} \frac{(-1)^m k^{2m} a^{4m}}{2^{4m}(2m)!} \sum_{\nu=0}^{m-1} \left[A_{4\nu+1}^{(2n+1)} \binom{2m}{m-1-\nu} + A_{4\nu+3}^{(2n+1)} \binom{2m}{m-\nu} \right]$$

$$\Sigma_n^{(2)}(k) = \sum_{m=0}^{\infty} \frac{(-1)^m k^{2m+1} a^{4m+2}}{2^{4m+3}(2m+1)!} \sum_{\nu=0}^{m} \left(A_{4\nu+3}^{(2n+1)} \left[\binom{2m+1}{m-\nu-1} + \binom{2m+1}{m+2+\nu} \right] \right.$$

$$\left. - A_{4\nu+5}^{(2n+1)} \left[\binom{2m+1}{m-\nu} + \binom{2m+1}{m+1+\nu} \right] \right)$$

$$\Sigma_n^{(3)}(k) = \sum_{m=1}^{\infty} \frac{(-1)^{m+1} k^{2m} a^{4m}}{2^{4m}(2m)!} \left[A_1^{(2n+1)} \binom{2m}{m} \right.$$

$$\left. + \sum_{\nu=1}^{m} \left(A_{4\nu-1}^{(2n+1)} + A_{4\nu+1}^{(2n+1)} \right) \binom{2m}{m-\nu} \right]$$

$$\Sigma_n^{(4)}(k) = \sum_{m=0}^{\infty} \frac{(-1)^m k^{2m+1} a^{4m+2}}{2^{4m+3}(2m+1)!} \sum_{\nu=0}^{m} \left(A_{4\nu+1}^{(2n+1)} + A_{4\nu+3}^{(2n+1)} \right)$$

$$\times \left[\binom{2m+1}{m-\nu} + \binom{2m+1}{m+1+\nu} \right]$$

Then, using these four functions, the expressions for $\mathcal{S}_{2n+1}$ and $\mathcal{G}_{2n+1}$ can be written in a more succinct way. The results are summarized in the following lemma.

Lemma 5.3.

$$
\begin{aligned}
&\mathcal{S}_{2n+1}(u,z,k)\\
&\quad = \Sigma_n^{(1)}(k)\int_0^u \cos k\left[z-\frac{a^2}{2}\cosh 2u'\right]\sinh u'\,\mathrm{seh}_{2n+1}(u',k)\,du'\\
&\quad - \Sigma_n^{(2)}(k)\int_0^u \sin k\left[z-\frac{a^2}{2}\cosh 2u'\right]\sinh u'\,\mathrm{seh}_{2n+1}(u',k)\,du'\\
&\quad + \Sigma_n^{(3)}(k)\int_0^u \cos k\left[z-\frac{a^2}{2}\cosh 2u'\right]\sinh 3u'\,\mathrm{seh}_{2n+1}(u',k)\,du'\\
&\quad - \Sigma_n^{(4)}(k)\int_0^u \sin k\left[z-\frac{a^2}{2}\cosh 2u'\right]\sinh 3u'\,\mathrm{seh}_{2n+1}(u',k)\,du'
\end{aligned}
$$

and

$$
\begin{aligned}
&\mathcal{G}_{2n+1}(u,z,k;u_0)\\
&\quad = \Sigma_n^{(1)}(k)\int_u^{u_0} \cos k\left[z-\frac{a^2}{2}\cosh 2u'\right]\sinh u'\,\mathrm{gekh}_{2n+1}(u',k)\,du'\\
&\quad - \Sigma_n^{(2)}(k)\int_u^{u_0} \sin k\left[z-\frac{a^2}{2}\cosh 2u'\right]\sinh u'\,\mathrm{gekh}_{2n+1}(u',k)\,du'\\
&\quad + \Sigma_n^{(3)}(k)\int_u^{u_0} \cos k\left[z-\frac{a^2}{2}\cosh 2u'\right]\sinh 3u'\,\mathrm{gekh}_{2n+1}(u',k)\,du'\\
&\quad - \Sigma_n^{(4)}(k)\int_u^{u_0} \sin k\left[z-\frac{a^2}{2}\cosh 2u'\right]\sinh 3u'\,\mathrm{gekh}_{2n+1}(u',k)\,du'
\end{aligned}
$$

Proof. The formula for $\mathcal{G}_{2n+1}$ is essentially given by $\mathcal{S}_{2n+1}$ with seh_{2n+1} replaced by gekh_{2n+1}. □

In summary,

Theorem 5.5. *Suppose that V is elliptic parabolic of length ℓ, width w, and depth d. Then the magnetic scalar potential generated by V is given by*

$$
\begin{aligned}
\Psi(u,v,z) = \frac{2\pi a^5 M_0}{3}\sum_{n=0}^{\infty}\int_0^{\infty} & C_n(k)[\mathcal{S}_{2n+1}(u,z,k)\,\mathrm{gekh}_{2n+1}(u)\\
&+ \mathcal{G}_{2n+1}(u,z,k;u_0)\,\mathrm{seh}_{2n+1}(u)]\,\mathrm{se}_{2n+1}(v)\,dk.
\end{aligned}
$$

where $u_0 = \operatorname{arctanh}(\ell/w)$, $a = \sqrt{d(w^2-\ell^2)}$ *and*

$$
C_n(k) = \frac{k^2 A_1^{(2n+1)} B_1^{(2n+1)}}{\mathrm{se}_{2n+1}^2(\pi/2)\,\dot{\mathrm{se}}_{2n+1}(0)\,\dot{\mathrm{ce}}_{2n+1}(\pi/2)}
$$

The components (B_u, B_v, B_z) of the magnetic flux leakage field $\mathbf{B}_\Sigma$ generated by this defect are obtained as usual via differentiation.

Theorem 5.6. *The magnetic flux leakage field* $\mathbf{B}_\Sigma$ *generated by an elliptic parabolic V of length* ℓ*, width* w*, and depth* d *is given by* $\mathbf{B}_\Sigma = (B_u, B_v, B_z)$ *where*

$$B_u(u,v,z) = -\frac{2\sqrt{2}\pi a^4 M_0}{3\sqrt{\cosh 2u - \cos 2v}} \sum_{n=0}^{\infty} \int_0^{\infty} C_n(k)\, [\mathcal{DS}_{2n+1}(u,z,k) + \mathcal{DG}_{2n+1}(u,z,k;u_0)]\, \mathrm{se}_{2n+1}(v,k)\, dk$$

with

$$\mathcal{DS}_{2n+1}(u,z,k) = \frac{\partial}{\partial u}\Big(\mathcal{S}_{2n+1}(u,z,k)\, \mathrm{gekh}_{2n+1}(u)\Big)$$

and

$$\mathcal{DG}_{2n+1}(u,z,k;u_0) = \frac{\partial}{\partial u}\Big(\mathcal{S}_{2n+1}(u,z,k)\, \mathrm{gekh}_{2n+1}(u)\Big)$$

$$B_v(u,v,z) = \frac{2\sqrt{2}\pi a^4 M_0}{3\sqrt{\cosh 2u - \cos 2v}} \sum_{n=0}^{\infty} \int_0^{\infty} C_n(k)\, [\mathcal{S}_{2n+1}(u,z,k)\, \mathrm{gekh}_{2n+1}(u,k) + \mathcal{G}_{2n+1}(u,z,k;u_0)\, \mathrm{seh}_{2n+1}(u,k)]\, \mathrm{dse}_{2n+1}(v,k) dk,$$

where

$$\mathrm{dse}_{2n+1}(v,k) = \frac{\partial \mathrm{se}_{2n+1}}{\partial v} = (-1)^n \sum_{\nu=0}^{\infty} (-1)^\nu (2\nu+1) A_{2\nu+1}^{(2n+1)} \cos(2\nu+1)u.$$

$$B_z(u,v,z) = -\frac{2\pi a^5 M_0}{3} \sum_{n=0}^{\infty} \int_0^{\infty} C_n(k)\, [\tilde{\mathcal{S}}_{2n+1}(u,z,k)\, \mathrm{gekh}_{2n+1}(u) + \tilde{\mathcal{G}}_{2n+1}(u,z,k;u_0)\, \mathrm{seh}_{2n+1}(u)] \mathrm{se}_{2n+1}(v) dk$$

with

$$\begin{aligned}
&\tilde{\mathcal{S}}_{2n+1}(u,z,k) \\
&= k\Bigg(\Sigma_n^{(1)}(k) \int_0^u \sin k\left[z - \frac{a^2}{2}\cosh 2u'\right] \sinh u'\, \mathrm{seh}_{2n+1}(u',k)\, du' \\
&+ \Sigma_n^{(2)}(k) \int_0^u \cos k\left[z - \frac{a^2}{2}\cosh 2u'\right] \sinh u'\, \mathrm{seh}_{2n+1}(u',k)\, du' \\
&+ \Sigma_n^{(3)}(k) \int_0^u \sin k\left[z - \frac{a^2}{2}\cosh 2u'\right] \sinh 3u'\, \mathrm{seh}_{2n+1}(u',k)\, du' \\
&+ \Sigma_n^{(4)}(k) \int_0^u \cos k\left[z - \frac{a^2}{2}\cosh 2u'\right] \sinh 3u'\, \mathrm{seh}_{2n+1}(u',k)\, du'\Bigg)
\end{aligned}$$

and

$$\begin{aligned}
&\tilde{\mathcal{G}}_{2n+1}(u,z,k;u_0)\\
&\quad = k\left(\Sigma_n^{(1)}(k)\int_u^{u_0}\sin k\left[z-\frac{a^2}{2}\cosh 2u'\right]\sinh u'\,\mathrm{gekh}_{2n+1}(u',k)\,du'\right.\\
&\qquad + \Sigma_n^{(2)}(k)\int_u^{u_0}\cos k\left[z-\frac{a^2}{2}\cosh 2u'\right]\sinh u'\,\mathrm{gekh}_{2n+1}(u',k)\,du'\\
&\qquad + \Sigma_n^{(3)}(k)\int_u^{u_0}\sin k\left[z-\frac{a^2}{2}\cosh 2u'\right]\sinh 3u'\,\mathrm{gekh}_{2n+1}(u',k)\,du'\\
&\qquad \left. +\Sigma_n^{(4)}(k)\int_u^{u_0}\cos k\left[z-\frac{a^2}{2}\cosh 2u'\right]\sinh 3u'\,\mathrm{gekh}_{2n+1}(u',k)\,du'\right)
\end{aligned}$$

The cartesian components (B_x, B_y, B_z) of $\mathbf{B}_\Sigma$ are obtained from the elliptic parabolic components (B_u, B_v, B_z) via the transformation matrix

$$T = \frac{1}{a(\cosh 2u - \cos 2v)}\begin{pmatrix} 2\sinh u\cos v & -2\cosh u\sin v & 0\\ 2\cosh u\sin v & 2\sinh u\cos v & 0\\ 0 & 0 & a(\cosh 2u-\cos 2v)\end{pmatrix}$$

as

$$\begin{pmatrix} B_x\\ B_y\\ B_z\end{pmatrix} = T\begin{pmatrix} B_u\\ B_v\\ B_z\end{pmatrix},$$

with the right-hand side completely expressed in cartesian coordinates.

References

1. M. Abramowitz and I. Stegun, *Handbook of Mathematical Functions*, Dover Publications Inc., (New York, 1972).
2. R. F. Arenstorf, *Lectures on Special Functions (Bessel Functions, Spherical Harmonics, Elliptic Functions)*, Lecture Notes, (Vanderbilt University, 1973).
3. G. Arfken, *Mathematical Methods for Physicists*, Academic Press, (New York, 1966).
4. R. Campbell, *Théorie Générale de L'Équation de Mathieu et de Quelques Autres Équations Différentielles de la Mécanique*, Masson et Cie Éditeurs, (Paris, 1955).
5. R. Courant and D. Hilbert, *Methoden der Mathematischen Physik I,II*, (Springer Verlag, Heidelberg, 1968).
6. E. W. Hobson, *The Theory of Spherical and Ellipsoidal Harmonics*, (Chelsea Publishing Company, New York, 1955).
7. G. Hämmerlin and K. H. Hoffmann, *Numerical Mathematics*, (Springer Verlag, New York, 1991).
8. J. D. Jackson, *Classical Electrodynamics*, (John Wiley & Sons, 2nd ed., New York, 1975).

9. N. N. Lebedev, *Special Functions and their Applications*, (Dover Publications Inc., New York, 1972).
10. N. W. McLachlan, *Theory and Applications of Mathieu Functions*, (Oxford Clarendon Press, London, 1947).
11. F. W. J. Olver, *Asymptotics and Special Functions*, (Academic Press, New York, 1974).
12. J. Stoer and R. Bulirsch, *Introduction to Numerical Analysis*, 2nd, ed., Texts in Applied Mathematics 12, (Springer Verlag, New York, 1993).
13. B. Schutz, *Geometrical Methods of Mathematical Physics*, (Cambridge University Press, UK, 1980).
14. A. Terras, *Harmonic Analysis on Symmetric Spaces and Applications I*, (Springer Verlag, New York 1985).
15. W. Walter, *Potentialtheorie*, BI Hochschulskripten, 765/765a, (Bibliographisches Institut, Mannheim, 1971).
16. G. N. Watson, *A Treatise on the Theory of Bessel Functions*, (Cambridge University Press, London, 1922).
17. K. Warnick, R. Selfridge and D. Arnold, *IEEE Trans. on Education*, **40**, 53 (1997).

Advances in Deterministic and Stochastic Analysis
Eds. N. M. Chuong *et al.* (pp. 177–196)

§10. ON THE CAUCHY PROBLEM FOR A QUASILINEAR WEAKLY HYPERBOLIC SYSTEM IN TWO VARIABLES AND APPLICATIONS TO THAT FOR WEAKLY HYPERBOLIC CLASSICAL MONGE-AMPÈRE EQUATIONS

HA TIEN NGOAN* and NGUYEN THI NGA*

Institute of Mathematics
18 Hoang Quoc Viet Road, 10307 Hanoi, Vietnam
E-mail: htngoan@math.ac.vn

The Cauchy problem for a normal quasilinear weakly hyperbolic system in two variables is considered. Sufficient conditions for its diagonalization are given. The local solvability of the noncharacteristic Cauchy problem for some weakly hyperbolic classical Monge-Ampère equations is proved.

1. Introduction

The classical hyperbolic Monge-Ampère equation with two variables is that of the form

$$F(x_1, x_2, z, p_1, p_2, r, s, t) = Ar + Bs + Ct + (rt - s^2) - E = 0, \quad (1)$$

where $z = z(x_1, x_2)$ is an unknown function defined for $(x_1, x_2) \in R^2, p_1 = z_{x_1}, p_2 = z_{x_2}, r = z_{x_1x_1}, s = z_{x_1x_2}$ and $t = z_{x_2x_2}$. The coefficients A, B, C and E are real C^2-functions of (x_1, x_2, z, p_1, p_2) and satisfy the condition of hyperbolicity:

$$\Delta := B^2 - 4(AC + E) > 0.$$

In this case the characteristic equation of the Eq. (1)

$$\lambda^2 + B\lambda + (AC + E) = 0 \quad (2)$$

has two different real roots

$$\lambda_1 = \lambda_1(x_1, x_2, z, p_1, p_2), \ \lambda_2 = \lambda_2(x_1, x_2, z, p_1, p_2).$$

*Supported in part by the National Basic Research Program in Natural Science, Vietnam

In this paper we consider the case $\Delta \geq 0$, i.e. the characteristic roots λ_1, λ_2 may coincide at some poits. In this case the Eq. (1) is said to be *weakly hyperbolic* and it can be written in the following equivalent form

$$\begin{vmatrix} z_{x_1x_1} + C & z_{x_1x_2} + \lambda_1 \\ z_{x_2x_1} + \lambda_2 & z_{x_2x_2} + A \end{vmatrix} = 0. \tag{3}$$

We are interested in looking for local classical solution to the Eq. (1). So we suppose that we are given C^2-functions $X_1^0(\alpha_1)$, $X_2^0(\alpha_1)$, $Z^0(\alpha_1)$, $P_1^0(\alpha_1)$, $P_2^0(\alpha_1)$, $\alpha_1 \in I$, where I is an interval $(0, \delta), \delta$ is some sufficiently small positive number.

The Cauchy problem for the Eq. (1) consists in looking for $z(x) \in C^2$ which is a solution of (1) such that

$$\begin{cases} z(x)\big|_{x=X^0(\alpha_1)} & = Z^0(\alpha_1), \\ z_{x_j}(x)\big|_{x=X^0(\alpha_1)} & = P_j^0(\alpha_1), j = 1, 2, \end{cases} \tag{4}$$

where $X^0(\alpha_1) \equiv (X_1^0(\alpha_1), X_2^0(\alpha_1))$.

From (4) we have the following necessary condition for the initial Cauchy data

$$\frac{\partial Z^0(\alpha_1)}{\partial \alpha_1} = P_1^0(\alpha_1)\frac{\partial X_1^0(\alpha_1)}{\partial \alpha_1} + P_2^0(\alpha_1)\frac{\partial X_2^0(\alpha_1)}{\partial \alpha_1}, \tag{5}$$

which is assumed to be fulfilled for all $\alpha_1 \in I$.

We assume that the Cauchy problem (1)-(4) is not charateristic (see Refs. 6,8), i.e.

$$\begin{aligned} &C\big(X_1^{\prime 0}(\alpha_1)\big)^2 + A\big(X_2^{0'}(\alpha_1)\big)^2 - BX_1^{0'}(\alpha_1)X_2^{0'}(\alpha_1) + \\ &\big(X_1^{0'}(\alpha_1)P_1^{0'}(\alpha_1) + X_2^{0'}(\alpha_1)P_2^0(\alpha_1)\big) \neq 0, \forall \alpha_1 \in I \end{aligned} \tag{6}$$

where the coefficients A, B, C are calculated at $(X_1^0(\alpha_1)$, $X_2^0(\alpha_1)$, $Z^0(\alpha_1)$, $P_1^0(\alpha_1)$, $P_2^0(\alpha_1))$.

Equation (1) was investigated in Refs. 1,2 by G. Darboux and E. Goursat under the assumptions that $\Delta > 0$ and there exist two independent first integrals for the Eq. (1). We recall that a function $\varphi(x_1, x_2, z, p_1, p_2)$ is said to be a *first integral* for the Eq. (1) if it satisfies the following system of equations

$$\begin{cases} \dfrac{\partial \varphi}{\partial x_1} + p_1\dfrac{\partial \varphi}{\partial z} - C\dfrac{\partial \varphi}{\partial p_1} - \lambda_1\dfrac{\partial \varphi}{\partial p_2} & = 0, \\ \dfrac{\partial \varphi}{\partial x_2} + p_2\dfrac{\partial \varphi}{\partial z} - \lambda_2\dfrac{\partial \varphi}{\partial p_1} - A\dfrac{\partial \varphi}{\partial p_2} & = 0. \end{cases}$$

However, there exist Monge-Ampère Eq. (1) which do not possess two independent first integrals. For example, as it has been proved in Ref. 7 that the following equation

$$rt - s^2 + c^2(x_1, x_2, z, p_1, p_2) = 0$$

has two independent first integrals if and only if

$$c(x_1, x_2, z, p_1, p_2) \equiv \text{constant} \neq 0.$$

In the case $\Delta > 0$, but without the assumption on existence of two independent first integrals, the Eq. (1) had been also considered in Refs. 3, 5–9 by reducing it to a hyperbolic quasilinear system of first-order partial differential equations with two variables.

In this paper when considering the case $\Delta \geq 0$ we do not assume existence of two independent first integrals for the Eq. (1). To avoid this difficulty in Ref. 8 we have proposed another solving method for the Cauchy problem (3)-(4) which reduces it to that for the following normal quasilinear first-order system of equations in two variables

$$\begin{cases} \dfrac{\partial X_1}{\partial \alpha_2} &= (\lambda_1 - 1)\dfrac{\partial X_1}{\partial \alpha_1} + A\dfrac{\partial X_2}{\partial \alpha_1} + \dfrac{\partial P_2}{\partial \alpha_1} \\ \dfrac{\partial X_2}{\partial \alpha_2} &= -C\dfrac{\partial X_1}{\partial \alpha_1} - (\lambda_2 + 1)\dfrac{\partial X_2}{\partial \alpha_1} - \dfrac{\partial P_1}{\partial \alpha_1} \\ \dfrac{\partial Z}{\partial \alpha_2} &= (\lambda_1 P_1 - CP_2)\dfrac{\partial X_1}{\partial \alpha_1} + (AP_1 - \lambda_2 P_2)\dfrac{\partial X_2}{\partial \alpha_1} - \dfrac{\partial Z}{\partial \alpha_1} \\ & \quad -P_2\dfrac{\partial P_1}{\partial \alpha_1} + P_1\dfrac{\partial P_2}{\partial \alpha_1} \\ \dfrac{\partial P_1}{\partial \alpha_2} &= (-CA + \lambda_1\lambda_2)\dfrac{\partial X_2}{\partial \alpha_1} + (\lambda_1 - 1)\dfrac{\partial P_1}{\partial \alpha_1} - C\dfrac{\partial P_2}{\partial \alpha_1} \\ \dfrac{\partial P_2}{\partial \alpha_2} &= (CA - \lambda_1\lambda_2)\dfrac{\partial X_1}{\partial \alpha_1} + A\dfrac{\partial P_1}{\partial \alpha_1} - (\lambda_2 + 1)\dfrac{\partial P_2}{\partial \alpha_1} \end{cases} \tag{7}$$

with the following initial conditions

$$\begin{cases} X(\alpha_1, \alpha_2)\big|_{\alpha_2 = 0} &= X^0(\alpha_1), \\ Z(\alpha_1, \alpha_2)\big|_{\alpha_2 = 0} &= Z^0(\alpha_1), \\ P(\alpha_1, \alpha_2)\big|_{\alpha_2 = 0} &= P^0(\alpha_1). \end{cases} \tag{8}$$

Here (X_1, X_2, Z, P_1, P_2) are unknown functions of two variables α_1, α_2 and the functions $X_1^0(\alpha_1)$, $X_2^0(\alpha_1)$, $Z^0(\alpha_1)$, $P_1^0(\alpha_1)$, $P_2^0(\alpha_1)$ are given as the same as above, which satisfy the conditions (5), (6).

We assume that (X_1, X_2, Z, P_1, P_2) is a C^2-solution to the Cauchy problem (7)-(8) satisfying the following condition

$$\det \frac{D(X_1, X_2)}{D(\alpha_1, \alpha_2)}\Big|_{\alpha_2=0} = \begin{vmatrix} \frac{\partial X_1}{\partial \alpha_1} & \frac{\partial X_1}{\partial \alpha_2} \\ \frac{\partial X_2}{\partial \alpha_1} & \frac{\partial X_2}{\partial \alpha_2} \end{vmatrix}_{\alpha_2=0} \neq 0. \tag{9}$$

From the implicit function theorem we can locally solve the following system of equations

$$\begin{cases} X_1(\alpha_1, \alpha_2) & = x_1 \\ X_2(\alpha_1, \alpha_2) & = x_2 \end{cases} \tag{10}$$

to obtain $\alpha_1 = \psi_1(x_1, x_2), \alpha_2 = \psi_2(x_1, x_2)$.

In Ref. 10 we have proved the following theorem.

Theorem 1.1. *Suppose that the Cauchy problem* (7)-(8), *satisfying the conditions* (5), (6), *possesses a* C^2*-solution* (X_1, X_2, Z, P_1, P_2). *Then the condition* (9) *is satisfied and a local* C^2*-solution* $z(x)$ *to the Cauchy problem* (1)-(4) *is given by the following formular*

$$z(x_1, x_2) = Z(\psi_1(x_1, x_2), \psi_2(x_1, x_2)).$$

Moreever,

$$z_{x_1}(x_1, x_2) = P_1(\psi_1(x_1, x_2), \psi_2(x_1, x_2)),$$
$$z_{x_2}(x_1, x_2) = P_2(\psi_1(x_1, x_2), \psi_2(x_1, x_2)).$$

So, to study the Cauchy problem (1)-(4) we first concentrate to investigate the solvability of the Cauchy problem (7)-(8). To do this in Sec. 2 we recall definition on hyperbolicity for systems (7) and some results on solvability of Cauchy problem for them. In Sec. 3 we first reduce the system (7) to an intermediate one (25). Provided some restrictions on coefficients a_{ij} and initial Cauchy date, the last system will be in Sec. 4 reduced in Theorem 4.1 to a diagonal one (42), the Cauchy prblem for which is well solved locally according to the Theorem 2.3. In Sec. 5 we describe in Theorem 5.1 a class of weakly hyperbolic classical Monge-Ampère equations, Cauchy problem for which is well solved locally. In Sec. 6 we give some examples of weakly hyperbolic classical Monge-Ampère equations that satisfy conditions of Theorem 5.1.

2. Hyperbolicity

From (3) we denote C by a_{11}, λ_1 by a_{12}, λ_2 by a_{21} and A by a_{22}. From now we consider instead of (7) the following quasilinear system in two variables

$$\begin{cases} \dfrac{\partial X_1}{\partial \alpha_2} & = (a_{12}-1)\dfrac{\partial X_1}{\partial \alpha_1} + a_{22}\dfrac{\partial X_2}{\partial \alpha_1} + \dfrac{\partial P_2}{\partial \alpha_1} \\ \dfrac{\partial X_2}{\partial \alpha_2} & = -a_{11}\dfrac{\partial X_1}{\partial \alpha_1} - (a_{21}+1)\dfrac{\partial X_2}{\partial \alpha_1} - \dfrac{\partial P_1}{\partial \alpha_1} \\ \dfrac{\partial Z}{\partial \alpha_2} & = (a_{12}P_1 - a_{11}P_2)\dfrac{\partial X_1}{\partial \alpha_1} \\ & \quad +(a_{22}P_1 - a_{21}P_2)\dfrac{\partial X_2}{\partial \alpha_1} - \dfrac{\partial Z}{\partial \alpha_1} - P_2\dfrac{\partial P_1}{\partial \alpha_1} + P_1\dfrac{\partial P_2}{\partial \alpha_1} \\ \dfrac{\partial P_1}{\partial \alpha_2} & = (-a_{11}a_{22} + a_{12}a_{21})\dfrac{\partial X_2}{\partial \alpha_1} + (a_{12}-1)\dfrac{\partial P_1}{\partial \alpha_1} - a_{11}\dfrac{\partial P_2}{\partial \alpha_1} \\ \dfrac{\partial P_2}{\partial \alpha_2} & = (a_{11}a_{22} - a_{12}a_{21})\dfrac{\partial X_1}{\partial \alpha_1} + a_{22}\dfrac{\partial P_1}{\partial \alpha_1} - (a_{21}+1)\dfrac{\partial P_2}{\partial \alpha_1} \end{cases} \tag{11}$$

Here (X_1, X_2, Z, P_1, P_2) are unknown functions of the variables α_1 and α_2; a_{ij} are C^2-functions of (X_1, X_2, Z, P_1, P_2).

We set

$$U = (X_1, X_2, Z, P_1, P_2)^T,$$

$$\mathcal{A}(U) = \begin{bmatrix} a_{12}-1 & a_{22} & 0 & 0 & 1 \\ -a_{11} & -a_{21}-1 & 0 & -1 & 0 \\ a_{12}P_1 - a_{11}P_2 & a_{22}P_1 - a_{21}P_2 & -1 & -P_2 & P_1 \\ 0 & -a_{11}a_{22} + a_{12}a_{21} & 0 & a_{12}-1 & -a_{11} \\ a_{11}a_{22} - a_{12}a_{21} & 0 & 0 & a_{22} & -a_{21}-1 \end{bmatrix}. \tag{12}$$

We write the system (11) in the matrix form

$$\frac{\partial U}{\partial \alpha_2} = \mathcal{A}(U)\frac{\partial U}{\partial \alpha_1}. \tag{13}$$

Now we recall some definitions and results on quasilinear hyperbolic systems. To do this we will consider the following more general normal system in two variables

$$\frac{\partial U}{\partial \alpha_2} = H(U)\frac{\partial U}{\partial \alpha_1} + G(U), \tag{14}$$

where U and $G(U)$ are m-dimensional columm-vectors, $H(U)$ is a matrix of size $m \times m$. If the matrix $H(U)$ is diagonal, the system (14) is said to be a diagonal one.

The Cauchy problem for system (14) consists in looking for $U(\alpha_1, \alpha_2) \in C^1$ such that

$$U(\alpha_1, \alpha_2)\big|_{\alpha_2=0} = U^0(\alpha_1), \tag{15}$$

where $U^0(\alpha_1)$ is a given m-dimensional vector function.

Definition 2.1. (Ref. 4) The system (14) is said to be hyperbolic if for any $U \in R^m$ the following conditions hold:

1) All the eigenvalues of the matrix $H(U)$ are real;

2) There exists a basis in R^m consisting of its corresponding smooth with respect to U left eigenvectors.

Theorem 2.1. *(Ref. 11) If $a_{12} \neq a_{21}$ for any $(X_1, X_2, Z, P_1, P_2) \in R^5$, then the system* (14) *is hyperbolic.*

Theorem 2.2. *(Ref. 4) Suppose that the system* (14) *is hyperbolic. Then it can be reduced to a diagonal system of* $2m$ *quasilinear equations with respect to* $2m$ *unknowns.*

Theorem 2.3. *(Ref. 4) Suppose the system* (14) *is diagonal. Then there exists locally a unique smooth solution to the Cauchy problem* (14)-(15).

For the system (11) we do not assume that $a_{12} \neq a_{21}$. This means that these functions may coincide at some points (X_1, X_2, Z, P_1, P_2). In this case only condition 1) in Definition 2.1 is valid and the system (11) is said to be *weakly hyperbolic.* We show below in Theorem 5.1 that under some restrictions on coefficients $a_{ij}(X, Z, P)$, the weakly hyperbolic system (11) can be still reduced to a diagonal one of 7 quasilinear equations with respect to 7 unknowns. From the Theorem 2.3 it follows that there exists locally a unique smooth solution for the Cauchy problem (7)-(8).

3. Reduced System

Set

$$C(X_1, X_2, Z, P_1, P_2) = \begin{bmatrix} 1 & 0 & 0\,0\,0 \\ 0 & 1 & 0\,0\,0 \\ 0 & 0 & 1\,0\,0 \\ -a_{11} & -a_{21} & 0\,1\,0 \\ -a_{12} & -a_{22} & 0\,0\,1 \end{bmatrix}, \tag{16}$$

where a_{ij} are the same functions of X_1, X_2, Z, P_1, P_2 as in (11). Then it is easy to verify that

$$C^{-1}(X_1, X_2, Z, P_1, P_2) = \begin{bmatrix} 1 & 0 & 0 & 0 & 0 \\ 0 & 1 & 0 & 0 & 0 \\ 0 & 0 & 1 & 0 & 0 \\ a_{11} & a_{21} & 0 & 1 & 0 \\ a_{12} & a_{22} & 0 & 0 & 1 \end{bmatrix}. \tag{17}$$

Proposition 3.1. *Suppose that matrix $\mathcal{A}$ is defined by* (12) *and*

$$\tilde{\mathcal{A}} = \begin{bmatrix} -1 & 0 & 0 & 0 & 1 \\ 0 & -1 & 0 & -1 & 0 \\ 0 & 0 & -1 & -P_2 & P_1 \\ 0 & 0 & 0 & a_{12} - a_{21} - 1 & 0 \\ 0 & 0 & 0 & 0 & a_{12} - a_{21} - 1 \end{bmatrix}. \tag{18}$$

Then

$$C^{-1}\mathcal{A}C = \tilde{\mathcal{A}}. \tag{19}$$

Proof. The proof is obvious by matrix multiplication. □

Set

$$\tilde{U} \equiv (\tilde{X}_1, \tilde{X}_2, \tilde{Z}, \tilde{P}_1, \tilde{P}_2)^T = C^{-1}U.$$

That means

$$\begin{cases} \tilde{X}_1 &= X_1, \\ \tilde{X}_2 &= X_2, \\ \tilde{Z} &= Z, \\ \tilde{P}_1 &= P_1 + a_{11}(X, Z, P)X_1 + a_{21}(X, Z, P)X_2, \\ \tilde{P}_2 &= P_2 + a_{12}(X, Z, P)X_1 + a_{22}(X, Z, P)X_2. \end{cases} \tag{20}$$

We introduce now the following condition for the Cauchy problem (11)-(8).

(C_1): *For the coefficients $a_{ij}(X, Z, P)$ and the initial functions $X_1^0(\alpha_1), X_2^0(\alpha_1), Z^0(\alpha_1), P_1^0(\alpha_1)$, $P_2^0(\alpha_1)$ the following inequality holds for* *all* $\alpha_1 \in I$:

$$\det \frac{D(a_{11}, a_{12})}{D(P_1, P_2)}(X_1^0(\alpha_1))^2 + \det \frac{D(a_{21}, a_{22})}{D(P_1, P_2)}(X_2^0(\alpha_1))^2 +$$
$$\Big[\det \frac{D(a_{11}, a_{22})}{D(P_1, P_2)} + \det \frac{D(a_{21}, a_{12})}{D(P_1, P_2)}\Big] X_1^0(\alpha_1) X_2^0(\alpha_1) + \tag{21}$$
$$+ \Big(\frac{\partial a_{11}}{\partial P_1} + \frac{\partial a_{12}}{\partial P_2}\Big) X_1^0(\alpha_1) + \Big(\frac{\partial a_{21}}{\partial P_1} + \frac{\partial a_{22}}{\partial P_2}\Big) X_2^0(\alpha_1) + 1 \neq 0,$$

where the derivatives of functions a_{ij} *are calculated at* $(X_1^0(\alpha_1), X_2^0(\alpha_1), Z^0(\alpha_1), P_1^0(\alpha_1), P_2^0(\alpha_1))$.

Proposition 3.2. *Suppose that the condition* (C_1) *holds. Then for any* $(X, Z, \tilde{P}) \in R^5$, *which are sufficiently closed to* $(X^0(\alpha_1), Z^0(\alpha_1), \tilde{P}^0(\alpha_1))$, *the following system of equations with respect to* P_1, P_2

$$\begin{cases} P_1 + a_{11}(X, Z, P)X_1 + a_{21}(X, Z, P)X_2 &= \tilde{P}_1, \\ \\ P_2 + a_{12}(X, Z, P)X_1 + a_{22}(X, Z, P)X_2 &= \tilde{P}_2 \end{cases} \tag{22}$$

has a unique smooth solution

$$\begin{cases} P_1 &= f(X_1, X_2, Z, \tilde{P}_1, \tilde{P}_2), \\ \\ P_2 &= g(X_1, X_2, Z, \tilde{P}_1, \tilde{P}_2), \end{cases} \tag{23}$$

where

$$\tilde{P}^0(\alpha_1) = P^0(\alpha_1) + X^0(\alpha_1) A(X^0(\alpha_1), Z^0(\alpha_1), P^0(\alpha_1))$$

and

$$A(X, Z, P) = \begin{bmatrix} a_{11}(X, Z, P) & a_{12}(X, Z, P) \\ \\ a_{21}(X, Z, P) & a_{22}(X, Z, P) \end{bmatrix}.$$

Proof. Set

$$\begin{cases} f_1 \equiv P_1 + a_{11}(X, Z, P)X_1 + a_{21}(X, Z, P)X_2 - \tilde{P}_1, \\ \\ f_2 \equiv P_2 + a_{12}(X, Z, P)X_1 + a_{22}(X, Z, P)X_2 - \tilde{P}_2. \end{cases}$$

We have

$$
\begin{aligned}
\det \frac{Df}{DP} &= \begin{vmatrix} \frac{\partial f_1}{\partial P_1} & \frac{\partial f_1}{\partial P_2} \\ \frac{\partial f_2}{\partial P_1} & \frac{\partial f_2}{\partial P_2} \end{vmatrix} \\
&= \begin{vmatrix} 1 + \frac{\partial a_{11}}{\partial P_1} X_1 + \frac{\partial a_{21}}{\partial P_1} X_2 & \frac{\partial a_{11}}{\partial P_2} X_1 + \frac{\partial a_{21}}{\partial P_2} X_2 \\ \frac{\partial a_{12}}{\partial P_1} X_1 + \frac{\partial a_{22}}{\partial P_1} X_2 & 1 + \frac{\partial a_{12}}{\partial P_2} X_1 + \frac{\partial a_{22}}{\partial P_2} X_2 \end{vmatrix} \\
&= \Big(1 + \frac{\partial a_{11}}{\partial P_1} X_1 + \frac{\partial a_{21}}{\partial P_1} X_2\Big)\Big(1 + \frac{\partial a_{12}}{\partial P_2} X_1 + \frac{\partial a_{22}}{\partial P_2} X_2\Big) \\
&\quad - \Big(\frac{\partial a_{11}}{\partial P_2} X_1 + \frac{\partial a_{21}}{\partial P_2} X_2\Big)\Big(\frac{\partial a_{12}}{\partial P_1} X_1 + \frac{\partial a_{22}}{\partial P_1} X_2\Big) \\
&= 1 + \frac{\partial a_{12}}{\partial P_2} X_1 + \frac{\partial a_{22}}{\partial P_2} X_2 + \frac{\partial a_{11}}{\partial P_1} X_1 + \frac{\partial a_{11}}{\partial P_1}\frac{\partial a_{12}}{\partial P_2} X_1^2 + \frac{\partial a_{11}}{\partial P_1}\frac{\partial a_{22}}{\partial P_2} X_1 X_2 \\
&\quad + \frac{\partial a_{21}}{\partial P_1} X_2 + \frac{\partial a_{12}}{\partial P_2}\frac{\partial a_{21}}{\partial P_1} X_1 X_2 + \frac{\partial a_{21}}{\partial P_1}\frac{\partial a_{22}}{\partial P_2} X_2^2 - \frac{\partial a_{12}}{\partial P_1}\frac{\partial a_{11}}{\partial P_2} X_1^2 \\
&\quad - \frac{\partial a_{12}}{\partial P_1}\frac{\partial a_{21}}{\partial P_2} X_1 X_2 - \frac{\partial a_{22}}{\partial P_1}\frac{\partial a_{11}}{\partial P_2} X_1 X_2 - \frac{\partial a_{22}}{\partial P_1}\frac{\partial a_{21}}{\partial P_2} X_2^2 \\
&= \det \frac{D(a_{11}, a_{12})}{D(P_1, P_2)} X_1^2 + \det \frac{D(a_{21}, a_{22})}{D(P_1, P_2)} X_2^2 \\
&\quad + \Big[\det \frac{D(a_{11}, a_{22})}{D(P_1, P_2)} + \det \frac{D(a_{21}, a_{12})}{D(P_1, P_2)}\Big] X_1 X_2 \\
&\quad + \Big(\frac{\partial a_{12}}{\partial P_2} + \frac{\partial a_{11}}{\partial P_1}\Big) X_1 + \Big(\frac{\partial a_{22}}{\partial P_2} + \frac{\partial a_{21}}{\partial P_1}\Big) X_2 + 1.
\end{aligned} \tag{24}
$$

Since $a_{ij}(X, Z, P)$ and X, Z, P are C^2-functions, from (21), (24) and from the implicit function theorem the assertion of the proposition follows. □

Theorem 3.1. *The vector function $U(\alpha_1, \alpha_2)$ satisfies the system* (13) *if and only if the vector function $\tilde{U}(\alpha_1, \alpha_2)$ satisfies the following system*

$$
\frac{\partial \tilde{U}}{\partial \alpha_2} = \tilde{\mathcal{A}}(\tilde{U}) \frac{\partial \tilde{U}}{\partial \alpha_1} + \mathcal{B}(\tilde{U})\tilde{U}. \tag{25}
$$

Here $\tilde{\mathcal{A}}$ is defined by (18) *and*

$$
\mathcal{B} = C^{-1}\Big(\mathcal{A}\frac{\partial C}{\partial \alpha_1} - \frac{\partial C}{\partial \alpha_2}\Big)
$$

$$= \begin{bmatrix} -\dfrac{\partial a_{12}}{\partial \alpha_1} & -\dfrac{\partial a_{22}}{\partial \alpha_1} & 0\,0\,0 \\ \dfrac{\partial a_{11}}{\partial \alpha_1} & \dfrac{\partial a_{21}}{\partial \alpha_1} & 0\,0\,0 \\ \dfrac{\partial a_{11}}{\partial \alpha_1}P_2 - \dfrac{\partial a_{12}}{\partial \alpha_1}P_1 & \dfrac{\partial a_{21}}{\partial \alpha_1}P_2 - \dfrac{\partial a_{22}}{\partial \alpha_1}P_1 & 0\,0\,0 \\ -(a_{12}-a_{21}-1)\frac{\partial a_{11}}{\partial \alpha_1} + \dfrac{\partial a_{11}}{\partial \alpha_2} & -(a_{12}-a_{21}-1)\dfrac{\partial a_{21}}{\partial \alpha_1} + \dfrac{\partial a_{21}}{\partial \alpha_2} & 0\,0\,0 \\ -(a_{12}-a_{21}-1)\dfrac{\partial a_{12}}{\partial \alpha_1} + \dfrac{\partial a_{12}}{\partial \alpha_2} & -(a_{12}-a_{21}-1)\frac{\partial a_{22}}{\partial \alpha_1} + \dfrac{\partial a_{22}}{\partial \alpha_2} & 0\,0\,0 \end{bmatrix}, \tag{26}$$

where in (26) *the variables* P_1, P_2 *are replaced respectively by the functions* $f(X_1, X_2, Z, \tilde{P}_1, \tilde{P}_2)$, $g(X_1, X_2, Z, \tilde{P}_1, \tilde{P}_2)$, *which are defined in* (23).

Proof.

1) Suppose U is a solution of (13). Since $\tilde{U} = C^{-1}U$ we have

$$U = C\tilde{U}.$$

Therefore

$$\frac{\partial U}{\partial \alpha_2} = C\frac{\partial \tilde{U}}{\partial \alpha_2} + \frac{\partial C}{\partial \alpha_2}\tilde{U}, \tag{27}$$

$$\frac{\partial U}{\partial \alpha_1} = C\frac{\partial \tilde{U}}{\partial \alpha_1} + \frac{\partial C}{\partial \alpha_1}\tilde{U}. \tag{28}$$

By (13), (27) and (28)

$$C\frac{\partial \tilde{U}}{\partial \alpha_2} + \frac{\partial C}{\partial \alpha_2}\tilde{U} = \mathcal{A}(C\frac{\partial \tilde{U}}{\partial \alpha_1} + \frac{\partial C}{\partial \alpha_1}\tilde{U}).$$

Hence,

$$\frac{\partial \tilde{U}}{\partial \alpha_2} = C^{-1}\mathcal{A}C\frac{\partial \tilde{U}}{\partial \alpha_1} + C^{-1}(\mathcal{A}\frac{\partial C}{\partial \alpha_1} - \frac{\partial C}{\partial \alpha_2})\tilde{U}.$$

So we have

$$C^{-1}\mathcal{A}C = \tilde{\mathcal{A}} = \begin{bmatrix} -1 & 0 & 0 & 0 & 1 \\ 0 & -1 & 0 & -1 & 0 \\ 0 & 0 & -1 & -P_2 & P_1 \\ 0 & 0 & 0 & a_{12}-a_{21}-1 & 0 \\ 0 & 0 & 0 & 0 & a_{12}-a_{21}-1 \end{bmatrix}$$

and

$$\mathcal{B} = C^{-1}\Big(\mathcal{A}(C\tilde{U})\frac{\partial C}{\partial \alpha_1} - \frac{\partial C}{\partial \alpha_2}\Big)$$

$$= \begin{bmatrix} -\frac{\partial a_{12}}{\partial \alpha_1} & -\frac{\partial a_{22}}{\partial \alpha_1} & 0\,0\,0 \\ \frac{\partial a_{11}}{\partial \alpha_1} & \frac{\partial a_{21}}{\partial \alpha_1} & 0\,0\,0 \\ \frac{\partial a_{11}}{\partial \alpha_1}P_2 - \frac{\partial a_{12}}{\partial \alpha_1}P_1 & \frac{\partial a_{21}}{\partial \alpha_1}P_2 - \frac{\partial a_{22}}{\partial \alpha_1}P_1 & 0\,0\,0 \\ -(a_{12}-a_{21}-1)\frac{\partial a_{11}}{\partial \alpha_1} + \frac{\partial a_{11}}{\partial \alpha_2} & -(a_{12}-a_{21}-1)\frac{\partial a_{21}}{\partial \alpha_1} + \frac{\partial a_{21}}{\partial \alpha_2} & 0\,0\,0 \\ -(a_{12}-a_{21}-1)\frac{\partial a_{12}}{\partial \alpha_1} + \frac{\partial a_{12}}{\partial \alpha_2} & -(a_{12}-a_{21}-1)\frac{\partial a_{22}}{\partial \alpha_1} + \frac{\partial a_{22}}{\partial \alpha_2} & 0\,0\,0 \end{bmatrix}.$$

2) The proof of the reverse part is obvious. The theorem is proved. □

4. Diagonalization

It is clear from (18) that the system (25) is still not diagonal. We give now some sufficient conditions under which the system (25) can be reduced to a diagonal quasilinear one.

We introduce now another condition for the system (11).

(C_2) : *Each function $a_{ij}(X, Z, P)$ satisfies the conditions*

$$\begin{cases} \frac{\partial a_{ij}}{\partial Z} & = 0, \\ \frac{\partial a_{ij}}{\partial X_1} - a_{11}\frac{\partial a_{ij}}{\partial P_1} - a_{12}\frac{\partial a_{ij}}{\partial P_2} & = 0, \\ \frac{\partial a_{ij}}{\partial X_2} - a_{21}\frac{\partial a_{ij}}{\partial P_1} - a_{22}\frac{\partial a_{ij}}{\partial P_2} & = 0. \end{cases} \tag{29}$$

We set

$$\overset{\approx}{P}_1 = \frac{\partial \tilde{P}_1}{\partial \alpha_1}, \overset{\approx}{P}_2 = \frac{\partial \tilde{P}_2}{\partial \alpha_1}. \tag{30}$$

Proposition 4.1.

Suppose (X_1, X_2, Z, P_1, P_2) is a solution of (13). *Under conditions (C_1) and (C_2) we have*

1)

$$\big(a_{12}(X,Z,P) - a_{21}(X,Z,P) - 1\big)\frac{\partial a_{ij}(X,Z,P)}{\partial \alpha_1} - \frac{\partial a_{ij}(X,Z,P)}{\partial \alpha_2} = 0. \tag{31}$$

2) Suppose $(X, \tilde{P})$ are sufficiently closed to $(X^0, \tilde{P}^0)$. Then there exist smooth functions $b_{ij}(X, \tilde{P})$ and $c_{ij}(X, \tilde{P})$ such that

$$\frac{\partial a_{ij}(X,Z,P)}{\partial \alpha_1} = b_{ij}(X,\tilde{P})\overset{\approx}{P}_1 + c_{ij}(X,\tilde{P})\overset{\approx}{P}_2, \forall i,j = 1,2. \tag{32}$$

Proof. 1) Since

$$\frac{\partial a_{ij}}{\partial \alpha_1} = \frac{\partial a_{ij}}{\partial X_1}\frac{\partial X_1}{\partial \alpha_1} + \frac{\partial a_{ij}}{\partial X_2}\frac{\partial X_2}{\partial \alpha_1} + \frac{\partial a_{ij}}{\partial Z}\frac{\partial Z}{\partial \alpha_1} + \frac{\partial a_{ij}}{\partial P_1}\frac{\partial P_1}{\partial \alpha_1} + \frac{\partial a_{ij}}{\partial P_2}\frac{\partial P_2}{\partial \alpha_1}, \tag{33}$$

$$\frac{\partial a_{ij}}{\partial \alpha_2} = \frac{\partial a_{ij}}{\partial X_1}\frac{\partial X_1}{\partial \alpha_2} + \frac{\partial a_{ij}}{\partial X_2}\frac{\partial X_2}{\partial \alpha_2} + \frac{\partial a_{ij}}{\partial Z}\frac{\partial Z}{\partial \alpha_2} + \frac{\partial a_{ij}}{\partial P_1}\frac{\partial P_1}{\partial \alpha_2} + \frac{\partial a_{ij}}{\partial P_2}\frac{\partial P_2}{\partial \alpha_2} \tag{34}$$

from (33), (34) and (11), we have

$$\begin{aligned}
&(a_{12}-a_{21}-1)\frac{\partial a_{ij}}{\partial \alpha_1}-\frac{\partial a_{ij}}{\partial \alpha_2}\\
&=(a_{12}-a_{21}-1)(\frac{\partial a_{ij}}{\partial X_1}\frac{\partial X_1}{\partial \alpha_1}+\frac{\partial a_{ij}}{\partial X_2}\frac{\partial X_2}{\partial \alpha_1}+\frac{\partial a_{ij}}{\partial P_1}\frac{\partial P_1}{\partial \alpha_1}+\frac{\partial a_{ij}}{\partial P_2}\frac{\partial P_2}{\partial \alpha_1})\\
&-\frac{\partial a_{ij}}{\partial X_1}[(a_{12}-1)\frac{\partial X_1}{\partial \alpha_1}+a_{22}\frac{\partial X_2}{\partial \alpha_1}+\frac{\partial P_2}{\partial \alpha_1}]\\
&-\frac{\partial a_{ij}}{\partial X_2}[-a_{11}\frac{\partial X_1}{\partial \alpha_1}-(a_{21}+1)\frac{\partial X_2}{\partial \alpha_1}-\frac{\partial P_1}{\partial \alpha_1}]\\
&-\frac{\partial a_{ij}}{\partial P_1}[(-a_{11}a_{22}+a_{12}a_{21})\frac{\partial X_2}{\partial \alpha_1}+(a_{12}-1)\frac{\partial P_1}{\partial \alpha_1}-a_{11}\frac{\partial P_2}{\partial \alpha_1}]\\
&-\frac{\partial a_{ij}}{\partial P_2}[(a_{11}a_{22}-a_{12}a_{21})\frac{\partial X_1}{\partial \alpha_1}+a_{22}\frac{\partial P_1}{\partial \alpha_1}-(a_{21}+1)\frac{\partial P_2}{\partial \alpha_1}]\\
&=[-a_{21}\frac{\partial a_{ij}}{\partial X_1}+a_{11}\frac{\partial a_{ij}}{\partial X_2}-(a_{11}a_{22}-a_{12}a_{21})\frac{\partial a_{ij}}{\partial P_2}]\frac{\partial X_1}{\partial \alpha_1}\\
&+[a_{12}\frac{\partial a_{ij}}{\partial X_2}-a_{22}\frac{\partial a_{ij}}{\partial X_1}+(a_{11}a_{22}-a_{12}a_{21})\frac{\partial a_{ij}}{\partial P_1}]\frac{\partial X_2}{\partial \alpha_1}\\
&+[-a_{21}\frac{\partial a_{ij}}{\partial P_1}+\frac{\partial a_{ij}}{\partial X_2}-a_{22}\frac{\partial a_{ij}}{\partial P_2}]\frac{\partial P_1}{\partial \alpha_1}\\
&+[a_{12}\frac{\partial a_{ij}}{\partial P_2}-\frac{\partial a_{ij}}{\partial X_1}+a_{11}\frac{\partial a_{ij}}{\partial P_1}]\frac{\partial P_2}{\partial \alpha_1}.
\end{aligned} \tag{35}$$

It follows from (29) and (35) that

$$(a_{12}-a_{21}-1)\frac{\partial a_{ij}}{\partial \alpha_1}-\frac{\partial a_{ij}}{\partial \alpha_2}=0.\frac{\partial X_1}{\partial \alpha_1}+0.\frac{\partial X_2}{\partial \alpha_1}+0.\frac{\partial P_1}{\partial \alpha_1}+0.\frac{\partial P_2}{\partial \alpha_1}=0.$$

2)
From (22) we have

$$\frac{\partial P_1}{\partial \alpha_1}=-\frac{\partial a_{11}}{\partial \alpha_1}X_1-a_{11}\frac{\partial X_1}{\partial \alpha_1}-a_{21}\frac{\partial X_2}{\partial \alpha_1}-\frac{\partial a_{21}}{\partial \alpha_1}X_2+\frac{\partial \tilde{P}_1}{\partial \alpha_1} \tag{36}$$

$$\frac{\partial P_2}{\partial \alpha_1}=-\frac{\partial a_{12}}{\partial \alpha_1}X_1-a_{12}\frac{\partial X_1}{\partial \alpha_1}-a_{22}\frac{\partial X_2}{\partial \alpha_1}-\frac{\partial a_{22}}{\partial \alpha_1}X_2+\frac{\partial \tilde{P}_2}{\partial \alpha_1}. \tag{37}$$

Applying (29), (36), (37) to (33) we get

$$\begin{aligned}
\frac{\partial a_{ij}}{\partial \alpha_1}&=\frac{\partial a_{ij}}{\partial X_1}\frac{\partial X_1}{\partial \alpha_1}+\frac{\partial a_{ij}}{\partial X_2}\frac{\partial X_2}{\partial \alpha_1}\\
&+\frac{\partial a_{ij}}{\partial P_1}\Big(-\frac{\partial a_{11}}{\partial \alpha_1}X_1-a_{11}\frac{\partial X_1}{\partial \alpha_1}-a_{21}\frac{\partial X_2}{\partial \alpha_1}-\frac{\partial a_{21}}{\partial \alpha_1}X_2+\overset{\approx}{P}_1\Big)\\
&+\frac{\partial a_{ij}}{\partial P_2}\Big(-\frac{\partial a_{12}}{\partial \alpha_1}X_1-a_{12}\frac{\partial X_1}{\partial \alpha_1}-a_{22}\frac{\partial X_2}{\partial \alpha_1}-\frac{\partial a_{22}}{\partial \alpha_1}X_2+\overset{\approx}{P}_2\Big)
\end{aligned}$$

$$= \Big(\frac{\partial a_{ij}}{\partial X_1} - a_{11}\frac{\partial a_{ij}}{\partial P_1} - a_{12}\frac{\partial a_{ij}}{\partial P_2}\Big)\frac{\partial X_1}{\partial \alpha_1} + \Big(\frac{\partial a_{ij}}{\partial X_2} - a_{21}\frac{\partial a_{ij}}{\partial P_1} - a_{22}\frac{\partial a_{ij}}{\partial P_2}\Big)\frac{\partial X_2}{\partial \alpha_1}$$
$$+ \Big(-\frac{\partial a_{11}}{\partial \alpha_1}X_1 - \frac{\partial a_{21}}{\partial \alpha_1}X_2 + \approx{P}_1\Big)\frac{\partial a_{ij}}{\partial P_1} + \Big(-\frac{\partial a_{12}}{\partial \alpha_1}X_1 - \frac{\partial a_{22}}{\partial \alpha_1}X_2 + \approx{P}_2\Big)\frac{\partial a_{ij}}{\partial P_2}$$
$$= 0.\frac{\partial X_1}{\partial \alpha_1} + 0.\frac{\partial X_2}{\partial \alpha_1} + \Big(-\frac{\partial a_{11}}{\partial \alpha_1}X_1 - \frac{\partial a_{21}}{\partial \alpha_1}X_2 + \approx{P}_1\Big)\frac{\partial a_{ij}}{\partial P_1}$$
$$+ \Big(-\frac{\partial a_{12}}{\partial \alpha_1}X_1 - \frac{\partial a_{22}}{\partial \alpha_1}X_2 + \approx{P}_2\Big)\frac{\partial a_{ij}}{\partial P_2}.$$

Hence,

$$\frac{\partial a_{ij}}{\partial \alpha_1} = \Big(-\frac{\partial a_{11}}{\partial \alpha_1}X_1 - \frac{\partial a_{21}}{\partial \alpha_1}X_2 + \approx{P}_1\Big)\frac{\partial a_{ij}}{\partial P_1} + \Big(-\frac{\partial a_{12}}{\partial \alpha_1}X_1 - \frac{\partial a_{22}}{\partial \alpha_1}X_2 + \approx{P}_2\Big)\frac{\partial a_{ij}}{\partial P_2}. \tag{38}$$

Letting i and j vary from 1 to 2 in (38), we obtain the system

$$\begin{cases} (1+\frac{\partial a_{11}}{\partial P_1}X_1)\frac{\partial a_{11}}{\partial \alpha_1} + (\frac{\partial a_{11}}{\partial P_2}X_1)\frac{\partial a_{12}}{\partial \alpha_1} + (\frac{\partial a_{11}}{\partial P_1}X_2)\frac{\partial a_{21}}{\partial \alpha_1} + (\frac{\partial a_{11}}{\partial P_2}X_2)\frac{\partial a_{22}}{\partial \alpha_1} \\ \qquad = \frac{\partial a_{11}}{\partial P_1}\approx{P}_1 + \frac{\partial a_{11}}{\partial P_2}\approx{P}_2 \\ (\frac{\partial a_{12}}{\partial P_1}X_1)\frac{\partial a_{11}}{\partial \alpha_1} + (1+\frac{\partial a_{12}}{\partial P_2}X_1)\frac{\partial a_{12}}{\partial \alpha_1} + (\frac{\partial a_{12}}{\partial P_1}X_2)\frac{\partial a_{21}}{\partial \alpha_1} + (\frac{\partial a_{12}}{\partial P_2}X_2)\frac{\partial a_{22}}{\partial \alpha_1} \\ \qquad = \frac{\partial a_{12}}{\partial P_1}\approx{P}_1 + \frac{\partial a_{12}}{\partial P_2}\approx{P}_2 \\ (\frac{\partial a_{21}}{\partial P_1}X_1)\frac{\partial a_{11}}{\partial \alpha_1} + (\frac{\partial a_{21}}{\partial P_2}X_1)\frac{\partial a_{12}}{\partial \alpha_1} + (1+\frac{\partial a_{21}}{\partial P_1}X_2)\frac{\partial a_{21}}{\partial \alpha_1} + (\frac{\partial a_{21}}{\partial P_2}X_2)\frac{\partial a_{22}}{\partial \alpha_1} \\ \qquad = \frac{\partial a_{21}}{\partial P_1}\approx{P}_1 + \frac{\partial a_{21}}{\partial P_2}\approx{P}_2 \\ (\frac{\partial a_{22}}{\partial P_1}X_1)\frac{\partial a_{11}}{\partial \alpha_1} + (\frac{\partial a_{22}}{\partial P_2}X_1)\frac{\partial a_{12}}{\partial \alpha_1} + (\frac{\partial a_{22}}{\partial P_1}X_2)\frac{\partial a_{21}}{\partial \alpha_1} + (1+\frac{\partial a_{22}}{\partial P_2}X_2)\frac{\partial a_{22}}{\partial \alpha_1} \\ \qquad = \frac{\partial a_{22}}{\partial P_1}\approx{P}_1 + \frac{\partial a_{22}}{\partial P_2}\approx{P}_2. \end{cases} \tag{39}$$

This is a linear system with $\frac{\partial a_{ij}}{\partial \alpha_1}, i, j = 1, 2$ as unknown functions.

Set

$$d = \begin{vmatrix} 1+\frac{\partial a_{11}}{\partial P_1}X_1 & \frac{\partial a_{11}}{\partial P_2}X_1 & \frac{\partial a_{11}}{\partial P_1}X_2 & \frac{\partial a_{11}}{\partial P_2}X_2 \\ \frac{\partial a_{12}}{\partial P_1}X_1 & 1+\frac{\partial a_{12}}{\partial P_2}X_1 & \frac{\partial a_{12}}{\partial P_1}X_2 & \frac{\partial a_{12}}{\partial P_2}X_2 \\ \frac{\partial a_{21}}{\partial P_1}X_1 & \frac{\partial a_{21}}{\partial P_2}X_1 & 1+\frac{\partial a_{21}}{\partial P_1}X_2 & +\frac{\partial a_{21}}{\partial P_2}X_2 \\ \frac{\partial a_{22}}{\partial P_1}X_1 & \frac{\partial a_{22}}{\partial P_2}X_1 & \frac{\partial a_{22}}{\partial P_1}X_2 & 1+\frac{\partial a_{22}}{\partial P_2}X_2 \end{vmatrix}. \tag{40}$$

Elementary calculations yield

$$
\begin{aligned}
d = 1 &+ (\frac{\partial a_{21}}{\partial P_1}\frac{\partial a_{22}}{\partial P_2} - \frac{\partial a_{21}}{\partial P_2}\frac{\partial a_{22}}{\partial P_1})X_2^2 + (\frac{\partial a_1}{\partial P_1}\frac{\partial a_{12}}{\partial P_2} - \frac{\partial a_{12}}{\partial P_1}\frac{\partial a_{11}}{\partial P_2})X_1^2 \\
&+ (\frac{\partial a_{12}}{\partial P_2}\frac{\partial a_{21}}{\partial P_1} - \frac{\partial a_{21}}{\partial P_2}\frac{\partial a_{12}}{\partial P_1} + \frac{\partial a_{11}}{\partial P_1}\frac{\partial a_{22}}{\partial P_2} - \frac{\partial a_{22}}{\partial P_1}\frac{\partial a_{11}}{\partial P_2}X_1X_2) \\
&+ (\frac{\partial a_{12}}{\partial P_2} + \frac{\partial a_{11}}{\partial P_1})X_1 + (\frac{\partial a_{21}}{\partial P_1} + \frac{\partial a_{22}}{\partial P_2})X_2 \\
&= \det \frac{D(a_{11}, a_{12})}{D(P_1, P_2)}(X_1(\alpha_1))^2 + \det \frac{D(a_{21}, a_{22})}{D(P_1, P_2)}(X_2(\alpha_1))^2 + \\
&\left[\det \frac{D(a_{11}, a_{22})}{D(P_1, P_2)} + \det \frac{D(a_{21}, a_{12})}{D(P_1, P_2)}\right] X_1(\alpha_1)X_2(\alpha_1) + \\
&+ (\frac{\partial a_{11}}{\partial P_1} + \frac{\partial a_{12}}{\partial P_2})X_1(\alpha_1) + (\frac{\partial a_{21}}{\partial P_1} + \frac{\partial a_{22}}{\partial P_2})X_2(\alpha_1) + 1.
\end{aligned} \tag{41}
$$

Since $(X, \tilde{P})$ are sufficiently closed to $(X^0, \tilde{P}^0)$, from (41) and (21) it follows that $d \neq 0$. The assertion 2) follows now. □

We introduce new dependent variables

$$
W = (X_1, X_2, Z, \tilde{P}_1, \tilde{P}_2, \approx{P}_1, \approx{P}_2)^T.
$$

Theorem 4.1. *Assume that the conditions (C_1) and (C_2) hold. Then the system (25) can be diagonalized, i.e. it may be reduced to the following diagonal one:*

$$
\frac{\partial W}{\partial \alpha_2} = \approx{\mathcal{A}}(W)\frac{\partial W}{\partial \alpha_1} + F(W), \tag{42}
$$

here

$$
\approx{\mathcal{A}} = \begin{bmatrix}
-1 & 0 & 0 & 0 & 0 & 0 & 0 \\
0 & -1 & 0 & 0 & 0 & 0 & 0 \\
0 & 0 & -1 & 0 & 0 & 0 & 0 \\
0 & 0 & 0 & a_{12} - a_{21} - 1 & 0 & 0 & 0 \\
0 & 0 & 0 & 0 & a_{12} - a_{21} - 1 & 0 & 0 \\
0 & 0 & 0 & 0 & 0 & a_{12} - a_{21} - 1 & 0 \\
0 & 0 & 0 & 0 & 0 & 0 & a_{12} - a_{21} - 1
\end{bmatrix}, \tag{43}
$$

and

$$
F(W) = F_1(W) + F_2(W),
$$

with

$$F_1(W) = \begin{bmatrix} \tilde{\tilde{P}}_2 - (b_{12}\tilde{\tilde{P}}_1 + c_{12}\tilde{\tilde{P}}_2)X_1 - (b_{22}\tilde{\tilde{P}}_1 + c_{22}\tilde{\tilde{P}}_2)X_2 \\ -\tilde{\tilde{P}}_1 + (b_{11}\tilde{\tilde{P}}_1 + c_{11}\tilde{\tilde{P}}_2)X_1 + (b_{21}\tilde{\tilde{P}}_1 + c_{21}\tilde{\tilde{P}}_2)X_2 \\ -P_2\tilde{\tilde{P}}_1 + P_1\tilde{\tilde{P}}_2 + ((b_{11}\tilde{\tilde{P}}_1 + c_{11}\tilde{\tilde{P}}_2)P_2 - (b_{12}\tilde{\tilde{P}}_1 + c_{12}\tilde{\tilde{P}}_2)P_1)X_1 \\ 0 \\ 0 \\ ((b_{12} - b_{21})\tilde{\tilde{P}}_1 + (c_{12} - c_{21})\tilde{\tilde{P}}_2)\tilde{\tilde{P}}_1 \\ ((b_{12} - b_{21})\tilde{\tilde{P}}_1 + (c_{12} - c_{21})\tilde{\tilde{P}}_2)\tilde{\tilde{P}}_2 \end{bmatrix},$$

$$F_2(W) = \begin{bmatrix} 0 \\ 0 \\ ((b_{21}\tilde{\tilde{P}}_1 + c_{21}\tilde{\tilde{P}}_2)P_2 - (b_{22}\tilde{\tilde{P}}_1 + c_{22}\tilde{\tilde{P}}_2)P_1)X_2 \\ 0 \\ 0 \\ 0 \\ 0 \end{bmatrix}.$$

In the representations for $F_1(W), F_2(W)$ the variables P_1, P_2 must be replaced by $f(X_1, X_2, Z, \tilde{P}_1, \tilde{P}_2)$ and $g(X_1, X_2, Z, \tilde{P}_1, \tilde{P}_2)$ respectively.

Proof. From (25) we have

$$\begin{cases} \dfrac{\partial X_1}{\partial \alpha_2} = -\dfrac{\partial X_1}{\partial \alpha_1} + \dfrac{\partial \tilde{P}_2}{\partial \alpha_1} - \dfrac{\partial a_{12}}{\partial \alpha_1}X_1 - \dfrac{\partial a_{22}}{\partial \alpha_1}X_2 \\ \dfrac{\partial X_2}{\partial \alpha_2} = -\dfrac{\partial X_2}{\partial \alpha_1} - \dfrac{\partial \tilde{P}_1}{\partial \alpha_1} + \dfrac{\partial a_{11}}{\partial \alpha_1}X_1 + \dfrac{\partial a_{21}}{\partial \alpha_1}X_2 \\ \dfrac{\partial Z}{\partial \alpha_2} = -\dfrac{\partial Z}{\partial \alpha_1} - P_2\dfrac{\partial \tilde{P}_1}{\partial \alpha_1} \\ +P_1\dfrac{\partial \tilde{P}_2}{\partial \alpha_1} + \left[\dfrac{\partial a_{11}}{\partial \alpha_1}P_2 - \dfrac{\partial a_{12}}{\partial \alpha_1}P_1\right]X_1 + \left[\dfrac{\partial a_{21}}{\partial \alpha_1}P_2 - \dfrac{\partial a_{22}}{\partial \alpha_1}P_1\right]X_2 \\ \dfrac{\partial \tilde{P}_1}{\partial \alpha_2} = (a_{12} - a_{21} - 1)\dfrac{\partial \tilde{P}_1}{\partial \alpha_1} \\ \dfrac{\partial \tilde{P}_2}{\partial \alpha_2} = (a_{12} - a_{21} - 1)\dfrac{\partial \tilde{P}_2}{\partial \alpha_1} \end{cases} \tag{44}$$

So from last two equations of (44) it follows

$$\frac{\partial \tilde{\tilde{P}}_1}{\partial \alpha_2} = \frac{\partial}{\partial \alpha_2}(\frac{\partial \tilde{P}_1}{\partial \alpha_1}) = \frac{\partial}{\partial \alpha_1}(\frac{\partial \tilde{P}_1}{\partial \alpha_2})$$
$$= \frac{\partial}{\partial \alpha_1}[(a_{12} - a_{21} - 1)(\frac{\partial \tilde{P}_1}{\partial \alpha_1})]$$

$$
\begin{aligned}
&= \frac{\partial(a_{12}-a_{21}-1)}{\partial\alpha_1}\frac{\partial\tilde{P}_1}{\partial\alpha_1} + (a_{12}-a_{21}-1)\frac{\partial\overset{\approx}{P}_1}{\partial\alpha_1} \\
&= \overset{\approx}{P}_1(\frac{\partial a_{12}}{\partial\alpha_1} - \frac{\partial a_{21}}{\partial\alpha_1}) + (a_{12}-a_{21}-1)\frac{\partial\overset{\approx}{P}_1}{\partial\alpha_1}
\end{aligned} \tag{45}
$$

and

$$
\begin{aligned}
\frac{\partial\overset{\approx}{P}_2}{\partial\alpha_2} &= \frac{\partial}{\partial\alpha_2}(\frac{\partial\tilde{P}_2}{\partial\alpha_1}) = \frac{\partial}{\partial\alpha_1}(\frac{\partial\tilde{P}_2}{\partial\alpha_2}) \\
&= \frac{\partial}{\partial\alpha_1}[(a_{12}-a_{21}-1)(\frac{\partial\tilde{P}_2}{\partial\alpha_1})] \\
&= \frac{\partial(a_{12}-a_{21}-1)}{\partial\alpha_1}\frac{\partial\tilde{P}_2}{\partial\alpha_1} + (a_{12}-a_{21}-1)\frac{\partial\overset{\approx}{P}_2}{\partial\alpha_1} \\
&= \overset{\approx}{P}_2(\frac{\partial a_{12}}{\partial\alpha_1} - \frac{\partial a_{21}}{\partial\alpha_1}) + (a_{12}-a_{21}-1)\frac{\partial\overset{\approx}{P}_2}{\partial\alpha_1}.
\end{aligned} \tag{46}
$$

Hence from (44)- (46) and (30) we obtain

$$
\begin{cases}
\frac{\partial X_1}{\partial\alpha_2} = -\frac{\partial X_1}{\partial\alpha_1} + \overset{\approx}{P}_2 - \frac{\partial a_{12}}{\partial\alpha_1}X_1 - \frac{\partial a_{22}}{\partial\alpha_1}X_2 \\
\frac{\partial X_2}{\partial\alpha_2} = -\frac{\partial X_2}{\partial\alpha_1} - \overset{\approx}{P}_1 + \frac{\partial a_{11}}{\partial\alpha_1}X_1 + \frac{\partial a_{21}}{\partial\alpha_1}X_2 \\
\frac{\partial Z}{\partial\alpha_2} = -\frac{\partial Z}{\partial\alpha_1} - P_2\overset{\approx}{P}_1 + P_1\overset{\approx}{P}_2 + \left[\frac{\partial a_{11}}{\partial\alpha_1}P_2 - \frac{\partial a_{12}}{\partial\alpha_1}P_1\right]X_1 \\
\qquad + \left[\frac{\partial a_{21}}{\partial\alpha_1}P_2 - \frac{\partial a_{22}}{\partial\alpha_1}P_1\right]X_2 \\
\frac{\partial\tilde{P}_1}{\partial\alpha_2} = (a_{12}-a_{21}-1)\frac{\partial\tilde{P}_1}{\partial\alpha_1} \\
\frac{\partial\tilde{P}_2}{\partial\alpha_2} = (a_{12}-a_{21}-1)\frac{\partial\tilde{P}_2}{\partial\alpha_1} \\
\frac{\partial\overset{\approx}{P}_1}{\partial\alpha_2} = (a_{12}-a_{21}-1)\frac{\partial\overset{\approx}{P}_1}{\partial\alpha_1} + \overset{\approx}{P}_1\Big(\frac{\partial a_{12}}{\partial\alpha_1} - \frac{\partial a_{21}}{\partial\alpha_1}\Big) \\
\frac{\partial\overset{\approx}{P}_2}{\partial\alpha_2} = (a_{12}-a_{21}-1)\frac{\partial\overset{\approx}{P}_2}{\partial\alpha_1} + \overset{\approx}{P}_2\Big(\frac{\partial a_{12}}{\partial\alpha_1} - \frac{\partial a_{21}}{\partial\alpha_1}\Big)
\end{cases} \tag{47}
$$

From (47) and (32) the theorem immediately follows. □

The following result is a direct consquence of the last theorem, theorem 2.3 and proposition 3.2.

Theorem 4.2. *Assume that the conditions* (C_1) *and* (C_2) *hold. Then there exists locally a unique* C^2*-solution to the Cauchy problem* (11)-(8).

5. Application to the Classical Weakly Hyperbolic Monge-Ampère Equation

Applying Theorems 4.1 and 1.1 we obtain the following theorem on solvability of the Cauchy problem (1)-(4) for weakly hyperbolic Monge-Ampère equations.

Theorem 5.1. *Suppose that besides the conditions* (5), (6) *the following conditions also hold:*

1) The coefficients A, B, C, E are C^2-functions and do not depend on z;

2) $\Delta \geq 0$;

3) For all $\alpha_1 \in I$ it holds

$$\det \frac{D(C,\lambda_1)}{D(p_1,p_2)}(X_1^0(\alpha_1))^2 + \det \frac{D(\lambda_2,A)}{D(p_1,p_2)}(X_2^0(\alpha_1))^2 + \left[\det \frac{D(C,A)}{D(p_1,p_2)} + \det \frac{D(\lambda_2,\lambda_1)}{D(p_1,p_2)}\right] X_1^0(\alpha_1)X_2^0(\alpha_1) + \left(\frac{\partial C}{\partial p_1} + \frac{\partial \lambda_1}{\partial p_2}\right)X_1^0(\alpha_1) + \left(\frac{\partial \lambda_2}{\partial p_1} + \frac{\partial A}{\partial p_2}\right)X_2^0(\alpha_1) + 1 \neq 0, \tag{48}$$

where the derivatives of functions $A, C, \lambda_1, \lambda_2$ are calculated at $(X_1^0(\alpha_1), X_2^0(\alpha_1), Z^0(\alpha_1), P_1^0(\alpha_1), P_2^0(\alpha_1))$;

4) Each function $A(x,z,p), C(x,z,p), \lambda_1(x,z,p), \lambda_2(x,z,p)$ is a first integral for the Eq. (1)*; i.e., due to the condition 1) above, each of them satisfies the following system*

$$\begin{cases} \dfrac{\partial \varphi}{\partial x_1} - C\dfrac{\partial \varphi}{\partial p_1} - \lambda_1\dfrac{\partial \varphi}{\partial p_2} = 0, \\ \dfrac{\partial \varphi}{\partial x_2} - \lambda_2\dfrac{\partial \varphi}{\partial p_1} - A\dfrac{\partial \varphi}{\partial p_2} = 0. \end{cases}$$

Then there exists locally a smooth solution $z(x)$ to the Cauchy problem (1)-(4).

6. Examples

Example 6.1. (*Ref. 5*) The coeffiients A, B, C, E are constants with $\Delta \geq 0$. It is easy to see that all the assumptions of Therem 5.1 are satisfied. From (26) it follows that $\mathcal{B} = 0$. From (25) and (22) we have the following quasilinear system

$$\begin{cases} \dfrac{\partial X_1}{\partial \alpha_2} &= -\dfrac{\partial X_1}{\partial \alpha_1} + \dfrac{\partial \tilde{P}_2}{\partial \alpha_1} \\ \dfrac{\partial X_2}{\partial \alpha_2} &= -\dfrac{\partial X_2}{\partial \alpha_1} - \dfrac{\partial \tilde{P}_1}{\partial \alpha_1} \\ \dfrac{\partial Z}{\partial \alpha_2} &= -\dfrac{\partial Z}{\partial \alpha_1} + (\lambda_1 X_1 + A X_2 - \tilde{P}_2)\dfrac{\partial \tilde{P}_1}{\partial \alpha_1} + (-C X_1 - \lambda_2 X_2 + \tilde{P}_1)\dfrac{\partial \tilde{P}_2}{\partial \alpha_1} \\ \dfrac{\partial \tilde{P}_1}{\partial \alpha_2} &= (\lambda_1 - \lambda_2 - 1)\dfrac{\partial \tilde{P}_1}{\partial \alpha_1} \\ \dfrac{\partial \tilde{P}_2}{\partial \alpha_2} &= (\lambda_1 - \lambda_2 - 1)\dfrac{\partial \tilde{P}_2}{\partial \alpha_1} \end{cases} \tag{49}$$

with the following initial conditions

$$\begin{cases} X(\alpha_1, \alpha_2)\big|_{\alpha_2 = 0} &= X^0(\alpha_1), \\ Z(\alpha_1, \alpha_2)\big|_{\alpha_2 = 0} &= Z^0(\alpha_1), \\ \tilde{P}(\alpha_1, \alpha_2)\big|_{\alpha_2 = 0} &= \tilde{P}^0(\alpha_1), \end{cases} \tag{50}$$

where $\tilde{P}^0(\alpha_1) = (\tilde{P}_1^0(\alpha_1), \tilde{P}_2^0(\alpha_1))$, and

$$\tilde{P}_1^0(\alpha_1) = P_1^0(\alpha_1) + C X_1^0(\alpha_1) + \lambda_2 X_2^0(\alpha_1)$$
$$\tilde{P}_2^0(\alpha_1) = P_2^0(\alpha_1) + \lambda_1 X_1^0(\alpha_1) + A X_2^0(\alpha_1).$$

From the two last independent linear equations of (49) with initial conditions (50) it is easy to solve $\tilde{P}_1(\alpha_1, \alpha_2), \tilde{P}_2(\alpha_1, \alpha_2)$. Then we substitute them into the first two equations of (49) to obtain independent linear equations with respect to $X_1(\alpha_1, \alpha_2)$ and $X_2(\alpha_1, \alpha_2)$. At the last, substituting $\tilde{P}_1(\alpha_1, \alpha_2), \tilde{P}_2(\alpha_1, \alpha_2), X_1(\alpha_1, \alpha_2), X_2(\alpha_1, \alpha_2)$, which have been found, into the third equation of (49) we can solve $Z(\alpha_1, \alpha_2)$. Then by the Theorem 1.1 the local solution $z(x)$ to the Cauchy problem (1)-(4) can be obtained from the functions $X_1(\alpha_1, \alpha_2), X_2(\alpha_1, \alpha_2)$ and $Z(\alpha_1, \alpha_2)$.

Example 6.2. Suppose that $v(y, t)$ is a solution of the *Burger equation*

$$v_t + v v_y = 0,$$

which is C^2-function in a neighbohood of the point $(c_1 P_1^0(0) - c_2 P_2^0(0), c_2 X_1^0(0) + c_1 X_2^0(0))$ and satisfies the condition $-v_y(c_1 P_1^0(\alpha_1) - c_2 P_2^0(\alpha_1), c_2 X_1^0(\alpha_1) + c_1 X_2^0(\alpha_1))(c_2 X_1^0(\alpha_1) + c_1 X_2^0(\alpha_1))\ +1 \neq 0$, for all $\alpha_1 \in I$ and where c_1, c_2 are some given constants, $c_1^2 + c_2^2 > 0$.
Then the Cauchy problem for the Monge-Ampère equation

$$rt - s^2 + v^2(c_1 z_{x_1} - c_2 z_{x_2}, c_2 x_1 + c_1 x_2) = 0$$

with $A = B = C = 0, E = -v^2(c_1p_1 - c_2p_2, c_2x_1 + c_1x_2), \Delta = 4v^2(c_1p_1 - c_2p_2, c_2x_1 + c_1x_2), \lambda_1 = -\lambda_2 = v(c_1p_1 - c_2p_2, c_2x_1 + c_1x_2)$ satisfies all conditions of Theorem 5.1.

Example 6.3. Suppose that $v(y,t)$ is a solution of the following equation

$$v_t + f(v)v_y = 0,$$

which is C^2-function in a neighbourhood of the point

$$(-c_1P_1^0(0) - c_2P_2^0(0), c_1X_1^0(0) + c_2X_2^0(0))$$

and satisfies the condition

$$\begin{aligned} &- f'(v(-c_1P_1^0(\alpha_1) - c_2P_2^0(\alpha_1), c_1X_1^0(\alpha_1) + c_2X_2^0(\alpha_1)) \\ &\times v_y(-c_1P_1^0(\alpha_1) - c_2P_2^0(\alpha_1), \\ &c_1X_1^0(\alpha_1) + c_2X_2^0(\alpha_1))(c_1X_1^0(\alpha_1) + c_2X_2^0(\alpha_1)) + 1 \neq 0, \end{aligned}$$

for all $\alpha_1 \in I$ and where $f(v)$ is a given C^2-function of one variable, c_1, c_2 are some given constants, $c_1^2 + c_2^2 > 0$. Then the Cauchy problem for the following equation

$$\begin{aligned} &f(v(-c_1z_{x_1} - c_2z_{x_2}, c_1x_1 + c_2x_2))(r+t) + (rt - s^2) \\ &+ f^2(v(-c_1z_{x_1} - c_2z_{x_2}, c_1x_1 + c_2x_2)) = 0 \end{aligned}$$

with $A = f(v(-c_1z_{x_1} - c_2z_{x_2}, c_1x_1 + c_2x_2))$, $C = f(v(-c_1z_{x_1} - c_2z_{x_2}, c_1x_1 + c_2x_2))$, $B = 0$, $E = -f^2(v(-c_1z_{x_1} - c_2z_{x_2}, c_1x_1 + c_2x_2))$, $\Delta \equiv 0, \lambda_1 = \lambda_2 = 0$ satisfies all conditions of the Theorem 5.1.

References

1. G. Darboux, *Leçons sur la théorie générale des surfaces*, tome 3 (Gauthier-Villars, Paris, 1894).
2. E. Goursat, *Leçons sur l'intégration des équations aux dérivées partielles du second ordre*, tome 1 (Hermann, Paris, 1896).
3. J. Hadamard, *Le problème de Cauchy et les équations aux dérivées partielles linéaires hyperboliques* (Hermann, Paris, 1932).
4. B. L. Rodgestvenski, N. N. Yanenko, *Quasilinear hyperbolic systems* (Nauka, Moscow, 1978).
5. M. Tsuji, *Bull. Sci. Math.*, 433 (1995).
6. M. Tsuji and H. T. Ngoan, *Integration of hyperbolic Monge-Ampère equations*, in *Proceedings of the Fifth Vietnamese Mathematical Conference* (Hanoi, 1997) pp. 205-212.
7. Ha Tien Ngoan, D. Kong and M. Tsuji, *Ann. Scuola Norm. Sup. Pisa Cl. Sci.*, **27**, 309 (1998).
8. M. Tsuji and N. D. Thai Son, *Acta Math. Viet.*, **27**, 97 (2002).

9. H. T. Ngoan and N. T. Nga, *On the Cauchy problem for hyperbolic Monge-Ampère equations*, in *Proceedings of the Conference on Partial Differential Equations and their Applications* (Hanoi, December 27-29, 1999) pp. 77–91.
10. Ha Tien Ngoan and Nguyen Thi Nga, *Acta Mathematica Vietnamica*, **29**, 281 (2005).
11. Ha Tien Ngoan and Nguyen Thi Nga, *Vietnam Journal of Mathematics*, **34**, 109 (2006).

Advances in Deterministic and Stochastic Analysis
Eds. N. M. Chuong *et al.* (pp. 197–227)

§11. SOME SINGULAR PERTURBATION PROBLEMS RELATED TO THE NAVIER-STOKES EQUATIONS

MAKRAM HAMOUDA

Laboratoire d'Analyse Numérique, Université de Paris–Sud, Orsay, France

ROGER TEMAM‡

The Institute for Scientific Computing and Applied Mathematics,
Indiana University, Bloomington, IN, USA
‡*E-mail: temam@indiana.edu*

In this article, we consider the asymptotic analysis of the solutions of the Navier-Stokes problem, when the viscosity goes to zero; we consider the flow in a channel of $\mathbb{R}^3$, in the case where the boundary is non-characteristic. More precisely, a complete asymptotic expansion, at all orders, is given in the linear case. For the full nonlinear Navier-Stokes solution, we give a convergence theorem up to order 1, thus improving and simplifying the results of Ref. 11.

1. Introduction

We consider the Navier-Stokes equations in a channel $\Omega_\infty = \mathbb{R}^2 \times (0,h)$ with a permeable boundary, making the boundaries $z = 0, h$, non-characteristic. More precisely we have

$$\begin{cases} \dfrac{\partial u^\varepsilon}{\partial t} - \varepsilon \Delta u^\varepsilon + (u^\varepsilon . \nabla)\, u^\varepsilon + \nabla p^\varepsilon = f, \quad \text{in } \Omega_\infty, \\ \text{div } u^\varepsilon = 0, \quad \text{in } \Omega_\infty, \\ u^\varepsilon = (0,0,-U), \quad \text{on } \Gamma_\infty, \\ u^\varepsilon \text{ is periodic in the } x \text{ and } y \text{ directions with periods } L_1, L_2, \\ u^\varepsilon\big|_{t=0} = u_0. \end{cases} \tag{1}$$

Here $\Gamma_\infty = \partial\Omega_\infty = \mathbb{R}^2 \times \{0,h\}$ and we introduce also Ω and Γ:

$$\Omega = (0,L_1) \times (0,L_2) \times (0,h), \quad \Gamma = (0,L_1) \times (0,L_2) \times \{0,h\}.$$

Throughout this article we assume that f and u_0 are given functions as regular as necessary in the channel Ω_∞, and that U is a given > 0 constant; at the price of long technicalities, we can also consider the case where U is nonconstant, $U > 0$ everywhere.
By setting $v^\varepsilon = u^\varepsilon - (0, 0, -U)$, we obtain the following problem for v^ε:

$$\begin{cases} \dfrac{\partial v^\varepsilon}{\partial t} - \varepsilon \Delta v^\varepsilon - U D_3 v^\varepsilon + (v^\varepsilon.\nabla)\, v^\varepsilon + \nabla p^\varepsilon = f, \quad \text{in } \Omega_\infty, \\ \text{div } v^\varepsilon = 0, \quad \text{in } \Omega_\infty, \\ v^\varepsilon = 0, \quad \text{on } \Gamma_\infty, \\ v^\varepsilon \text{ is periodic in the } x \text{ and } y \text{ directions with periods } L_1, L_2, \\ v^\varepsilon\big|_{t=0} = v_0. \end{cases} \tag{2}$$

Linearizing in v^ε we find:

$$\begin{cases} \dfrac{\partial v^\varepsilon}{\partial t} - \varepsilon \Delta v^\varepsilon - U D_3 v^\varepsilon + \nabla p^\varepsilon = f, \quad \text{in } \Omega_\infty, \\ \text{div } v^\varepsilon = 0, \quad \text{in } \Omega_\infty, \\ v^\varepsilon = 0, \quad \text{on } \Gamma_\infty, \\ v^\varepsilon \text{ is periodic in the } x \text{ and } y \text{ directions with periods } L_1, L_2, \\ v^0\big|_{t=0} = v_0. \end{cases} \tag{3}$$

In the linear case, the limit problem ($\varepsilon = 0$) corresponding to (3) is the Euler problem those solutions may not verify the boundary conditions $(3)_3$. More precisely it follows from this work that the limit solution, denoted by v^0, is solution (for all time) of the following system:

$$\begin{cases} \dfrac{\partial v^0}{\partial t} - UD_3 v^0 + \nabla p^0 = f, \quad \text{in } \Omega_\infty, \\ \operatorname{div} v^0 = 0, \quad \text{in } \Omega_\infty, \\ v_3^0 = 0, \quad \text{on } \Gamma_0, \\ v^0 = 0, \quad \text{on } \Gamma_h, \\ v^0 \text{ is periodic in the } x \text{ and } y, \text{ directions with periods } L_1, L_2, \\ v^0\big|_{t=0} = v_0, \end{cases} \tag{4}$$

where $\Gamma_0 = \mathbb{R}^2 \times \{0\}$ and $\Gamma_h = \mathbb{R}^2 \times \{h\}$.
Our aim in this article is to derive asymptotic expansions for v^ε, when the viscosity ε goes to zero, for either the linear case (3) or the nonlinear case (2).
It is obvious that we can not expect a convergence result of v^ε to v^0 in the uniform or H^1 norm spaces since v^ε and v^0 do not have the same traces on Γ_0. This question was addressed in Ref. 11 in which the authors gave a representation of the NS solution u^ε up to the boundary and proved convergence results in several Sobolev spaces.
In this article, we continue this work and we first derive the complete asymptotic expansion of the linearized NS solution v^ε up to the boundary at all orders. For this purpose, we will use the classical tools of the singular perturbation theory based on the use of correctors; see Refs. 4 and 3, or Refs. 12 and 7 for a related point of view. A similar work is performed in the nonlinear case but we then restrict ourselves to orders 0 and 1 (and to limited time), although we believe that higher orders can be treated in the same way.

This article is organized as follows. In the next section, we show how to choose and construct the correctors of orders zero and one for the linearized NS equations. The third section deals with the description of the correctors of order $N \geq 2$ and the fourth section is devoted to the convergence theorem which validates our (empirical) choice for the correctors (linear case). Finally, in the last section (Sec. 5), we treat the nonlinear problem at orders zero and one and give the adequate correctors by proving their validity; this section improves and simplifies the results of Ref. 11.

2. Correctors of Order 0 and 1

Our method to construct the corrector of order zero is based on the observation made in Ref. 11 that, unlike the characteristic case, the Prandtl equation for this problem is very simple, namely

$$\begin{cases} -\varepsilon D_3^2\theta^\varepsilon - UD_3\theta^\varepsilon = 0, \quad \text{in } \Omega_\infty, \\ \text{div } \theta^\varepsilon = 0, \quad \text{in } \Omega_\infty, \\ \theta^\varepsilon = -v^0, \quad \text{on } \Gamma_0, \\ \theta^\varepsilon = 0, \quad \text{on } \Gamma_h. \end{cases} \tag{5}$$

However our method here is quite different than in Ref. 11; the correctors' representation that we give hereafter is simpler than in Ref. 11 and this simplification allows us to continue our analysis at higher orders. We intend to give an asymptotic expansion of $v^\varepsilon, p^\varepsilon$ of the following form:

$$\begin{cases} v^\varepsilon \sim \sum_{j=0}^N \varepsilon^j \, (v^j + \theta^{j,\varepsilon}), \\ p^\varepsilon \sim \sum_{j=0}^N \varepsilon^j \, p^j. \end{cases} \tag{6}$$

To determine the asymptotic expansion above we are guided by the general techniques of boundary layer theory. Hence for the exterior expansion $v^\varepsilon \sim \sum_{j=0}^N \varepsilon^j \, v^j$,
$p^\varepsilon \sim \sum_{j=0}^N \varepsilon^j \, p^j$, we would have

$$\frac{\partial v^0}{\partial t} - UD_3 v^0 + \nabla p^0 = f, \tag{7}$$

and, for $j \geq 1$:

$$\frac{\partial v^j}{\partial t} - UD_3 v^j + \nabla p^j = \Delta v^{j-1}. \tag{8}$$

The interior asymptotic expansion $\sum_{j=0}^N \varepsilon^j \, \theta^{j,\varepsilon}$ is determined by the condition that $\theta^{j,\varepsilon} = -v^j$ on Γ_∞, at all orders j and, for the equation we use the ansatz $D_3 \sim \varepsilon^{-1}$ which leads to (5) in Ref. 11. Hence, formally

$$-\varepsilon D_3^2\theta^{0,\varepsilon} - UD_3\theta^{0,\varepsilon} = 0, \qquad -\varepsilon D_3^2\theta^{1,\varepsilon} - UD_3\theta^{1,\varepsilon} = -\frac{1}{\varepsilon}\frac{\partial\theta^{0,\varepsilon}}{\partial t}, \tag{9}$$

and at orders $j \geq 2$

$$-\varepsilon D_3^2\theta^{j,\varepsilon} - UD_3\theta^{j,\varepsilon} = -\frac{1}{\varepsilon}\left\{\frac{\partial\theta^{j-1,\varepsilon}}{\partial t} + \Delta_\tau\theta^{j-2,\varepsilon}\right\}. \tag{10}$$

However this general scheme has to be adapted to the present situation because of the intricacies introduced by the incompressibility condition (and the global nature of this equation); hence (7)-(10) will only serve as a general guideline.

We begin at order 0 by proposing a corrector solving (approximatively) this system

$$\begin{cases} -\varepsilon D_3^2 \theta^{0,\varepsilon} - U D_3 \theta^{0,\varepsilon} = 0, & \text{on } (0,h), \\ \operatorname{div} \theta^{0,\varepsilon} = 0, \quad \text{in } \Omega_\infty, \\ \theta^{0,\varepsilon} = -\gamma_0 v^0, \quad \text{on } \Gamma_0, \\ \theta^{0,\varepsilon} = 0, \quad \text{on } \Gamma_h, \end{cases} \tag{11}$$

where γ_0 (resp. γ_h) is the trace on $z=0$ (respectively on $z=h$).
Firstly, we derive the tangential component $\theta_\tau^{0,\varepsilon}$ of $\theta^{0,\varepsilon}$, by solving the system composed of the tangential projection of $(11)_1$ and $(11)_{3,4}$, and then the normal component (z-direction) $\theta_n^{0,\varepsilon}$ will be given by the incompressibility condition $(11)_2$.

Thus, an approximation denoted $\overline{\theta}_\tau^{0,\varepsilon}$ of the tangential component $\theta_\tau^{0,\varepsilon}$ up to an *e.s.t.* (exponentially small term) reads:

$$\begin{cases} \theta_\tau^{0,\varepsilon} = \overline{\theta}_\tau^{0,\varepsilon} + e.s.t., \\ \overline{\theta}_\tau^{0,\varepsilon} = a_0^0 \mathrm{e}^{-Uz/\varepsilon}, \\ a_0^0 = -\gamma_0 v_\tau^0. \end{cases} \tag{12}$$

Here and in the sequel an *e.s.t.* is a function those norm in all classical spaces H^m, C^r is exponentially small in ε.

Notice that $\overline{\theta}_\tau^{0,\varepsilon}$ still satisfies equations $(11)_1$ and $(11)_3$, and has an exponentially small value at $z=h$:

$$\overline{\theta}_\tau^{0,\varepsilon} = O(\mathrm{e}^{-Uh/\varepsilon}), \quad \text{at } z=h. \tag{13}$$

Then, thanks to $(11)_2$, we write

$$\theta_n^{0,\varepsilon} = \int_z^h \operatorname{div}_\tau \overline{\theta}_\tau^{0,\varepsilon}(x,y,\zeta,t)\mathrm{d}\zeta,$$

and we deduce easily an approximation of $\theta_n^{0,\varepsilon}$ up to an *e.s.t.*,

$$\begin{cases} \overline{\theta}_n^{0,\varepsilon} = \varepsilon b_0^0 \mathrm{e}^{-Uz/\varepsilon}, \\ b_0^0 = -\dfrac{1}{U}\operatorname{div}_\tau(\gamma_0 v_\tau^0). \end{cases} \tag{14}$$

We note that this approximate corrector $\overline{\theta}^{0,\varepsilon} = (\overline{\theta}_\tau^{0,\varepsilon}, \overline{\theta}_n^{0,\varepsilon})$ verifies the incompressibility condition $(11)_2$ and the equation $(11)_1$. However $\gamma_0 \overline{\theta}_n^{0,\varepsilon} = O(\varepsilon)$ and $\gamma_h \overline{\theta}_n^{0,\varepsilon} = O(\mathrm{e}^{-Uh/\varepsilon})$, these boundary values being different than the desired ones, namely $\overline{\theta}_n^{0,\varepsilon} = 0$ at $z=0,h$.

To correct the boundary conditions of $\overline{\theta}^{0,\varepsilon}$, we introduce an additional corrector denoted by $\varphi^{0,\varepsilon} = \varepsilon\varphi^0$, which is solution of the following system:

$$\begin{cases} -\Delta\varphi^0 + \nabla\pi^0 = 0, \quad \text{in } \Omega_\infty, \\ \text{div } \varphi^0 = 0, \quad \text{in } \Omega_\infty, \\ \varphi^0_\tau = 0, \quad \text{on } \Gamma_0 \cup \Gamma_h, \\ \varphi^0_n = -\dfrac{1}{\varepsilon}\gamma_0\overline{\theta}^{0,\varepsilon}_n = \dfrac{1}{U}\text{div}_\tau(\gamma_0 v^0_\tau), \quad \text{on } \Gamma_0, \\ \varphi^0_n = 0, \quad \text{on } \Gamma_h. \end{cases} \tag{15}$$

Remark 2.1. The role of the corrector $\varphi^{0,\varepsilon}$ is just to account for an $O(\varepsilon)$-error in the boundary value of $\overline{\theta}^{0,\varepsilon}$ and this "error" in the boundary condition will be actually corrected at the next order. There are also some boundary value "errors" of order $O(\varepsilon^{-\alpha}\mathrm{e}^{-Uh/\varepsilon})$; they will be corrected all at once later on (at the final order N).

We now proceed to the first order. The function v^1 appearing in (6), is required to satisfy theses equations and boundary conditions:

$$\begin{cases} \dfrac{\partial v^1}{\partial t} - UD_3v^1 + \nabla p^1 = -\dfrac{\partial\varphi^0}{\partial t} + UD_3\varphi^0 + \Delta v^0, \quad \text{in } \Omega_\infty, \\ \text{div } v^1 = 0, \quad \text{in } \Omega_\infty, \\ v^1_3 = 0, \quad \text{on } \Gamma_0, \\ v^1 = 0, \quad \text{on } \Gamma_h, \\ v^1 \text{ is periodic in } x \text{ and } y. \end{cases} \tag{16}$$

For the corrector $\theta^{1,\varepsilon}$, as suggested by (9), we would like to have:

$$\begin{cases} -\varepsilon D_3^2\theta^{1,\varepsilon} - UD_3\theta^{1,\varepsilon} = -\dfrac{1}{\varepsilon}\dfrac{\partial\overline{\theta}^{0,\varepsilon}}{\partial t}, \quad \text{in } \Omega_\infty, \\ \text{div } \theta^{1,\varepsilon} = 0, \quad \text{in } \Omega_\infty, \\ \theta^{1,\varepsilon} = -v^1, \quad \text{on } \Gamma_0 \cup \Gamma_h. \end{cases} \tag{17}$$

As at order zero, we instead construct an approximation for $\theta^{1,\varepsilon}$ as follows: $\theta^{1,\varepsilon} = \overline{\theta}^{1,\varepsilon} + e.s.t.$, with a tangential component

$$\begin{cases} \overline{\theta}^{1,\varepsilon}_\tau = a^1_0\mathrm{e}^{-Uz/\varepsilon} + a^1_1\dfrac{z}{\varepsilon U}\mathrm{e}^{-Uz/\varepsilon}, \\ a^1_0 = -\gamma_0 v^1_\tau, \quad a^1_1 = \dfrac{\partial}{\partial t}(\gamma_0 v^0_\tau) = -\dfrac{\partial}{\partial t}(a^0_0), \end{cases} \tag{18}$$

and, for the normal component,

$$\begin{cases} \overline{\theta}_n^{1,\varepsilon} = \varepsilon b_0^1 \mathrm{e}^{-Uz/\varepsilon} + \dfrac{b_1^1}{U^2} z\mathrm{e}^{-Uz/\varepsilon}, \\ b_0^1 = -\dfrac{1}{U}\mathrm{div}_\tau(\gamma_0 v_\tau^1) - \dfrac{1}{U^3}\dfrac{\partial(\mathrm{div}_\tau(\gamma_0 v_\tau^0))}{\partial t}, \quad b_1^1 = \dfrac{\partial(\mathrm{div}_\tau(\gamma_0 v_\tau^0))}{\partial t}. \end{cases} \tag{19}$$

As before $\overline{\theta}_\tau^{1,\varepsilon}$ satisfies the desired boundary value exactly at $z = 0$ and up to an *e.s.t.* at $z = h$, whereas $\overline{\theta}_n^{1,\varepsilon}$ satisfies the desired boundary value up to order $O(\varepsilon)$ at $z = 0$ and up to an *e.s.t.* at $z = h$. To correct the $O(\varepsilon)$ error we introduce an additional corrector $\varphi^{1,\varepsilon} = \varepsilon\varphi^1$ verifying

$$\begin{cases} -\Delta\varphi^1 + \nabla\pi^1 = 0, \quad \text{in } \Omega, \\ \mathrm{div}\,\varphi^1 = 0, \quad \text{in } \Omega, \\ \varphi_\tau^1 = 0, \quad \text{on } \Gamma_0 \cup \Gamma_h, \\ \varphi_n^1 = -\dfrac{1}{\varepsilon}\gamma_0\overline{\theta}_n^{1,\varepsilon} = -b_0^1, \quad \text{on } \Gamma_0, \\ \varphi_n^1 = 0, \quad \text{on } \Gamma_h. \end{cases} \tag{20}$$

Finally, let us summarize what we obtained at order 1. We set $\widetilde{v}^{1,\varepsilon} = v^1 + \overline{\theta}^{1,\varepsilon} + \varphi^{1,\varepsilon}$. Thanks to (16), and (18)–(20), the functions $\widetilde{v}^{1,\varepsilon}$ and $\widetilde{p}^1 = p^1 + \pi^1$ satisfy the following set of equations and boundary conditions

$$\begin{cases} \dfrac{\partial\widetilde{v}^{1,\varepsilon}}{\partial t} - \varepsilon\Delta\widetilde{v}^{1,\varepsilon} - UD_3\widetilde{v}^{1,\varepsilon} + \nabla\widetilde{p}^1 = \dfrac{\partial\overline{\theta}^{1,\varepsilon}}{\partial t} + \varepsilon\dfrac{\partial\varphi^1}{\partial t} - \varepsilon\Delta_\tau\overline{\theta}^{1,\varepsilon} - \\ \quad -\dfrac{\partial\varphi^0}{\partial t} + UD_3\varphi^0 + \Delta v^0 - \dfrac{1}{\varepsilon}\dfrac{\partial\overline{\theta}^{0,\varepsilon}}{\partial t} - \varepsilon\Delta v^1 - \varepsilon UD_3\varphi^1, \quad \text{in } \Omega_\infty, \\ \mathrm{div}\,\widetilde{v}^{1,\varepsilon} = 0, \quad \text{in } \Omega_\infty, \\ \widetilde{v}^{1,\varepsilon} = 0, \quad \text{on } \Gamma_0, \\ \widetilde{v}_\tau^{1,\varepsilon} = a_0^1\mathrm{e}^{-Uh/\varepsilon} + a_1^1\dfrac{h}{\varepsilon U}\mathrm{e}^{-Uh/\varepsilon} = O(\varepsilon^{-1}\mathrm{e}^{-Uh/\varepsilon}), \quad \text{on } \Gamma_h, \\ \widetilde{v}_n^{1,\varepsilon} = \varepsilon b_0^1\mathrm{e}^{-Uh/\varepsilon} + \dfrac{b_1^1}{U^2}h\mathrm{e}^{-Uh/\varepsilon} = O(\mathrm{e}^{-Uh/\varepsilon}), \quad \text{on } \Gamma_h, \\ \widetilde{v}^{1,\varepsilon} \text{ is periodic in } x \text{ and } y. \end{cases} \tag{21}$$

3. The Corrector of Order N $(N \geq 2)$

In this section, we will focus on the Nth corrector for all integers $N \geq 2$. Starting with order $N = 2$, the equations for the v^j and $\theta^{j,\varepsilon}$ are different as observed in (8) and (10). More precisely for $N \geq 2$, we propose to define the tangential component $\theta_\tau^{N,\varepsilon}$ of $\theta^{N,\varepsilon}$ by

$$\begin{cases} -\varepsilon D_3^2\theta_\tau^{N,\varepsilon} - UD_3\theta_\tau^{N,\varepsilon} = -\dfrac{1}{\varepsilon}\Big\{\dfrac{\partial\overline{\theta}_\tau^{N-1,\varepsilon}}{\partial t} - \Delta\overline{\theta}_\tau^{N-2,\varepsilon}\Big\}, & \text{on } (0,h), \\ \text{div } \theta^{N,\varepsilon} = 0, \quad \text{in } \Omega, \\ \theta_\tau^{N,\varepsilon} = -v_\tau^N, \quad \text{on } \Gamma_0 \cup \Gamma_h, \end{cases} \tag{22}$$

where v^N verifies for all $N \geq 1$ the following form of the system (8):

$$\begin{cases} \dfrac{\partial v^N}{\partial t} - UD_3 v^N + \nabla p^N \\ \quad = -\dfrac{\partial \varphi^{N-1}}{\partial t} + UD_3\varphi^{N-1} + \Delta v^{N-1}, \quad \text{in } \Omega, \\ \text{div } v^N = 0, \quad \text{in } \Omega, \\ v_3^N = 0, \quad \text{on } \Gamma_0, \\ v^N = 0, \quad \text{on } \Gamma_h, \\ v^N \text{ is periodic in } x \text{ and } y. \end{cases} \tag{23}$$

As for $N = 0, 1$, the φ^N to be subsequently defined are introduced as correctors of the boundary conditions.

To solve system (22), we need a sequence of lemmas.

3.1. *Preliminary results*

The first two lemmas are elementary, they will be helpful for computing the correctors.

Lemma 3.1. *We have, for any $j \in \mathbb{N}$,*

$$\int_h^z \xi^j e^{-U\xi/\varepsilon} d\xi = -\sum_{k=0}^{j} \Big(\frac{\varepsilon}{U}\Big)^{j-k+1} \frac{j!}{k!} [z^k e^{-Uz/\varepsilon} - h^k e^{-Uh/\varepsilon}]. \tag{24}$$

Lemma 3.2. *A particular solution of the differential equation*

$$-\varepsilon D_3^2 f - UD_3 f = \frac{z^j}{U^j\, j! \varepsilon^{j+1}} e^{-Uz/\varepsilon}, \quad j \in \mathbb{N}, \tag{25}$$

is given by

$$f = \sum_{k=0}^{j} \frac{z^{k+1}}{(k+1)!\, \varepsilon^{k+1} U^{2j+1-k}} e^{-Uz/\varepsilon}. \tag{26}$$

Both lemmas are proved by induction on j. Based on these lemmas, the next lemmas provide the explicit form of the corrector $\theta^{N,\varepsilon}$. We start by describing the tangential component and the normal component is then deduced from the incompressibility condition.

Lemma 3.3. *An approximation up to exponential order of the tangential component $\theta_\tau^{N,\varepsilon}$ of the corrector $\theta^{N,\varepsilon}$, solution of (22), is given as follows:*

$$\overline{\theta}_\tau^{N,\varepsilon} = \sum_{j=0}^{N} a_j^N \frac{z^j}{\varepsilon^j\, j!\, U^j} e^{-Uz/\varepsilon}, \quad \forall N \geq 2, \tag{27}$$

where

$$\begin{cases} a_0^N = -\gamma_0 v_\tau^N, \\ a_N^N = -\dfrac{\partial a_{N-1}^{N-1}}{\partial t} = (-1)^{N+1} \dfrac{\partial^N}{\partial t^N}(\gamma_0 v_\tau^0), \\ a_j^N = - \displaystyle\sum_{j-1 \leq i \leq N-2} \frac{1}{U^{2(i-j+1)}} \Big(\frac{\partial a_i^{N-1}}{\partial t} - \Delta_\tau a_i^{N-2}\Big) + \\ \qquad + \dfrac{(-1)^{N+1}}{U^{2(N-j)}} \dfrac{\partial^N}{\partial t^N}(\gamma_0 v_\tau^0), \quad \forall\, 1 \leq j \leq N-1. \end{cases} \tag{28}$$

Proof. By induction on N. For $N = 2$, (27) is verified thanks to Lemma 3.2 and the explicit form of the first order corrector $\overline{\theta}_\tau^{1,\varepsilon}$ which was obtained in (18).

The tangential component $\theta_\tau^{N+1,\varepsilon}$ satisfies the system (22) those first equation can be rewritten as follows:

$$\begin{aligned} -\varepsilon D_3^2 \theta_\tau^{N+1,\varepsilon} - U D_3 \theta_\tau^{N+1,\varepsilon} &= -\frac{1}{\varepsilon}\Big\{ \frac{\partial}{\partial t}(a_N^N) \frac{z^N}{\varepsilon^N\, N!\, U^N} \mathrm{e}^{-Uz/\varepsilon} + \\ &+ \sum_{j=0}^{N-1} \Big(\frac{\partial a_j^N}{\partial t} - \Delta_\tau a_j^{N-1}\Big) \frac{z^j}{\varepsilon^j\, j!\, U^j} \mathrm{e}^{-Uz/\varepsilon} \Big\}. \end{aligned} \tag{29}$$

Using Lemma 3.2, the resolution of this equation, taking into account the corresponding boundary condition as given by $(22)_3$, yields:

$$\overline{\theta}_\tau^{N+1,\varepsilon} = -\gamma_0 v_\tau^{N+1} \mathrm{e}^{-Uz/\varepsilon} + PS, \tag{30}$$

where PS denotes a particular solution of (29); based on Lemma 3.2, its expression reads:

$$PS = -\frac{\partial}{\partial t}(a_N^N)\sum_{j=0}^{N}\frac{z^{j+1}}{\varepsilon^{j+1}\,(j+1)!\,U^{2N+1-j}}\,\mathrm{e}^{-Uz/\varepsilon} - \\ -\sum_{j=0}^{N-1}(\frac{\partial a_j^N}{\partial t}+\Delta_\tau a_j^{N-1})\sum_{i=0}^{j}\frac{z^{i+1}}{\varepsilon^{i+1}\,(i+1)!\,U^{2j+1-i}}\,\mathrm{e}^{-Uz/\varepsilon}. \tag{31}$$

By identification of (30)-(31) and by the general form of the corrector $\overline{\theta}_\tau^{N+1,\varepsilon}$, *i.e.*,

$$\overline{\theta}_\tau^{N+1,\varepsilon} = \sum_{j=0}^{N+1} a_j^{N+1}\frac{z^j}{\varepsilon^j\,j!\,U^j}\mathrm{e}^{-Uz/\varepsilon},$$

we deduce that

$$\begin{cases} a_0^{N+1} = -\gamma_0 v_\tau^{N+1}, \\ a_{N+1}^{N+1} = -\dfrac{\partial a_N^N}{\partial t} = (-1)^{N+2}\dfrac{\partial^{N+1}}{\partial t^{N+1}}(\gamma_0 v_\tau^0). \end{cases}$$

Finally, after some simple transformations of (31) and by identification we obtain (28). This concludes the proof of Lemma 3.3. □

Thanks to the incompressibility condition and to Lemma 3.1, we have

Lemma 3.4. *An approximation of the normal component of the corrector* $\theta^{N,\varepsilon}$ *reads:*

$$\overline{\theta}_n^{N,\varepsilon} = \sum_{k=0}^{N} b_k^N\,\frac{z^k\,e^{-Uz/\varepsilon}}{\varepsilon^{k-1}\,k!\,U^k}, \tag{32}$$

where

$$b_k^N = \sum_{j=k}^{N}\frac{\mathrm{div}_\tau(a_j^N)}{U^{2(j-k)+1}}.$$

Remark 3.1. For all $N \in \mathbb{N}$ and $0 \le i \le N$, the quantities a_i^N, b_i^N are functions of t, x and y, but they are independent of z and ε. Thanks to the incompressibility condition $(22)_2$, the corrector $\overline{\theta}^{N,\varepsilon}$ still satisfies the same equation as $\theta_\tau^{N,\varepsilon}$ given by $(22)_1$. More precisely we have

$$\begin{cases} -\varepsilon D_3^2\overline{\theta}^{N,\varepsilon} - UD_3\overline{\theta}^{N,\varepsilon} = -\dfrac{1}{\varepsilon}\Big\{\dfrac{\partial\overline{\theta}^{N-1,\varepsilon}}{\partial t} - \Delta\overline{\theta}^{N-2,\varepsilon}\Big\}, \quad \text{on } (0,h), \\ \text{div } \overline{\theta}^{N,\varepsilon} = 0, \quad \text{in } \Omega, \\ \overline{\theta}_\tau^{N,\varepsilon} = -\gamma_0 v_\tau^N, \qquad \overline{\theta}_n^{N,\varepsilon} = \dfrac{\varepsilon}{U} b_0^N, \qquad \text{on } \Gamma_0, \\ \overline{\theta}_\tau^{N,\varepsilon} = O(\varepsilon^{-N}\mathrm{e}^{-Uh/\varepsilon}), \qquad \overline{\theta}_n^{N,\varepsilon} = O(\varepsilon^{-N+1}\mathrm{e}^{-Uh/\varepsilon}), \qquad \text{on } \Gamma_h. \end{cases} \tag{33}$$

Finally, to recover the desired boundary conditions, we introduce an additional corrector, $\varphi^{N,\varepsilon} = \varepsilon\varphi^N$, $\pi^{N,\varepsilon} = \varepsilon\pi^N$, defined by

$$\begin{cases} -\Delta\varphi^N + \nabla\pi^N = 0, \quad \text{in } \Omega, \\ \text{div } \varphi^N = 0, \quad \text{in } \Omega, \\ \varphi_\tau^N = 0, \quad \text{on } \Gamma_0 \cup \Gamma_h, \\ \varphi_n^N = -\dfrac{1}{\varepsilon}\gamma_0\overline{\theta}_n^{N,\varepsilon} = -b_0^N, \quad \text{on } \Gamma_0, \\ \varphi_n^N = 0, \quad \text{on } \Gamma_h. \end{cases} \tag{34}$$

Notice that φ^N is independent of ε.

We now define $\widetilde{v}^{N,\varepsilon} = v^N + \overline{\theta}^{N,\varepsilon} + \varphi^{N,\varepsilon}$ and $\widetilde{p}^N = p^N + \pi^{N,\varepsilon}$ which satisfy the following equations and boundary conditions:

$$\begin{cases} \dfrac{\partial\widetilde{v}^{N,\varepsilon}}{\partial t} - \varepsilon\Delta\widetilde{v}^{N,\varepsilon} - UD_3\widetilde{v}^{N,\varepsilon} + \nabla\widetilde{p}^N = \dfrac{\partial\overline{\theta}^{N,\varepsilon}}{\partial t} - \varepsilon\dfrac{\partial\varphi^N}{\partial t} - \varepsilon\Delta_\tau\overline{\theta}^{N,\varepsilon} - \dfrac{\partial\varphi^{N-1}}{\partial t} + \\ + UD_3\varphi^{N-1} + \Delta v^{N-1} - \dfrac{1}{\varepsilon}\Big(\dfrac{\partial\overline{\theta}^{N-1,\varepsilon}}{\partial t} - \Delta\overline{\theta}^{N-2,\varepsilon}\Big) - \varepsilon\Delta v^N - \varepsilon UD_3\varphi^N, \text{ in } \Omega_\infty, \\ \text{div } \widetilde{v}^{N,\varepsilon} = 0, \quad \text{in } \Omega_\infty, \\ \widetilde{v}^{N,\varepsilon} = 0, \quad \text{on } \Gamma_0, \\ \widetilde{v}_\tau^{N,\varepsilon} = O(\varepsilon^{-N}\mathrm{e}^{-Uh/\varepsilon}), \qquad \widetilde{v}_n^{N,\varepsilon} = O(\varepsilon^{-N+1}\mathrm{e}^{-Uh/\varepsilon}), \quad \text{on } \Gamma_h, \\ \widetilde{v}^{N,\varepsilon} \text{ is periodic in } x \text{ and } y. \end{cases} \tag{35}$$

4. Convergence Result

In this section, we will validate our choice for the correctors by proving a convergence result which concludes the study of the linear Navier-Stokes problem defined in Sec. 1.

We start by introducing the following expression:

$$\begin{cases} w^{N,\varepsilon} = v^\varepsilon - \displaystyle\sum_{k=0}^{N} \varepsilon^k (v^k + \overline{\theta}^{k,\varepsilon} + \varphi^{k,\varepsilon}), \\ p^\varepsilon = \displaystyle\sum_{k=0}^{N} \varepsilon^k (p^k + \pi^{k,\varepsilon}). \end{cases} \tag{36}$$

The correctors that we introduced contained exponentially small errors at the boundary $z = h$. To correct all these "errors", we now introduce an additional corrector which will account for all the *e.s.t.* at $z = h$. This corrector, denoted by $\widetilde{\varphi}^{N,\varepsilon}$, satisfies the following system of equations:

$$\begin{cases} -\varepsilon\, \Delta \widetilde{\varphi}^{N,\varepsilon} + \nabla \widetilde{\pi}^{N,\varepsilon} = 0, \quad \text{in } \Omega, \\ \operatorname{div} \widetilde{\varphi}^{N,\varepsilon} = 0, \quad \text{in } \Omega, \\ \widetilde{\varphi}^{N,\varepsilon}_\tau = 0, \quad \text{on } \Gamma_0, \\ \widetilde{\varphi}^{N,\varepsilon}_\tau = -\dfrac{1}{\varepsilon^N} \displaystyle\sum_{k=0}^{N} \varepsilon^k \gamma_h \overline{\theta}^{k,\varepsilon}_\tau = O(\varepsilon^{-2N}\, \mathrm{e}^{-Uh/\varepsilon}), \quad \text{on } \Gamma_h, \\ \widetilde{\varphi}^{N,\varepsilon}_n = 0, \quad \text{on } \Gamma_0, \\ \widetilde{\varphi}^{N,\varepsilon}_n = -\dfrac{1}{\varepsilon^N} \displaystyle\sum_{k=0}^{N} \varepsilon^k \gamma_h \overline{\theta}^{k,\varepsilon}_n = O(\varepsilon^{-2N+1}\, \mathrm{e}^{-Uh/\varepsilon}), \quad \text{on } \Gamma_h. \end{cases} \tag{37}$$

Note that by the classical results concerning the Stokes problem (see for instance[10]), all the Sobolev norms of $\widetilde{\varphi}^{N,\varepsilon}$ are exponentially small, namely $O(\varepsilon^{-2N}\, \mathrm{e}^{-Uh/\varepsilon})$.

We then define new quantities, which will allow us to prove the convergence result and give a complete asymptotic expansion of the linearized NS solution v^ε, namely

$$\widetilde{w}^{N,\varepsilon} = w^{N,\varepsilon} + \varepsilon^N \widetilde{\varphi}^{N,\varepsilon}, \tag{38}$$

$$\widetilde{p}^{\varepsilon} = p^\varepsilon + \varepsilon^N \widetilde{\pi}^{N,\varepsilon}.$$

Thanks to (3), (23), to our iterative construction for the correctors as described by the system (22) (with N replaced by k), (34) and (37), the equation verified by the pair $(\widetilde{w}^{N,\varepsilon}, \widetilde{p}^\varepsilon)$ reads:

$$\begin{aligned} \frac{\partial \widetilde{w}^{N,\varepsilon}}{\partial t} &- \varepsilon \Delta \widetilde{w}^{N,\varepsilon} - U D_3 \widetilde{w}^{N,\varepsilon} + \nabla(\widetilde{p}^\varepsilon - p^\varepsilon - \varepsilon^N \widetilde{p}^{N,\varepsilon}) = \\ &= -\varepsilon^N \left[\frac{\partial \overline{\theta}^{N,\varepsilon}}{\partial t} + \frac{\partial \varphi^{N,\varepsilon}}{\partial t} + \frac{\partial \widetilde{\varphi}^{N,\varepsilon}}{\partial t} \right] - \varepsilon^N \Delta_\tau \overline{\theta}^{N-1,\varepsilon} + \\ &+ \varepsilon^{N+1} \Delta v^N - \varepsilon^{N+1} \Delta_\tau \overline{\theta}^{N,\varepsilon} + \varepsilon^N U D_3 \varphi^{N,\varepsilon} + \varepsilon^N U D_3 \widetilde{\varphi}^{N,\varepsilon}, \quad \forall\, N \geq 1. \end{aligned} \tag{39}$$

Using equations $(3)_2$, $(22)_2$, $(23)_2$, $(34)_2$ and $(37)_2$, we deduce the incompressibility condition

$$\text{div}\ \widetilde{w}^{N,\varepsilon} = 0. \tag{40}$$

Moreover, our construction leads to the desired boundary conditions for $\widetilde{w}^{N,\varepsilon}$, namely the homogenous ones: from $(3)_3$, $(22)_3$, $(23)_{3,4}$, $(34)_{3-5}$ and $(37)_{3-6}$, we conclude that

$$\widetilde{w}^{N,\varepsilon} = 0, \quad \text{on } \Gamma_0 \cup \Gamma_h. \tag{41}$$

Also the periodicity condition is still satisfied for $\widetilde{w}^{N,\varepsilon}$:

$$\widetilde{w}^{N,\varepsilon} \text{ is periodic in the } x \text{ and } y \text{ directions with periods } L_1, L_2. \tag{42}$$

Finally, we state and prove the following theorem:

Theorem 4.1. *For each $N \geq 1$, the quantity $\widetilde{w}^{N,\varepsilon}$ defined by (38) tends to zero when $\varepsilon \to 0$. More precisely, we have the estimates:*

$$\|\widetilde{w}^{N,\varepsilon}\|_{L^\infty(0,T;\mathbf{L^2(\Omega)})} \leq C\ \varepsilon^{N+1}, \tag{43}$$

$$\|\widetilde{w}^{N,\varepsilon}\|_{L^2(0,T;\mathbf{H^1(\Omega)})} \leq C\ \varepsilon^{N+1/2}, \tag{44}$$

where $\mathbf{L^2(\Omega)} = (L^2(\Omega))^3, \mathbf{H^1(\Omega)} = (H^1(\Omega))^3$, and C denotes a constant which depends on the data (and N) but not on ε.

Since the $\widetilde{\varphi}^{N,\varepsilon}$ are exponentially small, the same results hold for $w^{N,\varepsilon}$.

Proof. We multiply (39) by $\widetilde{w}^{N,\varepsilon}$ and integrate over Ω. Thanks to (40), (41) and (42) we see that the last two terms in the left-hand side of the resulting equation vanish. Also all boundary terms vanish and we find:

$$\frac{1}{2}\frac{\mathrm{d}}{\mathrm{d}t}\|\widetilde{w}^{N,\varepsilon}\|^2_{\mathbf{L^2(\Omega)}} + \varepsilon\|\nabla\widetilde{w}^{N,\varepsilon}\|^2_{\mathbf{L^2(\Omega)}} = \varepsilon^{N+1} I_1^N + \varepsilon^N I_2^N + \varepsilon^N I_3^N, \tag{45}$$

where

$$\begin{cases} I_1^N = \displaystyle\int_\Omega \Big[- \frac{\partial \varphi^N}{\partial t} + \Delta v^N + \Delta_\tau \overline{\theta}^{N,\varepsilon} + U D_3 \varphi^N \Big] . \ \widetilde{w}^{N,\varepsilon} \mathrm{d}\Omega, \\ I_2^N = \displaystyle\int_\Omega \Big[- \frac{\partial \widetilde{\varphi}^{N,\varepsilon}}{\partial t} + U D_3 \widetilde{\varphi}^{N,\varepsilon} \Big] . \ \widetilde{w}^{N,\varepsilon} \mathrm{d}\Omega, \\ I_3^N = \displaystyle\int_\Omega \Big[- \frac{\partial \overline{\theta}^{N,\varepsilon}}{\partial t} + \Delta_\tau \overline{\theta}^{N-1,\varepsilon} \Big] . \ \widetilde{w}^{N,\varepsilon} \mathrm{d}\Omega. \end{cases}$$

Now, we proceed to estimate the terms I_1^N, I_2^N, I_3^N. Using the Cauchy-Schwarz inequality, we easily obtain that

$$|\varepsilon^{N+1} I_1^N| \leq C\,\varepsilon^{2N+2} + \frac{\varepsilon}{4}\|\nabla\widetilde{w}^{N,\varepsilon}\|^2_{\mathbf{L^2(\Omega)}}. \tag{46}$$

Thanks to the explicit expression of our corrector $\overline{\theta}^{N,\varepsilon}$, which is given by Lemma 3.3 and Lemma 3.4, we easily see that

$$\|\Delta_\tau \overline{\theta}^{N,\varepsilon}\|_{\mathbf{L}^2(\Omega)} \leq C\,\varepsilon^{1/2}.$$

Using equation (37) for $\widetilde{\varphi}^{N,\varepsilon}$ and the classical results on the Stokes problem, we deduce that

$$|\varepsilon^N I_2^N| \leq C\,\varepsilon^{-1}\,\mathrm{e}^{-Uh/\varepsilon} + \|\widetilde{w}^{N,\varepsilon}\|^2_{\mathbf{L}^2(\Omega)}. \tag{47}$$

We recall here that in order to obtain an estimate for the term $\partial\widetilde{\varphi}^{N,\varepsilon}/\partial t$, it suffices to differentiate the system (37) in time t, multiply by $\partial\widetilde{\varphi}^{N,\varepsilon}/\partial t$, and integrate over Ω.

Finally, we deal with the most difficult term I_3^N. For that purpose, we use again the explicit expression of the corrector $\overline{\theta}^{N,\varepsilon}$, for all $N \in \mathbb{N}$. Thus, we have

$$\Big\| z(-\frac{\partial\overline{\theta}^{N,\varepsilon}}{\partial t} + \Delta_\tau \overline{\theta}^{N-1,\varepsilon})\Big\|_{\mathbf{L}^2(\Omega)} \leq \varepsilon^{3/2}. \tag{48}$$

This yields

$$\begin{aligned}
|\,\varepsilon^N I_3^N\,| &\leq \varepsilon^N \Big\| z(-\frac{\partial\overline{\theta}^{N,\varepsilon}}{\partial t} + \Delta_\tau \overline{\theta}^{N-1,\varepsilon})\Big\|_{\mathbf{L}^2(\Omega)} \|\frac{\widetilde{w}^{N,\varepsilon}}{z}\|_{\mathbf{L}^2(\Omega)},\\
&\leq \text{(Thanks to the Hardy inequality)}\\
&\leq C\,\varepsilon^N \Big\| z(-\frac{\partial\overline{\theta}^{N,\varepsilon}}{\partial t} + \Delta_\tau \overline{\theta}^{N-1,\varepsilon})\Big\|_{\mathbf{L}^2(\Omega)} \|\nabla\widetilde{w}^{N,\varepsilon}\|_{\mathbf{L}^2(\Omega)},\\
&\leq C\,\varepsilon^{2N+2} + \frac{\varepsilon}{4}\,\|\nabla\widetilde{w}^{N,\varepsilon}\|^2_{\mathbf{L}^2(\Omega)}.
\end{aligned} \tag{49}$$

Combining (46), (47), and (49), we arrive at the following energy inequality

$$\frac{\mathrm{d}}{\mathrm{d}t}\|\widetilde{w}^{N,\varepsilon}\|^2_{\mathbf{L}^2(\Omega)} + \varepsilon\,\|\nabla\widetilde{w}^{N,\varepsilon}\|^2_{\mathbf{L}^2(\Omega)} \leq C\,\varepsilon^{2N+2} + 2\|\widetilde{w}^{N,\varepsilon}\|^2_{\mathbf{L}^2(\Omega)}. \tag{50}$$

Applying the Gronwall inequality to (50), we obtain (43) and (44). This concludes the proof of Theorem 4.1. □

5. The Nonlinear Case

This section is aimed at deriving a similar asymptotic expansion of the solutions of the full nonlinear Navier-Stokes equations; we believe that all orders can be handled in the same way, but because of the complexity of the calculations we restrict ourselves to order 1 thus recovering and improving the results of Ref. 11.

Proceeding as in the linear case, we consider $v^\varepsilon = u^\varepsilon - (0,0,-U)$ which is solution of the following system:

$$\begin{cases} \dfrac{\partial v^\varepsilon}{\partial t} - \varepsilon \Delta v^\varepsilon - U D_3 v^\varepsilon + (v^\varepsilon . \nabla)\, v^\varepsilon + \nabla p^\varepsilon = f, \quad \text{in } \Omega_\infty, \\ \text{div } v^\varepsilon = 0, \quad \text{in } \Omega_\infty, \\ v^\varepsilon = 0, \quad \text{on } \Gamma_\infty, \\ v^\varepsilon \ \text{ is periodic in the } x \text{ and } y \text{ directions with periods } L_1, L_2. \end{cases} \tag{51}$$

As in Ref. 11 our analysis shows that the leading terms remain the same as in the linear case and consequently the zeroth order corrector that we propose here is the same as in Sec. 2. Of course we should treat the nonlinear term appropriately. For that reason, and besides its importance, the derivation of a convergence result at order zero will be helpful for the choice of the corrector at order one.

5.1. *Corrector of order zero*

Let

$$w^{0,\varepsilon} = v^\varepsilon - v^0 - \overline{\theta}^{0,\varepsilon} - \varepsilon \varphi^0 - \widetilde{\varphi}^{0,\varepsilon}, \tag{52}$$

where, $\overline{\theta}^{0,\varepsilon}, \varphi^0, \widetilde{\varphi}^{0,\varepsilon}$ are respectively given by (12)-(14), (15), (37) with $N = 0$, and v^0 is now solution of the full nonlinear Euler equation

$$\begin{cases} \dfrac{\partial v^0}{\partial t} - U D_3 v^0 + (v^0 . \nabla)\, v^0 + \nabla p^0 = f, \quad \text{in } \Omega_\infty, \\ \text{div } v^0 = 0, \quad \text{in } \Omega_\infty, \\ v_3^0 = 0, \quad \text{on } \Gamma_0, \\ v^0 = 0, \quad \text{on } \Gamma_h, \\ v^0 \ \text{ is periodic in the } x \text{ and } y, \text{ directions with periods } L_1, L_2 \end{cases} \tag{53}$$

Remark 5.1. The existence of a smooth solution for the system (53) is proved in Ref. 8.

Thus, we deduce the following result:

Theorem 5.1. *There exists a time $T_* > 0$ such that we have*

$$\|v^\varepsilon - v^0 - \overline{\theta}^{0,\varepsilon}\|_{L^\infty(0,T_*;\mathbf{L}^2(\Omega))} \leq \kappa\ \varepsilon, \tag{54}$$

$$\|v^\varepsilon - v^0 - \overline{\theta}^{0,\varepsilon}\|_{L^2(0,T_*;\mathbf{H}^1(\Omega))} \leq \kappa\ \varepsilon^{1/2}, \tag{55}$$

where κ is a constant independent of ε.

Remark 5.2.

(1) The complete proof of Theorem 5.1 is similar to that of the corresponding theorem at order one (that is Theorem 5.2), so we skip it. However, it is helpful for the subsequent order to make explicit the form of the nonlinear term. Using (52) we see that

$$\begin{aligned}(v^\varepsilon.\nabla)\ v^\varepsilon = {} & (v^0.\nabla)\ v^0 + (v^\varepsilon.\nabla)\ w^{0,\varepsilon} + (w^{0,\varepsilon}.\nabla)\ (v^\varepsilon - w^{0,\varepsilon}) + \\ & + (\overline{\theta}^{0,\varepsilon}.\nabla)\ \overline{\theta}^{0,\varepsilon} + (v^0.\nabla)\ \overline{\theta}^{0,\varepsilon} + (\overline{\theta}^{0,\varepsilon}.\nabla)\ v^0 + \\ & + \varepsilon\ (\varphi^0.\nabla)\ \overline{\theta}^{0,\varepsilon} + \varepsilon\ (\overline{\theta}^{0,\varepsilon}.\nabla)\ \varphi^0 + \varepsilon\ (v^0.\nabla)\ \varphi^0 + \\ & + \varepsilon\ (\varphi^0.\nabla)\ v^0 + \varepsilon^2\ (\varphi^0.\nabla)\ \varphi^0 + (\overline{\theta}^{0,\varepsilon}.\nabla)\ \widetilde{\varphi}^{0,\varepsilon} + \\ & + (v^0.\nabla)\ \widetilde{\varphi}^{0,\varepsilon} + \varepsilon\ (\varphi^0.\nabla)\ \widetilde{\varphi}^{0,\varepsilon} + (\widetilde{\varphi}^{0,\varepsilon}.\nabla)\ \widetilde{\varphi}^{0,\varepsilon} + \\ & + (\widetilde{\varphi}^{0,\varepsilon}.\nabla)\ v^0 + (\widetilde{\varphi}^{0,\varepsilon}.\nabla)\ \overline{\theta}^{0,\varepsilon} + \varepsilon\ (\widetilde{\varphi}^{0,\varepsilon}.\nabla)\ \varphi^0.\end{aligned} \tag{56}$$

(2) In order to obtain (54)-(55), it is sufficient to obtain the same estimates for $w^{0,\varepsilon}$. Indeed, using the equations for φ^0 and $\widetilde{\varphi}^{0,\varepsilon}$ which are given respectively by (15) and (37) with $N = 0$, we have the following estimates:

$$\|\nabla\varphi^0\|_{\mathbf{L}^2(\Omega)} \leq C, \qquad \|\nabla\widetilde{\varphi}^{0,\varepsilon}\|_{\mathbf{L}^2(\Omega)} \leq C\ \varepsilon^{-2N-1/2}\mathrm{e}^{-Uh/\varepsilon}. \tag{57}$$

5.2. *Corrector of order one*

In this section, we are interested in the derivation of an asymptotic representation of v^ε up to order 1, that is to find an approximation of v^ε in Ω_∞ and up to the boundary as follows:

$$\begin{aligned} v^\varepsilon &\sim v^0 + \theta^{0,\varepsilon} + \varepsilon\varphi^0 + \varepsilon(v^1 + \theta^{1,\varepsilon}) + \widetilde{w}^{1,\varepsilon}, \\ p^\varepsilon &\sim p^0 + \varepsilon p^1. \end{aligned}$$

First, we infer from (56) the equations satisfied by the mode v^1 at order one. These are described by the following system:

$$\begin{cases} \dfrac{\partial v^1}{\partial t} - UD_3 v^1 + (v^1.\nabla)\, v^0 + (v^0.\nabla)\, v^1 + \nabla p^1 \\ \qquad = -\dfrac{\partial \varphi^0}{\partial t} + UD_3\varphi^0 - (v^0.\nabla)\,\varphi^0 - (\varphi^0.\nabla)\, v^0 + \Delta v^0, \quad \text{in } \Omega_\infty, \\ \text{div } v^1 = 0, \quad \text{in } \Omega_\infty, \\ v_3^1 = 0, \quad \text{on } \Gamma_0, \\ v^1 = 0, \quad \text{on } \Gamma_h, \\ v^1 \text{ is periodic in the } x \text{ and } y, \text{ directions with periods } L_1, L_2, \end{cases} \tag{58}$$

Remark 5.3. The existence,uniqueness and regularity of the solutions of system (58) is proved in the Appendix.

5.2.1. *Construction of the corrector $\theta^{1,\varepsilon}$*

We now present the equations that we propose for $\theta^{1,\varepsilon}$ by considering the next dominant terms in the nonlinear NS equation $(51)_1$ (of course without the source term and the pressure already taken into account). Hence, we obtain

$$\begin{cases} -\varepsilon D_3^2 \theta_\tau^{1,\varepsilon} - UD_3\theta_\tau^{1,\varepsilon} = -\dfrac{1}{\varepsilon}[\dfrac{\partial \overline{\theta}_\tau^{0,\varepsilon}}{\partial t} + (\overline{\theta}_\tau^{0,\varepsilon}.\nabla_\tau)\, v_\tau^0 + (v^0.\nabla)\overline{\theta}_\tau^{0,\varepsilon}] - \\ \qquad\qquad -\varphi_3^0 D_3 \overline{\theta}_\tau^{0,\varepsilon} - \dfrac{1}{\varepsilon}(\overline{\theta}^{0,\varepsilon}.\nabla)\, \overline{\theta}_\tau^{0,\varepsilon}, \quad \text{in } \Omega_\infty, \\ \text{div } \theta^{1,\varepsilon} = 0, \quad \text{in } \Omega_\infty, \\ \theta^{1,\varepsilon} = -v^1, \quad \text{on } \Gamma_0 \cup \Gamma_h. \end{cases} \tag{59}$$

Remark 5.4. Contrary to the linear case, it is important here to emphasize that equation $(59)_1$ is valid only for the tangential component of the corrector $\theta^{1,\varepsilon}$. The equation satisfied by the normal component $\theta_n^{1,\varepsilon}$ is different and it will be derived later on. It will be slightly different than for the tangential corrector equation since the nonlinear term is not divergence free.

We rewrite equation $(59)_1$ using the explicit form of $\overline{\theta}^{0,\varepsilon}$:

$$\begin{aligned}
&-\varepsilon D_3^2\theta_\tau^{1,\varepsilon} - UD_3\theta_\tau^{1,\varepsilon} \\
&\qquad = \frac{1}{\varepsilon}\,\frac{\partial(\gamma_0 v_\tau^0)}{\partial t}\,\mathrm{e}^{-Uz/\varepsilon} - \frac{1}{\varepsilon}\Big[- v_\tau^0.\nabla_\tau(\gamma_0 v_\tau^0) + U\,v_3^0\ \gamma_0 v_\tau^0 + \\
&\qquad + U\varphi_3^0(x,y,z)\gamma_0 v_\tau^0 - [(\gamma_0 v_\tau^0).\nabla_\tau]v_\tau^0\Big]\ \mathrm{e}^{-Uz/\varepsilon} - \\
&\qquad -\frac{1}{\varepsilon}\Big\{[(\gamma_0 v_\tau^0).\nabla_\tau](\gamma_0 v_\tau^0) + \mathrm{div}_\tau(\gamma_0 v_\tau^0)\gamma_0 v_\tau^0\Big\}\ \mathrm{e}^{-2Uz/\varepsilon}.
\end{aligned} \tag{60}$$

Thus, thanks to Lemma 3.2, we have an approximation, up to an *e.s.t.*, of the exact solution of (59)-(60) which is denoted, as in the linear case, by $\overline{\theta}_\tau^{1,\varepsilon}$ and given as follows:

$$\begin{aligned}
&\overline{\theta}_\tau^{1,\varepsilon} = -\gamma_0 v1_\tau\ \mathrm{e}^{-Uz/\varepsilon} + \frac{1}{\varepsilon U}\,\frac{\partial(\gamma_0 v0_\tau)}{\partial t}\,z\mathrm{e}^{-Uz/\varepsilon} - \\
&-\frac{1}{2U2}\Big\{[(\gamma_0 v0_\tau).\nabla_\tau](\gamma_0 v0_\tau) + \mathrm{div}_\tau(\gamma_0 v0_\tau)\gamma_0 v0_\tau\Big\}(\mathrm{e}^{-Uz/\varepsilon} - \mathrm{e}^{-2Uz/\varepsilon}) + \\
&+\frac{1}{\varepsilon 2}\,\mathrm{e}^{-Uz/\varepsilon}\int_0^z \mathrm{e}^{Us/\varepsilon}\Big(\int_h^s \mathrm{e}^{-U\zeta/\varepsilon}[-v0_\tau.\nabla_\tau(\gamma_0 v0_\tau) + U\,v0_3\ \gamma_0 v0_\tau + \\
&+U\varphi_3^0\gamma_0 v0_\tau - ((\gamma_0 v0_\tau).\nabla_\tau)v0_\tau](x,y,\zeta)\mathrm{d}\zeta\Big)\mathrm{d}s.
\end{aligned} \tag{61}$$

Remark 5.5. *1.* The approximation $\overline{\theta}_\tau^{1,\varepsilon}$ of $\theta_\tau^{1,\varepsilon}$ satisfies exactly the same equation (60) as $\theta_\tau^{1,\varepsilon}$.
2. It is easy to observe, but important to mention, that the boundary values of $\overline{\theta}_\tau^{1,\varepsilon}$ satisfy

$$\begin{cases} \overline{\theta}_\tau^{1,\varepsilon} = -\gamma_0 v_\tau^1, & \text{on } \Gamma_0, \\ |\gamma_h \overline{\theta}_\tau^{1,\varepsilon}| \le c\,\varepsilon^{-1}\,\mathrm{e}^{-Uh/\varepsilon}, & \text{on } \Gamma_h, \end{cases} \tag{62}$$

where $c > 0$ is constant depending on the data but independent of ε and z; (62) will be useful later on.

Using the incompressibility condition $(59)_2$ and Lemma 3.1, we deduce an explicit approximation $\overline{\theta}_n^{1,\varepsilon}$ of $\theta_n^{1,\varepsilon}$. Thus, we have

$$\begin{aligned}\overline{\theta}_n^{1,\varepsilon} =& -\frac{\varepsilon}{U}\,\mathrm{div}_\tau(\gamma_0 v_\tau^1)\ \mathrm{e}^{-Uz/\varepsilon} - \frac{\varepsilon}{2U^3}\,\mathrm{div}_\tau[((\gamma_0 v_\tau^0).\nabla_\tau)(\gamma_0 v_\tau^0)\\ &+ +\mathrm{div}_\tau(\gamma_0 v_\tau^0)\gamma_0 v_\tau^0]\ (\mathrm{e}^{-Uz/\varepsilon} - \mathrm{e}^{-2Uz/\varepsilon})\\ &+ \frac{1}{U^2}(\frac{\varepsilon}{U}+z)\ \mathrm{div}_\tau[\frac{\partial(\gamma_0 v_\tau^0)}{\partial t}]\ \mathrm{e}^{-Uz/\varepsilon}\\ &+ \frac{1}{\varepsilon^2}\int_z^h \mathrm{e}^{-Us/\varepsilon}\Big(\int_0^s \mathrm{e}^{U\xi/\varepsilon}\Big(\int_h^\xi \mathrm{e}^{-U\zeta/\varepsilon}\mathrm{div}_\tau[-v_\tau^0.\nabla_\tau(\gamma_0 v_\tau^0)\\ &+ U\,v_3^0\ \gamma_0 v_\tau^0 + U\varphi_3^0\gamma_0 v_\tau^0 - ((\gamma_0 v_\tau^0).\nabla_\tau)v_\tau^0]\mathrm{d}\zeta\Big)\mathrm{d}\xi\Big)\mathrm{d}s.\end{aligned} \tag{63}$$

Therefore, the equations and boundary conditions for $\overline{\theta}_n^{1,\varepsilon}$ read as follows:

$$\left\{\begin{aligned}&-\varepsilon D_3^2\overline{\theta}_n^{1,\varepsilon} - U D_3\overline{\theta}_n^{1,\varepsilon}\\ &\quad= -\frac{1}{2U}\mathrm{div}_\tau[((\gamma_0 v_\tau^0).\nabla_\tau)(\gamma_0 v_\tau^0) + \mathrm{div}_\tau(\gamma_0 v_\tau^0)\gamma_0 v_\tau^0]\ \mathrm{e}^{-2Uz/\varepsilon} +\\ &\quad+\frac{1}{\varepsilon}\int_h^z \mathrm{e}^{-U\zeta/\varepsilon}\mathrm{div}_\tau\Big[[U\varphi_3^0\gamma_0 v_\tau^0 - ((\gamma_0 v_\tau^0).\nabla_\tau)v_\tau^0](x,y,\zeta)\Big]\mathrm{d}\zeta -\\ &\quad-\frac{1}{\varepsilon}\frac{\partial\overline{\theta}_n^{0,\varepsilon}}{\partial t} - \frac{1}{\varepsilon}v_n^0 D_3\overline{\theta}^{0,\varepsilon},\\ &\mid \gamma_0\overline{\theta}_n^{1,\varepsilon}\mid \le c\,\varepsilon, \quad \text{on } \Gamma_0,\\ &\overline{\theta}_n^{1,\varepsilon} = O(\varepsilon\ \mathrm{e}^{-Uh/\varepsilon}), \quad \text{on } \Gamma_h.\end{aligned}\right. \tag{64}$$

Due to our choice of a simpler form for the corrector $\overline{\theta}^{1,\varepsilon}$, we observe that, in $(62)_2$ and $(64)_{2,3}$, we need to introduce an additional corrector in order to obtain the desired homogenous boundary conditions. For that purpose, we define $(\widetilde{\varphi}^{1,\varepsilon}, \widetilde{\pi}^{1,\varepsilon})$ as the solutions of the following system:

$$\left\{\begin{aligned}&-\varepsilon\,\Delta\widetilde{\varphi}^{1,\varepsilon} + \nabla\widetilde{\pi}^{1,\varepsilon} = 0, \quad \text{in } \Omega,\\ &\mathrm{div}\ \widetilde{\varphi}^{1,\varepsilon} = 0, \quad \text{in } \Omega,\\ &\widetilde{\varphi}_\tau^{1,\varepsilon} = 0, \quad \text{on } \Gamma_0,\\ &\widetilde{\varphi}_\tau^{1,\varepsilon} = -\varepsilon^{-1}\ \gamma_h\overline{\theta}_\tau^{0,\varepsilon} - \gamma_h\overline{\theta}_\tau^{1,\varepsilon} = O(\varepsilon^{-1}\,\mathrm{e}^{-Uh/\varepsilon}), \quad \text{on } \Gamma_h,\\ &\widetilde{\varphi}_n^{1,\varepsilon} = -\gamma_0\overline{\theta}_n^{1,\varepsilon} = O(\varepsilon), \quad \text{on } \Gamma_0,\\ &\widetilde{\varphi}_n^{1,\varepsilon} = -\varepsilon^{-1}\ \gamma_h\overline{\theta}_n^{0,\varepsilon} - \gamma_h\overline{\theta}_n^{1,\varepsilon} = O(\mathrm{e}^{-Uh/\varepsilon}), \quad \text{on } \Gamma_h.\end{aligned}\right. \tag{65}$$

By the properties of the Stokes operators, all the Sobolev norms of $\widetilde{\varphi}^{1,\varepsilon}$ are of order ε.

5.2.2. *Convergence result: the main theorem*

Finally, we state and prove the main result of this article which gives the asymptotic expansion at order one of v^ε:

Theorem 5.2. *For v^ε solution of the Navier-Stokes problem (51), there exist a time $T_* > 0$ and a constant $\kappa > 0$ depending on the data but not on ε, such that:*

$$\|v^\varepsilon - (v^0 + \overline{\theta}^{0,\varepsilon} + \varepsilon\varphi^0) - \varepsilon(v^1 + \overline{\theta}^{1,\varepsilon} + \widetilde{\varphi}^{1,\varepsilon})\|_{L^\infty(0,T_*;\mathbf{L^2(\Omega)})} \leq \kappa\, \varepsilon^2, \quad (66)$$

$$\|v^\varepsilon - (v^0 + \overline{\theta}^{0,\varepsilon} + \varepsilon\varphi^0) - \varepsilon(v^1 + \overline{\theta}^{1,\varepsilon} + \widetilde{\varphi}^{1,\varepsilon})\|_{L^2(0,T_*;\mathbf{H^1(\Omega)})} \leq \kappa\, \varepsilon^{3/2}. \quad (67)$$

Proof. Let

$$w^{1,\varepsilon} = v^\varepsilon - (v^0 + \overline{\theta}^{0,\varepsilon} + \varepsilon\varphi^0) - \varepsilon(v^1 + \overline{\theta}^{1,\varepsilon} + \widetilde{\varphi}^{1,\varepsilon})$$

Before writing the equations satisfied by $w^{1,\varepsilon}$, we start by expanding the nonlinear term:

$$\begin{aligned}
(v^\varepsilon.\nabla)\, v^\varepsilon &= (v^\varepsilon.\nabla)\, w^{1,\varepsilon} + (w^{1,\varepsilon}.\nabla)\, (v^\varepsilon - w^{1,\varepsilon}) + (v^0.\nabla)\, v^0 + \\
&+ \varepsilon \left[(v^0.\nabla)\, \varphi^0 + (\varphi^0.\nabla)\, v^0 + (v^0.\nabla)\, v^1 + (v^1.\nabla)\, v^0\right] \\
&+ (\overline{\theta}^{0,\varepsilon}.\nabla)\, \overline{\theta}^{0,\varepsilon} + \varepsilon\, (\varphi^0.\nabla)\, \overline{\theta}^{0,\varepsilon} + (\overline{\theta}^{0,\varepsilon}.\nabla)\, v^0 + (v^0.\nabla)\, \overline{\theta}^{0,\varepsilon} + \\
&+ \varepsilon\, (\overline{\theta}^{0,\varepsilon}.\nabla)\, (\varphi^0 + v^1) + \varepsilon\, (\overline{\theta}^{1,\varepsilon}.\nabla)\, (v^0 + \varepsilon\, v^1) + \\
&+ \varepsilon\, [(v^0 + \overline{\theta}^{0,\varepsilon} + \varepsilon\varphi^0 + \varepsilon\, v^1 + \varepsilon\overline{\theta}^{1,\varepsilon} + \varepsilon\widetilde{\varphi}^{1,\varepsilon}).\nabla]\, \widetilde{\varphi}^{1,\varepsilon} + \qquad (68)\\
&+ \varepsilon\, [(v^0 + \overline{\theta}^{0,\varepsilon} + \varepsilon\varphi^0 + \varepsilon\, v^1 + \varepsilon\overline{\theta}^{1,\varepsilon} + \varepsilon\widetilde{\varphi}^{1,\varepsilon}).\nabla]\, \overline{\theta}^{1,\varepsilon} + \\
&+ \varepsilon\, ((\widetilde{\varphi}^{1,\varepsilon} + v^1 + \overline{\theta}^{1,\varepsilon}).\nabla)\, \overline{\theta}^{0,\varepsilon} + \varepsilon\, (\overline{\theta}^{1,\varepsilon}.\nabla)\, \varphi^0 + \\
&+ \varepsilon^2 \left[(\varphi^0.\nabla)\, \varphi^0 + (\varphi^0.\nabla)\, v^1 + (v^1.\nabla)\, \varphi^0 + (v^1.\nabla)\, v^1\right] + \\
&+ \varepsilon\, (\widetilde{\varphi}^{1,\varepsilon}.\nabla)\, (v^0 + \varepsilon\, v^1 + \varepsilon\, \varphi^0).
\end{aligned}$$

Thanks to $(51)_1$, $(53)_1$, (9), $(15)_1$,$(58)_1$, (60)-$(64)_1$ and $(65)_1$, we deduce the main equation verified by $w^{1,\varepsilon}$ for which the boundary conditions are derived from $(51)_{3,4}$, $(53)_{3,4}$, $(15)_{3-5}$,$(58)_{3,4}$, (62)-$(64)_{2,3}$ and $(65)_{3-6}$. Notice that both the incompressibility condition and the periodicity property are conserved for $w^{1,\varepsilon}$. Hence, we arrive at the following system for $(w^{1,\varepsilon}, p^\varepsilon)$:

$$\begin{cases} \dfrac{\partial w^{1,\varepsilon}}{\partial t} - \varepsilon \Delta w^{1,\varepsilon} - U D_3 w^{1,\varepsilon} + \nabla[p^\varepsilon - (p^0 + \pi^0) - \varepsilon(p^1 + \widetilde{\pi}^{1,\varepsilon})] + \\ \qquad +(v^\varepsilon.\nabla)\, w^{1,\varepsilon} + (w^{1,\varepsilon}.\nabla)\,(v^\varepsilon - w^{1,\varepsilon}) = -\displaystyle\sum_{j=0}^{2} \varepsilon^j J^{j,\varepsilon}, \\ \operatorname{div} w^{1,\varepsilon} = 0, \\ w^{1,\varepsilon} = 0, \quad \text{on } \Gamma_0 \cup \Gamma_h, \\ w^{1,\varepsilon} \text{ is periodic in the } x \text{ and } y \text{ directions with periods } L_1, L_2, \end{cases} \tag{69}$$

where,

$$J^{0,\varepsilon} = \underline{(\overline{\theta}^{0,\varepsilon}.\nabla)\, v_n^0} + \varepsilon\, v_n^0 D_3 \overline{\theta}^{1,\varepsilon} + \varepsilon\, v_n^1 D_3 \overline{\theta}^{0,\varepsilon} + \varepsilon\, \overline{\theta}_n^{1,\varepsilon} D_3 \overline{\theta}^{0,\varepsilon}, \tag{70}$$

$$\begin{aligned} J^{1,\varepsilon} =& \Delta_\tau \overline{\theta}^{0,\varepsilon} - \frac{\partial \overline{\theta}^{1,\varepsilon}}{\partial t} + \frac{1}{\varepsilon}[\underline{(\overline{\theta}^{0,\varepsilon}.\nabla)} + v_n^0 D_3]\, \overline{\theta}_n^{0,\varepsilon} + (\varphi_\tau^0.\nabla_\tau)\, \overline{\theta}_\tau^{0,\varepsilon} + (\varphi^0.\nabla)\, \overline{\theta}_n^{0,\varepsilon} \\ &+ (\overline{\theta}^{0,\varepsilon}.\nabla)\,(\varphi^0 + v^1) + (\overline{\theta}^{1,\varepsilon}.\nabla)\,(v^0 + \varepsilon\, v^1) + \overline{\theta}_n^{0,\varepsilon} D_3 \overline{\theta}^{1,\varepsilon} + \overline{\theta}_n^{0,\varepsilon} D_3 v_\tau^0 \\ &+ [(v^0 + \overline{\theta}^{0,\varepsilon} + \varepsilon\varphi^0 + \varepsilon\, v^1 + \varepsilon\widetilde{\varphi}^{1,\varepsilon})_\tau.\nabla_\tau]\, \overline{\theta}^{1,\varepsilon} - \underline{(\overline{\theta}^{1,\varepsilon}.\nabla)\, \overline{\theta}^{1,\varepsilon}} \\ &+ \varepsilon\, (\varphi_n^0 + v_n^1) D_3 \overline{\theta}^{1,\varepsilon} + \underline{(\widetilde{\varphi}^{1,\varepsilon}.\nabla)\, \overline{\theta}^{0,\varepsilon}} + (v_\tau^1.\nabla_\tau)\overline{\theta}^{0,\varepsilon} + (\overline{\theta}^{1,\varepsilon}.\nabla)\, \varphi^0 \\ &+ (\overline{\theta}_\tau^{1,\varepsilon}.\nabla_\tau)\, \overline{\theta}^{0,\varepsilon} - \frac{1}{2U} \operatorname{div}_\tau[((\gamma_0 v_\tau^0).\nabla_\tau)(\gamma_0 v_\tau^0) + \operatorname{div}_\tau(\gamma_0 v_\tau^0)\gamma_0 v_\tau^0]\, \mathrm{e}^{-2Uz/\varepsilon} \\ &+ \frac{U}{\varepsilon} \int_h^z \mathrm{e}^{-U\zeta/\varepsilon} \operatorname{div}_\tau \Big[[\varphi_3^0 \gamma_0 v_\tau^0 - ((\gamma_0 v_\tau^0).\nabla_\tau) v_\tau^0](x, y, \zeta) \Big] \mathrm{d}\zeta, \end{aligned} \tag{71}$$

and

$$\begin{aligned} J^{2,\varepsilon} =& \frac{1}{\varepsilon}\Big(\frac{\partial \widetilde{\varphi}^{1,\varepsilon}}{\partial t} + U D_3 \widetilde{\varphi}^{1,\varepsilon}\Big) + \Delta v^1 + \Delta_\tau \overline{\theta}^{1,\varepsilon} + \frac{1}{\varepsilon}(\widetilde{\varphi}^{1,\varepsilon}.\nabla)\,(v^0 + \varepsilon\, v^1 + \varepsilon\varphi^0) \\ &+ [(v^0 + \overline{\theta}^{0,\varepsilon} + \varepsilon\varphi^0 + \varepsilon\, v^1 + \varepsilon\overline{\theta}^{1,\varepsilon} + \varepsilon\widetilde{\varphi}^{1,\varepsilon}).\nabla]\, \widetilde{\varphi}^{1,\varepsilon} + \varepsilon^2 \widetilde{\varphi}^{1,\varepsilon} D_3 \overline{\theta}^{1,\varepsilon} \\ &+ \Big[(\varphi^0.\nabla)\, \varphi^0 + (\varphi^0.\nabla)\, v^1 + (v^1.\nabla)\, \varphi^0 + (v^1.\nabla)\, v^1\Big]. \end{aligned} \tag{72}$$

Remark 5.6. The underlined terms in $J^{0,\varepsilon}$ and $J^{1,\varepsilon}$ need to be split as follows before estimating them later on:

$$(u.\nabla)\, v = (u_\tau.\nabla_\tau)\, v + u_3\, D_3 v.$$

Now, multiplying equation $(69)_1$ by $w^{1,\varepsilon}$ and integrating over Ω, we obtain

$$\frac{1}{2}\frac{\mathrm{d}}{\mathrm{d}t}\|w^{1,\varepsilon}\|^2_{\mathbf{L}^2(\Omega)} + \varepsilon\|\nabla w^{1,\varepsilon}\|^2_{\mathbf{L}^2(\Omega)} + \sum_{j=3}^{6} \mathcal{J}^{j,\varepsilon} = -\sum_{j=0}^{2} \varepsilon^j \mathcal{J}^{j,\varepsilon}, \tag{73}$$

where,

$$\mathcal{J}^{j,\,\varepsilon} = \int_\Omega J^{j,\,\varepsilon} \,.\, w^{1,\,\varepsilon} \mathrm{d}\Omega, \quad \forall j \in \{0,1,2\}.$$

Thanks to the incompressibility, boundary and periodicity conditions ($(69)_{2,3,4}$ and $(51)_{2,3,4}$), we deduce that the following terms vanish:

$$\begin{cases} \mathcal{J}^{3,\,\varepsilon} = -\displaystyle\int_\Omega U D_3 w^{1,\,\varepsilon} \,.\, w^{1,\,\varepsilon} \mathrm{d}\Omega = 0, \\ \mathcal{J}^{4,\,\varepsilon} = \displaystyle\int_\Omega \nabla[p^\varepsilon - (p^0 + \pi^0) - \varepsilon(p^1 + \widetilde{\pi}^{1,\,\varepsilon})] \,.\, w^{1,\,\varepsilon} \mathrm{d}\Omega = 0, \\ \mathcal{J}^{5,\,\varepsilon} = \displaystyle\int_\Omega [(v^\varepsilon.\nabla)\; w^{1,\,\varepsilon} \mathrm{d}\Omega = 0. \end{cases} \tag{74}$$

Thanks to the Hardy inequality, we have

$$\begin{aligned} |\mathcal{J}^{6,\,\varepsilon}| &= \Big| \int_\Omega (w^{1,\varepsilon}.\nabla)\; (v^\varepsilon - w^{1,\,\varepsilon}) \,.\, w^{1,\,\varepsilon} \mathrm{d}\Omega \Big| \\ &\leq \Big| \int_\Omega \Big[(w_\tau^{1,\varepsilon}.\nabla_\tau)\; (v^\varepsilon - w^{1,\,\varepsilon}).w^{1,\,\varepsilon} + w_n^{1,\varepsilon} D_3(v^\varepsilon - w^{1,\,\varepsilon} - \overline{\theta}^{0,\varepsilon}).w^{1,\,\varepsilon} \Big] \mathrm{d}\Omega \Big| + \\ &+ \Big| \int_\Omega w_n^{1,\varepsilon} D_3 \overline{\theta}_n^{0,\varepsilon} w^{1,\,\varepsilon} \mathrm{d}\Omega \Big| + \Big| \int_\Omega w_n^{1,\varepsilon} D_3 \overline{\theta}_\tau^{0,\varepsilon} w^{1,\,\varepsilon} \mathrm{d}\Omega \Big| \\ &\leq \| \nabla_\tau (v^\varepsilon - w^{1,\,\varepsilon}) + D_3(v^\varepsilon - w^{1,\,\varepsilon} - \overline{\theta}^{0,\varepsilon}) + D_3 \overline{\theta}_n^{0,\varepsilon} \|_{\mathbf{L}^\infty(\Omega)} \, \| w^{1,\varepsilon} \|^2_{\mathbf{L}^2(\Omega)} + \\ &+ \| z^2 D_3 \overline{\theta}_\tau^{0,\varepsilon} \|_{\mathbf{L}^\infty(\Omega)} \, \Big\| \frac{w^{1,\varepsilon}}{z} \Big\|^2_{\mathbf{L}^2(\Omega)} \\ &\leq c \, \| w^{1,\varepsilon} \|^2_{\mathbf{L}^2(\Omega)} + \frac{U}{\varepsilon} \| z^2 \, \mathrm{e}^{-Uz/\varepsilon} \|_{L^\infty(0,h)} \| \gamma_0 v_\tau^0 \|_{L^\infty_{x,y}} \| \nabla w^{1,\varepsilon} \|^2_{\mathbf{L}^2(\Omega)}. \end{aligned} \tag{75}$$

Finally, for the term $\mathcal{J}^{6,\,\varepsilon}$ we impose the following condition:

$$\frac{U}{\varepsilon} \| z^2 \, \mathrm{e}^{-Uz/\varepsilon} \|_{L^\infty(0,h)} \| \gamma_0 v_\tau^0 \|_{L^\infty_{x,y}} \leq \frac{4\varepsilon}{e^2 U} \| \gamma_0 v_\tau^0 \|_{L^\infty_{x,y}} \leq \frac{\varepsilon}{6}. \tag{76}$$

Since $\gamma_0 v_\tau^0$ vanishes at $t = 0$, $\| \gamma_0 v_\tau^0 \|_{L^\infty_{x,y}}$ remains initially small and therefore there exists a time $T_* > 0$ such that

$$\| \gamma_0 v_\tau^0 \|_{L^\infty_{x,y}} \leq \frac{Ue^2}{24}, \quad \forall\, t \in [0, T_*]. \tag{77}$$

Consequently, we obtain

$$|\mathcal{J}^{6,\,\varepsilon}| \leq c \, \| w^{1,\varepsilon} \|^2_{\mathbf{L}^2(\Omega)} + \frac{\varepsilon}{6} \| \nabla w^{1,\varepsilon} \|^2_{\mathbf{L}^2(\Omega)}. \tag{78}$$

We now estimate the right hand side of (73). We start by the easier terms included in $\mathcal{J}^{2,\,\varepsilon}$. First, by the properties of the solutions of the Stokes problem, the solutions of (65) satisfy:

$$\|\frac{\partial \widetilde{\varphi}^{1,\varepsilon}}{\partial t}\|_{\mathbf{L^2(\Omega)}} \leq c\,\varepsilon, \qquad \|\nabla \widetilde{\varphi}^{1,\varepsilon}\|_{\mathbf{L^2(\Omega)}} \leq c\,\varepsilon. \tag{79}$$

The same estimates are also valid for the tangential derivatives of $\widetilde{\varphi}^{1,\varepsilon}$. Hence, thanks to the Cauchy-Schwarz and Hölder inequalities, we have:

$$|\varepsilon^2\, \mathcal{J}^{2,\varepsilon}| \leq \kappa\varepsilon^4 + c\,\|w^{1,\varepsilon}\|^2_{\mathbf{L^2(\Omega)}}. \tag{80}$$

For $\mathcal{J}^{1,\varepsilon}$, we use the Hardy inequality * combined with the Cauchy-Schwarz and Hölder inequalities and we conclude that:

$$\begin{aligned}|\varepsilon\, \mathcal{J}^{1,\varepsilon}| &\leq \varepsilon\, \|z\, J^{1,\varepsilon}\| \left\|\frac{w^{1,\varepsilon}}{z}\right\|_{\mathbf{L^2(\Omega)}} \\ &\leq c\,\varepsilon^{5/2}\, \|\nabla w^{1,\varepsilon}\|_{\mathbf{L^2(\Omega)}} \\ &\leq c\,\varepsilon^4 + \frac{\varepsilon}{6}\|\nabla w^{1,\varepsilon}\|^2_{\mathbf{L^2(\Omega)}}.\end{aligned} \tag{81}$$

Finally, since $v_n^0 = v_n^1 = 0$ on $z = 0$, we have:

$$\begin{aligned}&|\,\varepsilon\, \mathcal{J}^{0,\varepsilon}| \leq \varepsilon\, \|z\, J^{0,\varepsilon}\| \left\|\frac{w^{1,\varepsilon}}{z}\right\|_{\mathbf{L^2(\Omega)}} \\ &\leq c\, \left\|z^2(\overline{\theta}_\tau^{0,\varepsilon} + \varepsilon\, D_3\overline{\theta}^{1,\varepsilon} + \varepsilon\, D_3\overline{\theta}^{0,\varepsilon})\right\|_{\mathbf{L^2(\Omega)}} \left\|\frac{\nabla_\tau v_n^0 + v_n^0 + v_n^1}{z}\right\|_{\mathbf{L^\infty(\Omega)}} + \\ &+\|z\overline{\theta}_n^{0,\varepsilon}\|_{\mathbf{L^2(\Omega)}}\|D_3 v_n^0\|_{\mathbf{L^\infty}} + \|\overline{\theta}_n^{1,\varepsilon}\|_{\mathbf{L^\infty}}\|\varepsilon\, D_3\overline{\theta}^{0,\varepsilon}\|_{\mathbf{L^2(\Omega)}}\Big\}\left\|\frac{w^{1,\varepsilon}}{z}\right\|_{\mathbf{L^2(\Omega)}}\end{aligned}$$

(thanks to the explicit expression of $\overline{\theta}_n^{1,\varepsilon}$ given by (61) – (63))

$$\leq c\,\varepsilon^{5/2}\, \|\nabla w^{1,\varepsilon}\|_{\mathbf{L^2(\Omega)}} \leq c\,\varepsilon^4 + \frac{\varepsilon}{6}\|\nabla w^{1,\varepsilon}\|^2_{\mathbf{L^2(\Omega)}}. \tag{82}$$

Hence, inserting all the results (74), (78), (80), (81) and (82) in (73), we deduce the following new energy estimate:

$$\frac{\mathrm{d}}{\mathrm{d}t}\|w^{1,\varepsilon}\|^2_{\mathbf{L^2(\Omega)}} + \varepsilon\|\nabla w^{1,\varepsilon}\|^2_{\mathbf{L^2(\Omega)}} \leq \ \leq \kappa\varepsilon^4 + c\,\|w^{1,\varepsilon}\|^2_{\mathbf{L^2(\Omega)}}, \tag{83}$$

to which we apply the Gronwall inequality to conclude the proof of Theorem 5.2. □

*Hardy's inequality states that $\|\frac{w}{z}\|_{\mathbf{L^2(\Omega)}} \leq 2\|\nabla w\|_{\mathbf{L^2(\Omega)}}$ provided $w(z=0)=0$.

Appendix

In this paragraph, we prove the existence and regularity of the solutions of the linearized Euler systems (4), (16) and (23). For the full nonlinear Euler system (53), we refer the reader to Ref. 8.
We first want to apply the Hille-Yosida theorem to prove the existence and uniqueness of the solution of (4). Thus, we start by introducing the adequate function spaces

$$H = \Big\{ v \in \mathbf{L^2}(\boldsymbol{\Omega});\ \text{div}\ v = 0,\ v_1|_{x=0} = v_1|_{x=L_1},$$

$$v_2|_{x=0} = v_2|_{x=L_2},\ \text{and}\ v_3 = 0\ \text{ en }\ z = 0, h \Big\},$$

$$D(A) = \Big\{ v \in H;\ z\frac{\partial v}{\partial z} \in \mathbf{L^2}(\boldsymbol{\Omega}),\ v = 0\ \text{ at }\ z = h\ \text{and}$$

$$\exists p \in \mathcal{D}'(\Omega)\ \text{ such that }\ -U\frac{\partial v}{\partial z} + \nabla p \in H \Big\}.$$

Then for $v \in D(A)$, we set $Av = -UD_3 v + \nabla p$, thus defining an unbounded linear operator $A : D(A) \subset H \to H$.

Remark 0.1. If $\int_\Omega p\ \mathrm{d}\Omega = 0$, then p is unique, otherwise it is unique up to an additive constant.

Let us now recall the statement of the Hille-Yosida theorem (for more details see for example Refs. 1,2,5 and 13):

Theorem 0.3. ***(Hille-Yosida Theorem).*** *Let H be a Hilbert space and let $B : D(B) \to H$ be a linear unbounded operator, with domain $D(B) \subset H$. Assume the following:*

(i) $D(B)$ is dense in H and B is closed,
(ii) B is a positive operator,
(iii) $\exists\, \mu_0 > 0$, such that $B + \mu_0 I$ is onto.

Then $(-B)$ is the infinitesimal generator of a semigroup of contractions $\{S(t)\}_{t\geq 0}$ in H. More precisely, the solution v of the system:

$$\begin{cases} \dfrac{\mathrm{d}v}{\mathrm{d}t} + Bv = f, \\ v_{|t=0} = v_0. \end{cases}$$

satisfies the following properties:
(H_0): If $v_0 \in H$ and $f \in L^1(0,T;H)$ then $v \in \mathcal{C}\ ([0,T];H)$, $\forall\ T > 0$,

and,
(H_1): *If* $v_0 \in D(B)$ *and* $f^{'} \in L^1(0,T;H)$ *then* $v \in \mathcal{C}([0,T];H) \cap L^\infty([0,T];D(B))$, *and* $\frac{\mathrm{d}v}{\mathrm{d}t} \in L^\infty([0,T];H), \forall\, T > 0$.

In order to apply the Hille-Yosida theorem to the system (4) with $B = A$, it is sufficient to prove the following proposition:

Proposition 0.1. *The operator* A *verifies*

(i) $A + \lambda I$ *is onto* $\forall\, \lambda > 0$,
(ii) $(A + \lambda I)^{-1} \in \mathcal{B}(H, D(A))$, $\forall\, \lambda \neq \{U|k|, k \in \mathbb{Z}^2\}$,
(iii) A *is closed.*

Proof. The proof of the property (i) is equivalent to finding a solution for the following system:

$$\begin{cases} -UD_3 u + \nabla p + \lambda u = f, \quad \text{in } \Omega_\infty, \\ \text{div } u = 0, \quad \text{in } \Omega_\infty, \\ u_3 = 0, \quad \text{on } \Gamma_0, \\ u = 0, \quad \text{on } \Gamma_h, \\ u \text{ is periodic in the } x \text{ and } y, \text{ directions with periods } L_1, L_2, \end{cases} \tag{84}$$

where f is a given function of H.
Now, we decompose the solution of (84) in the Fourier basis as follows:

$$u = \sum_{k=(k_1,k_2)\in\mathbb{Z}^2} u_k(z)\, \mathrm{e}^{k_1 x + k_2 y},$$
$$p = \sum_{k=(k_1,k_2)\in\mathbb{Z}^2} p_k(z)\, \mathrm{e}^{k_1 x + k_2 y},$$

and let

$$f = \sum_{k=(k_1,k_2)\in\mathbb{Z}^2} f_k(z)\, \mathrm{e}^{k_1 x + k_2 y}.$$

Hence, the system (84) is equivalent to the following system of equations:

$$\begin{cases} -UD_3 u_{1k} + ik_1 p_k + \lambda u_{1k} = f_{1k}, \\ -UD_3 u_{2k} + ik_2 p_k + \lambda u_{2k} = f_{2k}, \\ -UD_3 u_{3k} + p_k^{'} + \lambda u_{3k} = f_{3k}, \\ ik_1 u_{1k} + ik_2 u_{2k} + u_{3k}^{'} = 0, \\ u_{1k} = u_{2k} = 0, \quad \text{at } z = h, \\ u_{3k} = 0, \quad \text{at } z = 0, h. \end{cases} \tag{85}$$

We multiply $(85)_1$ by ik_1, $(85)_2$ by ik_2 and we differentiate $(85)_3$ with respect to z, then we sum up the equations that we obtained and find:

$$-|k|^2 p_k + p_k'' = \mathrm{div} f = 0.$$

Since $p_k'(h) = 0$ (from $(85)_3$ and $(85)_6$) we have

$$p_k(z) = \alpha_k[\mathrm{e}^{|k|z} + \mathrm{e}^{2h-z}], \tag{86}$$

where α_k is a constant to be determined later on.

Remark 0.2. The existence of α_k is guaranteed by $(85)_3$.
Indeed, we multiply $(85)_3$ by $(-\frac{1}{U}\mathrm{e}^{-\frac{\lambda}{U}z})$ and integrate the resulting equation on $(0,h)$; we deduce

$$\alpha_k = \frac{\int_0^h \mathrm{e}^{-\frac{\lambda}{U}z} f_{3k} \mathrm{d}z}{|k| \int_0^h [\mathrm{e}^{(|k|-\frac{\lambda}{U})z} - \mathrm{e}^{-(|k|+\frac{\lambda}{U})z}\mathrm{e}^{2|k|h}]\mathrm{d}z}, \tag{87}$$

for all $|k| \neq 0$. If $k = (0,0)$ then the pressure p_0 reads

$$p_0(z) = \int_z^h f_{30}(z)\mathrm{d}z. \tag{88}$$

From (87) we deduce the following estimate for α_k:

$$|\alpha_k| \leq \frac{||k|^2 - (\frac{\lambda}{U})^2|(\frac{U}{2\lambda}(1-\mathrm{e}^{-\frac{\lambda h}{U}}))^{1/2}}{|k|(|k|^2 - (\frac{\lambda}{U})^2)(\mathrm{e}^{(|k|-\frac{\lambda}{U})h} - 1)(\mathrm{e}^{(|k|-\frac{\lambda}{U})h} - \mathrm{e}^{2|k|h})}\left(\int_0^h (f_{3k})^2 \mathrm{d}z\right)^{1/2}, \tag{89}$$

where λ is chosen such that $\lambda \notin \{U|k|, k \in \mathbb{Z}^2\}$.

Lemma 0.1. *We have $p \in L^2(\Omega)$.*

Proof. By definition of p we have

$$\begin{aligned} \|p\|^2_{L^2(\Omega)} &= L_1 L_2 \sum_k \int_0^h (p_k(z))^2 \mathrm{d}z, \\ &= L_1 L_2 \sum_k (\alpha_k)^2 \int_0^h [\mathrm{e}^{|k|z} + \mathrm{e}^{|k|(2h-z)}]^2 \mathrm{d}z. \end{aligned} \tag{90}$$

There exits a constant C such that

$$|\alpha_k|^2 \leq C\mathrm{e}^{-2|k|h} \int_0^h (f_{3k})^2 \mathrm{d}z, \tag{91}$$

$$\int_0^h [\mathrm{e}^{|k|z} + \mathrm{e}^{|k|(2h-z)}]^2 \mathrm{d}z \leq C\mathrm{e}^{2|k|h}. \tag{92}$$

Hence, we conclude that

$$\|p\|^2_{L^2(\Omega)} \leq C \sum_k \int_0^h (f_k)^2 \mathrm{d}z = C(L_1, L_2, \lambda, h)\|f\|^2_{L^2(\Omega)} < +\infty. \tag{93}$$

This ends the proof of Lemma 0.1. □

Lemma 0.2. *The solution u of (84) satisfies:*

$$u \in \mathbf{L^2(\Omega)} \quad \text{and} \quad zD_3u \in \mathbf{L^2(\Omega)}. \tag{94}$$

Proof. We multiply $(85)_3$ by $(-\frac{1}{U}\mathrm{e}^{-\lambda z/U})$ and integrate on $(0, h)$; we deduce the following expression of u_{3k}:

$$u_{3k}(z) = -\frac{\mathrm{e}^{\lambda z/U}}{U} \int_0^z [f_{3k} - p_k']\mathrm{e}^{-\frac{\lambda}{U}s}\mathrm{d}s. \tag{95}$$

Thanks to (86) we have

$$\begin{aligned}
\sum_k \int_0^h (u_{3k})^2 \mathrm{d}s &\leq \frac{2}{U^2}\mathrm{e}^{\frac{2\lambda h}{U}} \sum_k \Big\{ \int_0^h (\int_0^z \mathrm{e}^{-\frac{\lambda}{U}s} f_{3k}\mathrm{d}s)^2 \mathrm{d}z + \\
&\quad + |k|^2\alpha_k^2 \int_0^h (\int_0^z [\mathrm{e}^{|k|s} + \mathrm{e}^{|k|(2h-s)}]\mathrm{d}s)^2 \Big\}, \\
&\leq \text{(thanks to (89))} \\
&\leq \frac{h}{\lambda U}(\mathrm{e}^{\frac{2\lambda h}{U}} - 1) \sum_k \int_0^h (f_{3k})^2 \mathrm{d}z + \frac{2h}{U^2}\mathrm{e}^{\frac{3\lambda h}{U}} \sum_k \int_0^h \mathrm{e}^{-\frac{\lambda}{U}z}|f_{3k}|\mathrm{d}z, \\
&\leq C(\lambda, U, h)\|f_3\|^2_{L^2(\Omega)}.
\end{aligned} \tag{96}$$

Hence $u_3 \in L^2(\Omega)$.

Now, since u_1 and u_2 satisfy similar equations ($(85)_1$ and $(85)_2$) it is sufficient to prove for example that $u_1 \in L^2(\Omega)$. For that purpose, we multiply $(85)_1$ by $(-\frac{1}{U}\,\mathrm{e}^{-\lambda z/U})$ and integrate on $(0, h)$, and we obtain the expression of u_{1k}:

$$u_{1k}(z) = -\frac{\mathrm{e}^{\lambda z/U}}{U} \int_h^z [f_{1k} - ik_1 p_k]\mathrm{e}^{-\frac{\lambda}{U}s}\mathrm{d}s. \tag{97}$$

Thus, we infer from (86) that

$$\sum_k \int_0^h (u_{3k}(z))^2 \mathrm{d}z \leq \frac{2}{U^2} \sum_k \Big\{ \int_0^h \mathrm{e}^{\frac{2\lambda}{U}z} (\int_h^z f_{1k}\mathrm{e}^{-\frac{\lambda}{U}s}\mathrm{d}s)^2 \mathrm{d}z + (k_1\alpha_k I_k)^2 \Big\}, \tag{98}$$

where

$$I_k = \int_0^h [\mathrm{e}^{(|k|-\frac{\lambda}{U})s} + \mathrm{e}^{-(|k|+\frac{\lambda}{U})s}\mathrm{e}^{2|k|h}]\mathrm{d}s.$$

Thanks to (89) and for $|k|$ large, we deduce the existence of a positive constant $C > 0$ such that:

$$|k_1\,\alpha_k\,I_k| \leq C \left(\int_0^h (f_{3k})^2 \mathrm{d}z \right)^{1/2}, \quad \forall\ k \in \mathbb{Z}^2. \tag{99}$$

Combining (98) and (99), we obtain

$$\sum_k \int_0^h (u_{3k}(z))^2\ \mathrm{d}z \leq C\ \|f\|^2_{L^2(\Omega)} < +\infty. \tag{100}$$

We conclude then that $u \in \mathbf{L^2(\Omega)}$.

Now, since we have the explicit expressions of u and p (given respectively by (97)-(95) and (86)), we prove exactly as below that $zD_3u_1, zD_3u_2 \in L^2(\Omega)$ and by the incompressibility condition $(85)_4$, we deduce that $zD_3u_3 \in L^2(\Omega)$.
This ends the proof of Lemma 0.2. □

Consequently, we have proved that for every $f \in H$ there exists $u \in D(A)$ such that $(A + \lambda I)(u) = f$. This proves (i). The property (ii) is obvious. To prove (iii), we consider a sequence $(u_m)_m \subset D(A)$ such that $u_m \to u$ and $Au_m \to \varphi$ in H. Thanks to (ii) we have

$$u_m = (A + \lambda I)^{-1}(A + \lambda I)u_m \to_{m \to +\infty} (A + \lambda I)^{-1}(\varphi + \lambda u). \tag{101}$$

By the uniqueness of the limit, we conclude that

$$(A + \lambda I)(u) = \varphi + \lambda u \Leftrightarrow u \in D(A) \text{ and } Au = \varphi. \tag{102}$$

This proves that A is closed and consequently ends the proof of the Proposition 0.1.

Theorem 0.4. *Under the following hypotheses:*

$$(H_2)\ :\ v_0 \in \mathcal{C}^\infty(\overline{\Omega}_\infty) \cap D(A) \quad \text{and} \quad f \in \mathcal{C}^\infty([0,T] \times \overline{\Omega}_\infty) \cap H,$$

the solution (v, p) of the system (4) satisfies:

$$v, p \in \mathcal{C}^\infty([0,T] \times \overline{\Omega}_\infty). \tag{103}$$

Remark 0.3. The hypothesis (H_2) on f can be replaced by $f \in \mathcal{C}^\infty([0,T] \times \overline{\Omega}_\infty)$. Indeed, we decompose f as follows:

$$f = P_H f + P_{H^\perp} f,$$

where P_H is the Leray-Hopf projector on H. Thus, only the pressure p is changed if we replace f by $P_H f$. We define a regular function θ such that $P_{H^\perp} f = \nabla\theta$, and then p is just replaced by $p - \theta$.

Let us state and prove the following Lemma in order to prove Theorem 0.4:

Lemma 0.3. *For $\tau \in \{x, y\}$ and $\sigma \in \{\tau, t\}$, we have:*

(i) *If Φ belongs to the space $\mathcal{C}^\infty(\overline{\Omega}_\infty) \cap H$ (respectively $\mathcal{C}^\infty([0,T] \times \overline{\Omega}_\infty) \cap H$ or $\mathcal{C}^\infty(\overline{\Omega}_\infty) \cap D(A)$) then $\partial\Phi/\partial\sigma$ belongs to the same space,*
(ii) *If $\Phi \in \mathcal{C}^\infty([0,T] \times \overline{\Omega}_\infty) \cap L^1(0,T;H)$ then $\partial\Phi/\partial\tau$ too.*

Proof. First, we notice that the periodicity is conserved after differentiation with respect to x, y, z or t.
To prove (i) it suffices to verify that: if $\Phi \in \mathcal{C}^\infty(\overline{\Omega}_\infty) \cap H$ (resp. $\mathcal{C}^\infty(\overline{\Omega}_\infty) \cap D(A)$) then $\partial\Phi/\partial\sigma \in H$ (resp. $D(A)$).
Hence, let $\Phi \in \mathcal{C}^\infty(\overline{\Omega}_\infty) \cap H$; by using the definition of H, we have easily $\partial\Phi/\partial\sigma \in H$ since we differentiate the boundary conditions only in the tangential direction on the boundaries $z = 0$ and $z = h$.
Now, let $\Phi \in \mathcal{C}^\infty(\overline{\Omega}_\infty) \cap D(A)$. Then, in particular $\Phi \in H$ and consequently $\partial\Phi/\partial\sigma \in H$. Moreover, since $\Phi = 0$ on $z = h$, and we have $\partial\Phi/\partial\sigma = 0$ on $z = h$. From the condition $\Phi \in D(A)$, we deduce the existence of a regular function p ($\nabla p = (I - P_H)(UD_3 v)$) such that $-UD_3 v + \nabla p \in H$. This implies that $-UD_3(\partial\Phi/\partial\sigma) + \nabla(\partial p/\partial\sigma) \in H$, and thus $\partial\Phi/\partial\sigma \in D(A)$.
The property (ii) is a consequence of (i). Indeed, if $\Phi \in \mathcal{C}^\infty([0,T] \times \overline{\Omega}_\infty) \cap L^1(0,T;H)$, then in particular $\Phi \in H$ for almost all $t \in [0,T]$. Thanks to (i) in Lemma 0.3, we have $\partial\Phi/\partial\sigma \in H$ for almost all $t \in [0,T]$.
This concludes the proof of Lemma 0.3. □

Now, we are able to prove Theorem 0.4:

Proof. of Theorem 0.4 We recall that under the hypothesis (H_2), the Hille-Yosida (Theorem 0.3) and more precisely (H_1) yields:

$$v \in \mathcal{C}([0,T]; \mathbf{L^2}(\boldsymbol{\Omega})) \text{ and } \frac{\mathrm{d}v}{\mathrm{d}t} \in L^\infty(0,T; \mathbf{L^2}(\boldsymbol{\Omega})), \quad \forall\, T > 0. \tag{104}$$

Thanks to Lemma 0.3, the linearity of the system (4) and after replacing the source function f and the initial data v_0 by their tangential derivatives, we infer that

$$\frac{\partial^m v}{\partial \tau^m} \in \mathcal{C}([0,T]; \mathbf{L^2}(\boldsymbol{\Omega})), \quad \forall\, \tau \in \{x, y\},\ \forall\, m \in \mathbb{N}. \tag{105}$$

On the other hand, since $v_0 \in D(A)$ and f is continuous at $t = 0$, then we deduce that $(\mathrm{d}v/\mathrm{d}t)(t = 0) \in H$. Now, if we differentiate the system (4) with respect to t, we obtain:

$$\frac{\partial v}{\partial t} \in \mathcal{C}([0,T]; \mathbf{L^2(\Omega)}). \tag{106}$$

From the incompressibility condition $(4)_2$, we deduce that

$$D_3 v_3 = -D_1 v_1 - D_2 v_2 \in \mathcal{C}([0,T]; L^2(\Omega)). \tag{107}$$

Using (106) and (107), the third component of the equation $(4)_1$ yields:

$$\frac{\partial p}{\partial z} \in \mathcal{C}([0,T]; L^2(\Omega)). \tag{108}$$

Again, the invariance of the system (4) by differentiation with respect to τ gives:

$$\frac{\partial^{m+1} p}{\partial z \partial \tau^m} \in \mathcal{C}([0,T]; L^2(\Omega)) \quad \forall\, \tau \in \{x, y\},\ \forall\, m \in \mathbb{N}. \tag{109}$$

In particular, we have

$$\frac{\partial p}{\partial x}, \frac{\partial p}{\partial y} \in \mathcal{C}([0,T]; L^2(\Omega)). \tag{110}$$

Hence, equation $(4)_1$ projected on the tangential directions x and y implies with (106) that:

$$D_3 v_1, D_3 v_2 \in \mathcal{C}([0,T]; L^2(\Omega)). \tag{111}$$

Gathering (105), (107) and (111), we conclude the following regularity for v:

$$v \in \mathcal{C}([0,T]; \mathbf{H^1(\Omega)}). \tag{112}$$

Now, we reiterate the procedure; from (112) and the invariance of (4) by differentiation in time t, we obtain

$$\frac{\partial v}{\partial t} \in \mathcal{C}([0,T]; \mathbf{H^1(\Omega)}) \ \ i.e.,\ \ v \in \mathcal{C}^1([0,T]; \mathbf{H^1(\Omega)}). \tag{113}$$

The invariance of (4) by differentiation with respect to x and y implies that:

$$D_3 v_3 = -D_1 v_1 - D_2 v_2 \in \mathcal{C}^1([0,T]; H^1(\Omega)). \tag{114}$$

Then equation $(4)_1$ and (113) imply that

$$\frac{\partial p}{\partial z} \in \mathcal{C}([0,T]; H^1(\Omega)). \tag{115}$$

Thus,

$$D_3 v_1, D_3 v_2 \in \mathcal{C}([0,T]; H^1(\Omega)), \tag{116}$$

and consequently,

$$v \in \mathcal{C}([0,T]; \mathbf{H}^2(\boldsymbol{\Omega})). \tag{117}$$

Finally, the invariance by differentiation in time gives

$$v \in \mathcal{C}^1([0,T]; \mathbf{H}^2(\boldsymbol{\Omega})). \tag{118}$$

This procedure allows us to prove a higher regularity for v and more precisely (103).

This concludes the proof of Theorem 0.4. □

Acknowledgements

This work was partially supported by the National Science Foundation under the grant NSF-DMS-0305110, and by the Research Fund of Indiana University.

References

1. H. Brézis,*Opérateurs maximaux monotones et semi-groupes de contractions dans les espaces de Hilbert* (North-Holland Publishing Co., Amesterdam, 1973).
2. N. Burq and P. Gérard, *Contrôle optimal des équations aux dérivées partielles* (Ecole Polytechnique, Palaiseau, France, 2003).
3. K. O. Friedrichs, *The mathematical strucure of the boundary layer problem in Fluid Dynamics*, eds., R. von Mises and K.O. Friedrichs, Brown Univ., Providence, RI (reprinted by Springer-Verlag, New York, 1971). pp. 171-174.
4. J. L. Lions,*Lectures Notes in Math*, **323** Springer-Verlag, New York, (1973).
5. J. L. Lions, *Les Presses de l'Université de Montréal* (Montreal, Que., 1965).
6. J. L. Lions, *Selected work,* Vol 1. (EDS Sciences, Paris, 2003).
7. R. E. O'Malley, *Communications of the Mathematical Institute, Rijksuniversiteit Utrecht, 5. Rijksuniversiteit Utrecht* (Mathematical Institute, Utrecht, 1977).
8. M. Petcu, *Advances in Differential Equations* (to appear).
9. L. Prandtl, *Veber Flüssigkeiten bei sehr kleiner Reibung*, in *Verh. III Intern. Math. Kongr. Heidelberg,* (Tuber, Leipzig, 1905) pp. 484-491.
10. R. Temam, *Navier-Stokes Equations* (North-Holland Pub. Company, 1977), Reedition in the AMS-Chelsea Series, AMS (Providence, 2001).
11. R. Temam and X. Wang,*J. Differential Equations* **179**, 647 (2002).
12. M. I. Vishik and L. A. Lyusternik,*Uspekki Mat. Nauk*, **12**, 3 (1957).
13. K. Yosida, *Functional analysis*, (Springer-Verlag, Berlin, 6th edition, 1980).

Chapter III

GEOMETRIC ANALYSIS

Advances in Deterministic and Stochastic Analysis
Eds. N. M. Chuong *et al.* (pp. 231–253)

§12. MONOTONE INVARIANTS AND EMBEDDINGS OF STATISTICAL MANIFOLDS

LE HONG VAN *

Max-Planck-Institute for Mathein Sciences
Inselstraße 22-26
D-04103 Leipzig
E-mail: hvle@mis.mpg.de

In this note we prove certain necessary and sufficient conditions for the existence of an embedding of statistical manifolds. In particular, we prove that any smooth (C^1 resp.) statistical manifold can be embedded into the space of probability measures on a finite set. As a result, we get positive answers to the Lauritzen question on a realization of smooth (C^1 resp.) statistical manifolds as statistical models.

Keywords: Fisher metric, Chentsov-Amari connections, statistical manifolds, statistical models.

1. Introduction

A *statistical model* is a family M of probability measures on a measurable space Ω. There are two natural geometrical structures on any statistical model equipped with a differentiable manifold structure. They are the Fisher tensor and the Chentsov-Amari tensor.

The Fisher tensor was given by Fisher in 1925 as an information characterization of a statistical model. Rao[13] proposed to consider this tensor as a Riemannian metric on the manifold of probability distributions. This Fisher metric has been systematically studied in Refs. 1,4,9 and others,[3,6,7,14] ect. in the field of geometric aspects of statistics and information theory.

Chentsov[4] and Amari[2] independently also discovered a natural structure on statistical models, namely a 1-parameter family of invariant connections, which includes the Levi-Civita connection of the Fisher metric. This family of invariant connections is defined by a 3-symmetric tensor T together with the Levi-Civita connection of the Fisher metric.

*Mathematical Institute of ASCR, Zitna 25, 11567 Praha 1, E-mail: hvle@math.cas.cz

Motivated by the question how much we can describe a statistical model via their Fisher metric and Chentsov-Amari tensor T, in 1987 Lauritzen proposed to call a Riemannian manifold (M, g) with a 3-symmetric tensor T *a statistical manifold*. Since two 3-symmetric tensors T and $k \cdot T$, $k \neq 0$, define the same family of Chentsov-Amari connections, we shall say that two statistical manifolds (M, g, T) and (M, g, kT) are conformal equivalent.

A natural and important question in the mathematical statistics is to understand, if a given family M of probability distributions can be considered as a subfamily of another given one N. In the language of statistical manifolds, this question can be formulated as a problem of isostatistical embedding of a statistical manifold (M, g, T) into another one (N, g', T'). Here we say that an immersion $f : (M, g, T) \to (\overline{M}, \overline{g}, \overline{T})$ is called *isostatistical*, if $f^*(\overline{g}) = g$ and $f^*(\overline{T}) = T$.

We shall see in Sec. 2 that the problem of the existence of an isostatistical embedding includes also the Lauritzen question in 1987, if any statistical manifold is a statistical model. It also concerns the following important problem posed by Amari in 1997, if any finite dimensional statistical model can be embedded into the space Cap^N of probability distributions of the sample space Ω^N of N elementary events for some finite N.

We shall construct a class of C^0 (and C^1) *monotone invariants* of statistical manifolds, which present obstructions to embedding of a given C^k statistical manifold M into another one N^n. Here a C^k statistical manifold (M, g, T) is a smooth differentiable manifold with C^k sections $g \in S^2T^*M$ and $T \in S^3T^*M$. These invariants measure certain relations between the metric tensor g and the 3-symmetric tensor T. In particular, using these invariants we show that no statistical manifold which is conformal equivalent to the space Cap^N can be embedded into the product of m copies of the normal Gaussian manifolds for any $N > 3$ and any finite m. In the Main Theorem (Sec. 5) we prove that any smooth (C^1 resp.) statistical manifold M^m can be isostatistically embedded to a the space Cap^N for some N big enough.

As a consequence we also get a new proof of Matumoto theorem on the existence of the contrast function for a statistical manifold (see subsection 2.6).

2. Statistical Models and Statistical Manifolds

In this section we recall the definitions of the Fisher metric and the Chentsov-Amari connections on statistical models. We introduce the notion of a weak Fisher metric and a weak potential function. At the end of the section we discuss the problem, if a given statistical manifold is a statistical model. Most of the facts in this section can be found in Ref. 1.

Suppose that M is a statistical model - a family of probability measures on a space Ω. We assume throughout this note that M and Ω are differentiable manifolds, and Ω is equipped with a fixed Borel measure $d\omega$. We also write

$$p(x,\omega) = p(x,\omega)d\omega, \tag{1}$$

where $p(x,\omega)$ in LHS of 1 is a Borel measure in M and $p(x,\omega)$ in the RHS of (1) is a non negative (density) function on $M \times \Omega$ which satisfies

$$\int_\Omega p(x,\omega)\, d\omega = 1\, \forall x \in M. \tag{2}$$

The *Fisher metric* $g^F(x)$ is defined on M as follows. For any $V, W \in T_xM$ we put

$$g^F(V,W)_x = \int_\Omega (\partial_V \ln p(x,\omega))(\partial_W \ln p(x,\omega))p(x,\omega). \tag{3}$$

The function under integral in Eq. (3) is well defined, if

$$p(x,\omega) > 0. \tag{4}$$

Denote by $Cap(\Omega)$ the space of all probability measures on Ω. Clearly we can consider the density function $p(x,\omega)$ as a mapping $M \to Cap(\Omega)$. Thus we shall call a function $p(x,\omega)$ *a probability potential of the metric* g_F, if $p(x,\omega)$ satisfies (2), (3), (4). It is known[1,4] that for a given Riemannian metric g_F on a smooth manifold M there exist many probability potentials $f(x,\omega)$ for g_F, even if we fix the space $(\Omega, d\omega)$.

Some time it is useful to consider functions $p(x,\omega)$ which satisfy (3) and (4) but not necessary (2). In this case, the Riemannian metric g^F will be called *weak Fisher metric*, and the function $p(x,\omega)$ will be called *a weak probability potential* of g^F.

2.1. *Example of a Weak Fisher Metric*

The standard Euclidean metric g^0 on the positive quadrant $\mathbb{R}^N_+(x_i > 0)$. It is straightforward to check that g_0 admits a weak probability potential $\{p_i(x) = \frac{1}{4}x_i^2, i = \overline{1,N}.\}$ Here $\Omega = \Omega^N$ - the sample space of N elementary events.

2.2. *The Fisher Metric on the Space $(Cap^N)_+$ of all positive probability distributions on Ω^N (see also Refs. 1,4,6)*

By definition we have

$$Cap_+^N := \{(p_1, \cdots, p_N) | p_i > 0 \text{ for } i = \overline{1, N} \,\&\, \sum p_i = 1\}.$$

We define the embedding map

$$f : Cap_+^N \to S^{N-1}(2),$$

$$(p_1, \cdots, p_N) \mapsto (q_1 = 2\sqrt{p_1}, \cdots, q_N = 2\sqrt{p_N}).$$

It is easy to see that the Fisher metric in the new coordinates (q_i) is the standard metric of constant positive curvature on the sphere $S^{N-1}(2)$.

2.3. *Divergence Potential (see 1,14)*

A function ρ on $M \times M$ with the following property

$$\rho(x, y) \geq 0 \text{ with equality iff } x = y \tag{5}$$

is called a divergence function. A divergence function ρ is called *a divergence potential* for a metric g on M, if

$$g(X, Y)_x = Hess(\rho)(i_1(X), i_1(Y)). \tag{6}$$

where

$$T_{(x,x)}(M, M) = (T_x M, 0) \oplus (0, T_x M) = (i_1(T_x M)) \oplus (i_2(T_x M)).$$

An example of a divergence potential for a Fisher metric is the Jensen function $J_H^{\lambda,\mu}(x, y)$ of the entropy function $H(x)$ on M, or a Kullback relative entropy function $K(x, y)$ on $M \times M$.

2.4. *Chentsov-Amari connections*

Let $p(x, \omega)$ be a probability potential for a Riemannian metric g. We define a symmetric 3-tensor T on M as follows

$$T(X, Y, Z) = \int (\partial_X \ln p(x, \omega))(\partial_Y \ln p(x, \omega))(\partial_Z \ln p(x, \omega))\, p(x, \omega). \tag{7}$$

We denote by ∇^F the Levi-Civita connection of the (weak) Fisher metric g^F. We define

$$< \nabla^t_X Y, Z > := < \nabla^F_X Y, Z > + t \cdot T(X, Y, Z). \tag{8}$$

The connections ∇^t are called the Chentsov-Amari connections.

Remark 2.1. (Refs. 1,8) Any divergence function $\rho(x,y)$ on $M \times M$ defines a tensor T on M via the following formula

$$T(X,Y,Z)_x = -\,\partial_{i_2(Z)} Hess(\rho)(i_1(X), i_1(Y))_{(x,x)} \\ +\,\partial_{i_1(Z)} Hess(\rho)(i_2(X), i_2(Y))_{(x,x)}.$$

If g and T are defined by the same divergence function $\rho(x,y)$, we shall call $\rho(x,y)$ *a divergence potential for the statistical manifold* (M,g,T). It is a known fact that the Kullback relative entropy function is a divergence potential for the associated statistical model.

2.5. *Statistical Submanifolds*

A submanifold N in a statistical manifold (M,g,T) with the induced Riemannian metric $g_{|N}$ and induced tensor $T_{|N}$ is called *statistical submanifold* of (M,g,T). Clearly, if $f(x,\omega)$ is a (weak) probability potential for (M,g,T), then its restriction to any submanifold $N \subset M$ is a (weak) probability potential of the induced statistical structure.

2.6. *Statistical Models and Statistical Manifolds*

Since any probability function $p(x,\omega)$ defines a map $M \to Cap(\Omega)$, we shall say that a statistical manifold (M,g,T) is a statistical model, if there probability potential $p(x,\omega)$ for g and T. By the remark in subsection 2.5, we get that a statistical submanifold of a statistical model is also a statistical model. Furthermore, if a statistical manifold (M,g,T) is a statistical model, then it must admit a divergence potential. Hence the following theorem of Matsumoto is a consequence of our Main Theorem in Sec. 5.

Theorem 2.1 (Ref. 8). *For any statistical manifold* (M,g,T) *there exists a divergence potential* ρ *for* g *and for* T.

3. Embeddings of Linear Statistical Spaces

An Euclidean space $(\mathbb{R}^n, g^0)$ equipped with a 3 -symmetric tensor T will be called *a linear statistical spaces.* We observe that the equivalence class of linear statistical spaces coincides with the orbit space of 3-symmetric tensors T under the action of the orthogonal group $O(n)$. In this section we discuss certain invariants of these orbits and we show several necessary and sufficient conditions for the existence of embedding of one linear statistical space into another linear statistical space by studying these invariants. A class of our necessary conditions consists of *monotone invariants* λ, i.e. we assign to any linear statistical space $(\mathbb{R}^n, g^0, T)$ a number $\lambda(\mathbb{R}^n, g^0, T)$ such that, if $(\mathbb{R}^n, g^0, T)$ is a statistical submanifold of $(\mathbb{R}^m, g^0, T')$, then we have

$$\lambda(\mathbb{R}^n, g^0, T) \leq \lambda(\mathbb{R}^m, g^0, T').$$

Since a tangent space of a statistical manifold is a linear statistical manifold, these invariants play important role in the problem of isostatistical immersion.

3.1. *Trace Type of a Symmetric 3-tensor*

Let us denote by $\mathcal{R}^n$ the subspace in $S^3(\mathbb{R}^n)$ consisting of the following 3-symmetric tensors

$$T^v(x, y, z) = <v, x><y, z> + <v, y><x, z> + <v, z><x, y>,$$

where $v \in \mathbb{R}^n$. Using the standard representation theory (see e.g. Ref. 12) we have the decomposition

$$S^3(\mathbb{R}^n) = \mathcal{R}(3\pi_1) \oplus \mathcal{R}^n. \tag{9}$$

To compute the orthogonal projection of a 3-symmetric tensor T on the space $\mathcal{R}^n$ in the decomposition (9) we can use the following Lemma. We denote by π_2 the orthogonal projection form $S^3(\mathbb{R}^n)$ to $\mathcal{R}^n$.

Lemma 3.1. *We have*

$$\pi_2(S) = \frac{1}{n+2} T^{Tr(S)}. \tag{10}$$

Here we identify the 1-form $Tr(S)$ with a vector in $\mathbb{R}^n$ by using the Euclidean metric g^0.

We omit the proof of Lemma 3.1 which is straightforward. In view of Lemma 3.1 we shall call any tensor $T \in \mathcal{R}^n$ of *trace type*.

We note that

$$\dim S^3(\mathbb{R}^n) = C_n^3 + 2C_n^2 + n = \frac{n(n+1)(n+2)}{6}.$$

Thus the dimension of the quotient $S^3(\mathbb{R}^n)/SO(n)$ is at least $C_n^3 + C_n^2 + n$. A direct computation shows that the dimension of the orbit $SO(n)([\sum_{i=1}^n a_i v_i^3])$ is $C_n^2 = \dim SO(n)$, if $\Pi a_i \neq 0$. Here $\{v_i\}$ is an orthonormal basis in $\mathbb{R}^n$. Hence the dimension of $S^3(\mathbb{R}^n)/O(n) = C_n^3 + C_n^2 + n$. This dimension is exactly the number of all complete invariants of pairs consisting of a positive definite bilinear form g and a 3-symmetric tensor T.

Since the dimension of $G_k(\mathbb{R}^n) = k(n-k)$, it follows that generically it is impossible to embed a linear statistical space (R^k, g^0, T) into a given statistical linear space (R^n, g^0, T), unless $k(n-k) \geq C_k^3 + C_k^2 + k$. Clearly the dimension condition is not sufficient as the following proposition shows.

Proposition 3.1. *A linear statistical space $(\mathbb{R}^k, g^0, T)$ can be embedded into a linear statistical space $(\mathbb{R}^N, g^0, T^v)$, if and only if $N \geq k$ and T is also a trace type: $T = T^w$ with $|w| \leq |v|$.*

Proof. The necessary condition follows from the fact that the restriction of T^v to $\mathbb{R}^k$ equals $T^{\overline{v}}$, where $\overline{v}$ is the orthogonal projection of v to $\mathbb{R}^k$. Conversely, if $|w| \leq |v|$ we can find an orthogonal transformation, such that w equals the orthogonal projection of v on $\mathbb{R}^k$. □

3.2. *Commasses as Monotone Invariants*

Since the metric g extends canonically on the space $S^3(\mathbb{R}^n)$, we can define the absolute norm

$$||T|| := \sqrt{<T,T>}.$$

Now we define *comasses* of a 3-symmetric tensor T as follows

$$\mathcal{M}^3(T) := \max_{|x|=1,|y|=1,|z|=1} T(x,y,z),$$

$$\mathcal{M}^2(T) := \max_{|x|=1,|y|=1} T(x,y,y),$$

$$\mathcal{M}^1(T) := \max_{|x|=1} T(x,x,x).$$

Clearly we have

$$0 \leq \mathcal{M}^1(T) \leq \mathcal{M}^2(T) \leq \mathcal{M}^3(T) \leq ||T||.$$

Proposition 3.2. *The comasses are positive functions which vanish at T, if and and only if T equals zero. They are monotone invariants of T, since if T is a restriction of 3-symmetric tensor $\overline{T}$ on $\mathbb{R}^N$, then*

$$||T|| \leq ||\overline{T}||,\ \mathcal{M}^i(T) \leq \mathcal{M}^i(\overline{T}),\ \forall i = 1,2,3. \tag{11}$$

Proof. To prove the first statement it suffices to show that $\mathcal{M}^1$ vanishes at only $T = 0$. To see this we use the identity

$$\begin{aligned}
&- 12T(x, y, z) = T(x + y + z, x + y + z, x + y + z)\\
&+ T(x + y - z, x + y - z, x + y - z) + T(x - y + z, x - y + z, x - y + z)\\
&+ T(-x + y + z, -x + y + z, -z + y + z)\\
&- 2(T(x, x, x) + T(y, y, y) + T(z, z, z)).
\end{aligned}$$

The second statement follows immediately from the definition. □

Now for a space $(\mathbb{R}^n, g^0, T)$ and for $1 \le k \le n$ we put

$$\lambda_k(T) := \min_{\mathbb{R}^k \subset \mathbb{R}^n} \mathcal{M}^1(T_{|\mathbb{R}^k}).$$

We can easily check that if $\overline{T}$ is a restriction of T to a subspace $\mathbb{R}^m \subset \mathbb{R}^n$, then

$$\lambda_k(\overline{T}) \ge \lambda_k(T) \ge 0 \text{ for all } k \le m.$$

Thus $\lambda_k(T)$ is a monotone invariant of linear statistical manifolds. These invariants are related by the following inequalities

$$\mathcal{M}^1(T) = \lambda_n(T) \ge \lambda_{n-1}(T) \cdots \ge \lambda_2(T) \ge \lambda_1(T) = 0.$$

The last equality follows from the fact, that the function $T(x, x, x)$ is anti-symmetric on $S^{n-1}(|x| = 1) \subset \mathbb{R}^n$ and S^{n-1} is connected. We observe that if T is of trace type, then $\lambda_{n-1}(T) = \cdots = \lambda_1(T) = 0$.

We are going to give a lower bound of the monotone invariant λ_{n-1} of a linear statistical space of certain type. The equality $\lambda_{n-1}(\mathbb{R}^n, g^0, T) \ge A$ means that no hyperplane with the norm $\mathcal{M}^1$ strictly less than A can be embedded in $(\mathbb{R}^{n-1}, g^0, T)$.

Lemma 3.2.

(a) *Let $T = \sum_{i=1}^n (N - \varepsilon_i)(x^i)^3$ be a 3-symmetric tensor on $\mathbb{R}^n$ with $n \ge 4$, $N \ge 4$ and $|\varepsilon_i| \le 1/4$. Then we have*

$$\lambda_{n-1}(T) \ge \frac{N}{\sqrt{10}} - 1/4.$$

(b) *Let $T = N \sum_{i=1}^n (x^i)^3$, and H be a hyperplane in $\mathbb{R}^n$ which is orthogonal to $(kn, 1, 1, \cdots, 1)$, and let $n \ge 5$, $k \ge 3$. Then we have*

$$\lambda_{n-2}(T_{|H}) \ge \frac{N}{5} - 1.$$

(c) *Let* $x = ((1-\varepsilon), \frac{1}{kn}, \cdots, \frac{1}{kn}) \in S^n(1) \subset \mathbb{R}^{n+1}$*, where* $n \geq 4, k \geq n$*. We denote by* H *the tangential plane* $T_x S^n$*, and by* T^0 *the following 3-symmetric tensor on* $\mathbb{R}^{n+1}_+$*:*

$$T^0_{ijk}(x_1, \cdots x_N) = \delta_{ijk}\frac{1}{x_i}. \tag{12}$$

Then we have

$$\lambda_{n-1}(T^0_{|H}) \geq \frac{kn}{5} - 1.$$

Remark 3.1. The tensor T^0 in (??) defines on $(\mathbb{R}^n, g^0)$ a statistical structure with a weak probability potential $\{\frac{1}{2}x_i^2, i = 1, n\}$.

Proof of Lemma 3.2. The reader shall see that a proof of Lemma 3.2 can be done in the same scheme of the proof of Sublemma 5.4. Therefore we do not repeat this argument here.

Remark 3.2. Lemma 3.2.a holds also for $n = 3$ but not for $n = 2$, Lemma 3.2.b holds also for $n = 4$, but not for $n = 3$, and Lemma 3.2.c holds also for $n = 3$ but not for $n = 2$.

There are also several obvious monotone invariants of T.

$$A^1(T) := \max_{|x|=|y|=|z|=1, <x,y>=<y,z>=<z,x>=0} T(x, y, z)$$

is well-defined for $n \geq 3$.

$$A^2(T) := \max_{|x|=|y|=1, <x,y>=0} T(x, y, y),$$

is well-defined for $n \geq 2$. We can check that

$$\ker A^1 = \mathcal{R}^n.$$

On the other hand we have

$$\ker A^2 \subset \mathcal{R}(3\pi_1).$$

Thus A^1 and A^2 are different invariants. □

Lemma 3.3. *Let* π_1 *be the first component of* T *in decomposition (9). Then* $\|T\|_1 := \|\pi_1(T)\|$ *is a monotone invariant of* T.

Proof. Let $\mathbb{R}^k$ be a subspace of $\mathbb{R}^n$. We denote by $\pi^n_k T$ the restriction of T to $\mathbb{R}^k$. Clearly

$$\pi^n_k(T) = \pi^n_k(\pi_1 T) + \pi^n_k(\pi_2 T).$$

We have noticed in Proposition 3.1 that the restriction of the trace form $\pi_2 T$ to any subspace is also a trace form. Thus $\pi^n_k(\pi_2)$ is an element in $\mathcal{R}^k \subset S^3(\mathbb{R}^k)$. Hence we have

$$\pi_1(\pi^n_k T) = \pi_1(\pi^n_k(\pi_1 T)). \tag{13}$$

Since all the projections π_1, π^n_k decrease the norm $||.||$, we get

$$||\pi^n_k T||_1 = ||\pi_1(\pi^n_k T)|| = ||\pi_1(\pi^n_k(\pi_1 T))|| \leq ||\pi_1(T)|| = ||T||_1. \qquad \square$$

Proposition 3.3. *A statistical line* $(\mathbb{R}, g^0, T)$ *can be embedded into* $(\mathbb{R}^N, g^0, T')$, *if and only if* $\mathcal{M}^1(T) \leq \mathcal{M}^1(T')$.

Proof. It suffices to show that we can embed $(\mathbb{R}, g^0, T)$ into $(\mathbb{R}^N, g^0, T')$, if we have $\mathcal{M}^1(T) \leq \mathcal{M}^1(T')$. We note that $T'(v, v, v)$ defines an anti-symmetric function on the sphere $S^{N-1}(|v| = 1) \subset \mathbb{R}^N$. Thus there is a point $v \in S^{N-1}$ such that $T'(v, v, v) = \mathcal{M}^1(T)$. Clearly the line $v \otimes \mathbb{R}$ defines the required embedding. $\square$

Let us consider the embedding problem for 2-dimensional linear statistical spaces. It is easy to see that

$$S^3(\mathbb{R}^2) = \mathbb{R}^2 \oplus \mathbb{R}^2.$$

Thus the quotient $S^3(\mathbb{R}^2)/SO(2)$ equals $(\mathbb{R}^2 \oplus \mathbb{R}^2)/S^1$. Geometrically there are several ways to see this. In the first way we denote components of $T \in S^3(\mathbb{R}^2)$ via $T_{111}, T_{112}, T_{122}, T_{222}$.

Lemma 3.4. *There exists an oriented orthonormal basic in* $\mathbb{R}^2$ *such that* $T_{111} = \mathcal{M}^1(T) > 0, T_{112} = 0$ *for all non-vanishing* T. *These numbers* $(T_{111}, T_{122}, T_{222})$ *are called canonical coordinates of* T. *Two tensors* T *and* T' *are equivalent, if and only if they have the same canonical coordinates.*

Proof. We choose an oriented orthonormal basis (v_1, v_2) by taking as v_1 a point on $S^1(|x| = 1)$, where the function $T(x, x, x)$ reaches the maximum. The first variation formula shows that in this case $T_{112} = 0$. This shows the existence of the canonical coordinates. Clearly, if two tensors have the same canonical coordinates, then they are equivalent. Next, if two tensors T and T' are equivalent, then their norms $\mathcal{M}^1$ are the same. We need to take care the case, when there are several points x at which $T(x, x, x)$ reaches the maximum. In any case, they have the same first coordinates. Next we note that

$$< Tr(T), Tr(T) >= (T_{111} + T_{122})^2 + T_{222}^2,$$

$$||T||^2 = T_{111}^2 + T_{122}^2 + T_{222}^2.$$

Thus if two tensors are equivalent and have the same first coordinates, they must have the same third coordinate T_{122}, and this third coordinate is uniquely defined up to sign. The condition on the orientation tells us that the sign must be $+$. This proves the second statement. □

Proposition 3.4. *We can always embed the 2-dimensional statistical space* $(\mathbb{R}^2, g^0, 0)$ *into any linear statistical space* $(\mathbb{R}^n, g^0, T)$, *if* $n \geq 7$.

Proof. It suffices to prove for $n = 7$. We denote by $\mathcal{O}(T)$ the set of of all unit vectors $v \in S^6$ such that $T(v, v, v) = 0$. Clearly $\mathcal{O}(T)$ is a set of dimension 5 in S^6. Since T is anti-symmetric, there exists a connected component $\mathcal{O}^0(T)$ of $\mathcal{O}(T)$ which is invariant under the anti-symmetry involution. Now we consider the following function f on $\mathcal{O}^0(T)$. For each $v \in \mathcal{O}^0(T)$ we denote by A^v the bilinear symmetric 2-form on the space $T_x\mathcal{O}^0(T)$ considered as a subspace in $\mathbb{R}^n$:

$$A^v(y, z) = T(v, y, z).$$

Then we define $f(v)$ equal to $\det(A^v)$. Since $\mathcal{O}(T)$ has dimension 5, the function $f(v)$ is anti-symmetric on $\mathcal{O}^0(T)$. Hence the set $\mathcal{O}_0^0(T)$ of all $v \in \mathcal{O}^0(T)$ with $f(v) = 0$ has dimension 4 and it contains a connected component which is also invariant under the anti-symmetric involution. For the simplicity we denote this connected component also by $\mathcal{O}_0^0(T)$. Now we consider the following two possible cases.

Case 1. We assume that there is a point $v \in \mathcal{O}_0^0(T)$ such that the nullity of A^v is at least 2. Then there are two linear independent vectors $y, z \in T_v$ such that the restriction of A^v on the plane $\mathbb{R}^2(y, z)$ vanishes. Since the set $\mathcal{O}^0(T)$ is connected and anti-symmetric and of co dimension 1 in S^{n-1}, the plane $\mathbb{R}(y, z)$ has a non-empty intersection with $\mathcal{O}^0(T)$ at a point w. Then the restriction of T on the plane $\mathbb{R}^2(v, w)$ is vanished, because

$$T(v, v, v) = T(w, w, w) = 0$$

$$T(v, w, w) = 0 \,(\text{since } A^v(w, w) = 0),$$

$$T(v, v, w) = 0 \,(\text{since } w \in T_v\mathcal{O}^0(T)).$$

Case 2. We assume that the nullity of A^v on $\mathcal{O}_0^0(T)$ is constantly 1. Using the anti-symmetric property of A^v we conclude that the restriction of A^v to the plane $\mathbb{R}^4(v)$ which is orthogonal to the kernel of A^v has index constantly 2. Thus there exists a vector z which is orthogonal to the kernel y of A^v such that $A^v(z,z)=0$. Clearly the restriction of A^v to the plane $\mathbb{R}^2(y,z)$ vanishes. Now we can repeat the argument in the case 1 to get a vector w such that the restriction of T to $\mathbb{R}^2(v,w)$ vanishes. □

Theorem 3.1.

a) Any statistical space $(\mathbb{R}^n, g^0, T)$ can be embedded in the statistical space $(\mathbb{R}^{n(n+1)}, g^0, T' = 2||T||\sum_{i=1}^{N(n)} x_i^3)$, where x_i are the canonical Euclidean coordinates on $\mathbb{R}^{n(n+1)}$.

b) The trivial space $(\mathbb{R}^n, g^0, 0)$ can be embedded into $(\mathbb{R}^{2n}, g^0, \sum_{i=1}^{2n}(dx^i)^3)$ for all n.

Proof.

a) We prove by induction. The statement for $n=1$ follows from Proposition 3.3. Suppose that the statement is valid for all $n \le k$.

Lemma 3.5. *Suppose that $T \in S^3(\mathbb{R}^{k+1})$. Then there are orthonormal coordinates $x_1, \cdots, x_k$ such that*

$$T = x_1 \sum_{i=1}^{k+1} a_i x_i^2 + \sum_{1<i,j,k} a_{ijk} x_i x_j x_k. \tag{14}$$

Proof of Lemma 3.5. We choose v_1 as the unit vector in $S^k \subset \mathbb{R}^{k+1}$, on which the function $T(v,v,v)$ reaches the maximum on the unit sphere S^k. The first variation formula shows that $T(v_1,v_1,w)=0$ for all w which is orthogonal to v_1. We denote by $\mathbb{R}^k$ the orthogonal complement to $\mathbb{R}\cdot v_1$. Now we consider a bilinear symmetric form A on $\mathbb{R}^k$ defined as follows

$$A(x,y) = S(v_1, x, y).$$

There is an orthonormal basis on $\mathbb{R}^k$, where we can write $A(x,y) = \sum_{i=2}^{k+1} a_i x_i^2$. Clearly in this orthonormal basis we can write T in the form in (14). □

Continuation of the proof of Theorem 3.1.a We shall show explicitly that that any statistical space $(\mathbb{R}^2, g^0, T = a_2 x_1 (x_2)^2)$ can be embedded in $(\mathbb{R}^4, g^0, \sum_{i=1}^4 (y_i)^3)$, if $0 \le |a_2| \le 1/2$. We put

$$L(v_1) := \pm(\frac{1}{2}, \frac{1}{2}, -\frac{1}{2}, -\frac{1}{2}) \tag{15}$$

$$L(v_2) := (\sqrt{\frac{1+2a_2}{2}}, -\sqrt{\frac{1+2a_2}{2}}, \sqrt{\frac{1-2a_2}{2}}, -\sqrt{\frac{1-2a_2}{2}}). \tag{16}$$

Here we take the sign+ in (15), if $a_2 > 0$, and we take the sign−, if $a_2 < 0$. Clearly, L defines the required embedding $\mathbb{R}^2 \to \mathbb{R}^4$.

This together with Proposition 3.3 and the induction assumption complete the proof of Theorem 3.1.a.

Proof of Theorem 3.1.b. We decompose the embedding $f : (\mathbb{R}^n, g^0, 0)$ to $(\mathbb{R}^{2n}, g^0, \sum_{i=1}^{2n}(x^i)^3)$ as follows

$$f(x_1, \cdots, x_n) = (f^1(x_1), \cdots, f^n(x_n))$$

where f^i embeds the line $(\mathbb{R}, (dx^i)^2, 0)$ into $(\mathbb{R}^2, (dx^{2i-1})^2 + (dx^{2i})^2, (dx^{2i-1})^3 + (dx^{2i})^3)$. Clearly, f is the required embedding. □

4. Monotone Invariants and Obstructions to Embeddings of Statistical Manifolds

Let $K(M, e)$ denote the category of statistical manifolds M with morphisms being embeddings. Functors of this category are called *monotone invariants* of statistical manifolds. Clearly any monotone invariant is an invariant of statistical manifolds.

4.1. *Examples*

There are many monotone invariants which arise from our analysis in Sec. 3.

a) *Trace type of a statistical manifold.* A statistical manifold (M, g, T) will be called of *trace type*, if for all $x \in M$ the form $T(x)$ is of trace type (see 3.1) It follows from Proposition 3.1 that any statistical submanifold of a statistical manifold of trace type is also of trace type. Thus the trace type is a monotone invariant. In particular we cannot embed the statistical space Cap^N and the normal Gaussian space into any statistical space of trace type. On the other hand, unlike the linear case, we cannot embed a statistical manifold of trace type into another one of trace type, even if the norm condition is satisfied. For example, if the trace form is closed (or exact), then the trace form of its submanifolds is also closed (resp. exact). Hence within a class of statistical manifolds of trace type we get a new monotone invariants which can be expressed via the closedness and the cohomology class of the corresponding trace form.

b) *Decomposability* of a statistical manifold. We note that the class of 3-symmetric tensors of trace form is a subclass of all *decomposable tensors* T^3 which are a symmetric product of 1-forms and symmetric 2-forms. Any statistical submanifold of a statistical manifold with a decomposable tensor T has also the (induced) decomposable tensor. Thus the decomposability is also a monotone invariant. The Gaussian normal 2-dimensional manifold is an example of decomposable type but not of trace type.

c) *Rank and comass.* We define for any statistical manifold (M, g, T) the following number

$$rank(T) = \sup rank(T(x))$$

$$||T||_0 = \sup_{x \in M} ||T(x)||.$$

$$\mathcal{M}^1(T)_0 = \sup_{x \in M} \mathcal{M}^1(T(x)).$$

$$||T||_{1,0} = \sup_{x \in M} ||T(x)||_1.$$

Clearly these four numbers are monotone invariants of statistical manifolds.

We recall that the normal Gaussian statistical manifold is the two dimensional statistical model which is upper half of the plane $\mathbb{R}^2(\mu, \sigma)$ with the potential

$$p(\mu, \sigma)(x) = \frac{1}{\sqrt{2\pi}\,\sigma} \exp(\frac{-(x-\mu)^2}{2\sigma^2}),$$

here $x \in \mathbb{R}$.

Proposition 4.1. *Any statistical manifold which is conformal equivalent to the space Cap^N cannot be embedded into the direct product of m copies of the normal Gaussian statistical manifold for any $N \geq 3$ and finite m.*

Proof. It is easy to check that $\mathcal{M}^1(Cap^N) = \infty$. Thus any statistical manifold which is conformal equivalent to Cap^N has also the infinite invariant $\mathcal{M}^1$. On the other hand, we compute easily that the norm $\mathcal{M}^1$ of the Gaussian normal manifold, as well as the norm $\mathcal{M}^1$ of a direct product of its finite copies, is finite. Namely the norm $\mathcal{M}^1(\mu, \sigma)$ is $\sqrt{2}$ for all (μ, σ). □

4.2. *Diameters of Statistical Manifolds*

For a positive number $\rho > 0$ and a statistical manifold (M, g, T) we set

$$d_\rho(M, g, T) := \sup\{l \in R^+ \cup \infty \mid \exists \text{ an embedding of } ([0, l], dx^2, \rho(dx)^3) \text{ to } (M, g, T).\}$$

We shall call $d_\rho(M, g, T)$ *the diameter with weight* ρ of (M, g, T). Clearly d_ρ are monotone invariants for all ρ.

To estimate the diameter with weight ρ of a given statistical manifold (M, g, T) we can proceed as follows. For each point $x \in M$ we denote by $D_\rho(x)$ the set of all unit tangential vector $v \in T_xM$ such that $T(v, v, v) = \rho$. We denote by $D^i_\rho(x)$ the connected components of $D^i_\rho(x)$. We say that a unite vector v in T_xM is *ρ-characteristic* with weight $c(x)$, if there exists i such that we have

$$c(x) = \min_{w \in D^i_\rho(x)} < v, w >> 0.$$

We shall say that a point $x \in M$ is *ρ-regular*, if there is an open neighborhood $U_\varepsilon(x) \subset M$ such that $D_\rho(U_\varepsilon) = U_\varepsilon \times D_\rho(x)$. It is easy to see that the set of all ρ-regular points is open and dense in M for any given ρ.

Proposition 4.2. *The diameter* d_ρ *of* (M^m, g, T) *is infinite, if* $m \geq 3$ *and there exists a number* $\varepsilon > 0$ *such that one of the following 2 conditions holds:*
a)There exists a $(\rho + \varepsilon)$*-regular point* $x \in M$ *such that the convex hull* $Cov(D^i_{\rho+\varepsilon}(x))$ *of one of connected components* $D^i_{\rho+\varepsilon}(x)$ *contains the origin point* $0 \in T_xM^m$ *as it interior point.*
b) (M^m, g, T) *has a complete Riemannian submanifold* $(N, \overline{g})$ *such that there exists a smooth section* $x \mapsto (D_{\rho+\varepsilon}(x) \cap TN)$ *over* N.

Proof. The statement under the first condition a) is based on the fundamental Lemma of the convex integration technique of Gromov. Namely Gromov proved that (2.4.1.A, Ref. 5), if the convex hull of some path connected subset $A_0 \subset \mathbb{R}^q$ contains a small neighborhood of the origin, then there exists a map $f : S^1 \to \mathbb{R}^q$ whose derivative sends S^1 into A_0. □

Lemma 4.1. *Under the condition in Proposition 4.2.a there exists a small neighborhood* $U_\delta(x)$ *in* M *and an embedded oriented curve* $S^1 \subset U_\delta(x)$ *such that for all point* $s(t) \in S^1$ *we have* $\mathcal{M}^1(T_{s(t)}S^1) \geq \rho + (\varepsilon/2)$.

Proof of Lemma 4.1. We denote by Exp the exponential map $T_xM^m \to M^m$ and by $DExp$ the differential of this exponential map restricted to $S^{m-1} \times T_xM^m \subset T(T_xM^m)$. Here S^{m-1} is the unit sphere in T_xM^m. The space T_xM^m is a linear statistical space, so we denote by $\mathcal{M}^1_x$ the induced norm-function on $S^{m-1} \times T_xM^m$ as follows:

$$\mathcal{M}^1_x(l) = T_x(l, l, l).$$

Since $DExp$ is a continuous function, whose restriction to $S^{m-1} \times 0$ is the identity, there exists a ball $B(0, \delta)$ with center in $0 \in T_x M$ such that

$$\mathcal{M}^1(DExp(l)) - \mathcal{M}^1_x(l)) < \varepsilon/4 \tag{17}$$

for all $l \in S^{m-1} \times B(\delta) \subset T(T_x M^m)$. We can assume that δ is so small such that $DExp$ is a homeomorphism on $S^{m-1} \times B(0, \delta)$.

Now we apply the above mentioned Gromov Lemma (2.4.1.A, Ref. 5) to get a oriented curve $S^1(t)$ in the linear space $T_x M$ such that

$$T(\frac{(\partial/\partial t)S^1(t)}{|(\partial/\partial t)S^1(t)|}) = \rho + \varepsilon \tag{18}$$

for all t. Next we observe that for all $\alpha > 0$ the curve $\alpha \cdot S^1(t)$ has the same norm as $S^1(t)$, i.e.

$$\mathcal{M}^1_x(T_{|(\alpha \cdot S^1)}(t)) = \mathcal{M}^1_x(T_{|(S^1)}(t)) = \rho + \varepsilon.$$

Thus we can assume that our curve $S^1(t)$, which satisfies (18), lies in the ball $B(0, \delta)$. By our choice of δ (see (17)), we get from (18)

$$\rho + \frac{3}{4}\varepsilon \leq \mathcal{M}^1(Exp(S^1(t))) \leq \rho + \frac{5}{4}\varepsilon, \tag{19}$$

for all t. This curve $Exp(S^1(t))$ is an immersed curve. To get an embedded curve we perturb the immersed curve such that the condition of Lemma 4.1 is satisfied. This is possible, since $m \geq 3$.

Now let us to continue the proof of Proposition 4.2.a. We denote by $S^1(t)$ the embedded curve in Lemma 4.1. Next by choosing a tubular neighborhood of $S^1(t)$ we can get a (small, thin) oriented embedded solid torus $T^3(t, s, r) = S^1(t) \times S^1(s) \times [0, R]$ in M^m such that our embedded curve is exactly the mean curve $S^1(t) \times \{0\} \times \{0\}$ on the solid torus. We can choose this torus T^3 so thin, such that for all s, t, r we have

$$\mathcal{M}^1(T^2_r(t, s)) \geq \rho + \frac{\varepsilon}{4}. \tag{20}$$

Using (20) we choose a smooth unit vector field $V(t, s)$ on the torus $T^3(t, s, r)$ which is tangential to each torus $T^2_r(t, s)$ such that $T(V, V, V) = \rho$. The integral curve of this vector field is either a circle or an curve of infinite length. If there exists an integral curve of infinite length, then this curve is our desired curve for the Proposition 4.2.a. Assume now that all the integral curves are circles. Then there exist an embedding $S^1(t) \times [0, \mu] \times [0, \mu]$ such that for all $(s, r) \in [0, \mu] \times [0, \mu]$ the circle $S^1(t) \times \{s\} \times \{r\}$ is an integral curve of V. Now we perturb V in a neighborhood $[0, \alpha] \times [0, \mu] \times [0, \alpha]$ with a very small α such that the perturbed unit vector field V' satisfies

$T(V', V', V') = \rho$ and the integral curve of vector field V' is not any more periodic. This completes the proof of the first part in Proposition 4.2.

Using the same argument we can prove the second part b) of Proposition 4.2. First we get the existence of an embedded curve $S^1(t)$ of arbitrary length on M such that $\mathcal{M}^1(T_{|S^1}(t)) \geq \rho + (1/4)\varepsilon$. Now we consider a torus tubular neighborhood of this curve in M and apply the same argument in the first part, namely we get on each torus $T^2(t, s)$ an integral curve whose unit tangential vector $V = (\partial/\partial t)S^1(t; s, r)$ satisfies the condition:

$$T(V, V, V) = \rho.$$

If there exists an infinite integral curve, then we are done. If not, that means all integral curve are circles, then we apply the perturbation method in the proof of the first part and get our desired curve. □

5. Existence of Isostatistical Embeddings into Cap^N

Main Theorem. *Any smooth (C^1 resp.) statistical manifold (M^n, g, T) can be immersed into the statistical manifold (Cap_+^N, g^F, T^{A-C}) for some finite number N. Hence any statistical manifold is a statistical model.*

We first deduce our Main Theorem for compact statistical manifolds (M^n, g, T) from Theorem 5.1 and Theorem 5.4.

Theorem 5.1. *Let (M^m, g, T) be a compact smooth (C^1 resp.) statistical manifold. Then there exist numbers $N \in \mathbb{N}^+$ and $A \geq 0$ as well as a smooth (C^1 resp.) embedding $f : (M^m, g, T) \to (\mathbb{R}^N, g_0, A \cdot T_0)$ such that $f^*(g_0) = g$ and $f^*(A \cdot T_0) = T$.*

Our proof of Theorem 5.1 uses the Nash embedding theorem, the Gromov embedding theorem and an algebraic trick. The existence of monotone invariants prevents us extend Theorem 5.1 for non-compact case (in contrast to the Riemannian case.)

Theorem 5.2 (The Nash embedding theorem[10,11]). *Any smooth (C^1 resp.) -Riemannian manifold (M^n, g) can be isometrically embedded into $(\mathbb{R}^N, g_0)$ for some N depending on M^n.*

We denote by T_0 the "standard" 3-tensor on $\mathbb{R}^n$:

$$T_0 = \sum_{i=1}^{n} dx_i^3.$$

Theorem 5.3 (The Gromov immersion theorem[5]). *Suppose that M^m is given with a smooth (C^1 resp.) symmetric 3-form T. Then there exists an embedding $f : M^m \to \mathbb{R}^{N_1(m)}$ with $N_1(m) = 3(n + \binom{n+1}{2} + \binom{n+2}{3})$ such that $f^*(T_0) = T$.*

Proof of Theorem 5.1. First we shall take an immersion $f_1 : (M^m, g, T) \to (\mathbb{R}^{N_1(m)}, g_0, T_0)$ such that

$$f_1^*(T_0) = T.$$

The existence of f_1 follows from the Gromov immersion theorem.

Then we choose a positive number A^{-1} such that

$$g - A^{-1}(f_1^*(g_0)) = g_1$$

is a Riemannian metric on M, i.e. g_1 is a positive symmetric bi-linear form. Such a number A exists, since M is compact.

Now we shall choose an isometric immersion $f_2 : (M^m, g_1) \to (\mathbb{R}^N, g_0)$. The existence of f_2 follows from the Nash isometric immersion theorem.

Lemma 5.1. *There is a linear isometric embedding $L_{m+1} : \mathbb{R}^{m+1} \to \mathbb{R}^{2m+2}$ such that $L_{n+1}(T_0) = 0$.*

Proof. We put

$$L_{m+1}(x_1, \cdots, x_{m+1}) = (f^1(x_1), \cdots, f^{m+1}(x_{m+1}))$$

where f^i embeds the line $(\mathbb{R}, (dx^i)^2, 0)$ into $(\mathbb{R}^2, (dx^{2i-1})^2 + (dx^{2i})^2, (dx^{2i-1})^3 + (dx^{2i})^3)$:

$$f^i(x_i) = \frac{1}{\sqrt{2}}(x_{2i-1} - x_{2i}).$$

Clearly, L_{m+1} is the required embedding. □

Completion of the proof of Theorem 5.1. Finally we take an embedding

$$f_3 : M^m \to \mathbb{R}^{(m+1)(m+2)+m} \oplus \mathbb{R}^{2m+2}$$

as follows.

$$f_3(x) = A^{-1} \cdot f_1(x) \oplus (L_{n+1} \circ f_2).$$

Since f_2 is an embedding, f_3 is the required embedding map for Theorem 5.1.

Theorem 5.4. *Suppose that C is a compact subset in Cap_+^{4n}. Then any bounded domain D in a linear statistical manifold $(\mathbb{R}^n, g_0, A \cdot T_0)$ can be realized as an embedded statistical submanifold of $Cap_+^{4n} \setminus C$.*

We denote by $Cap^3_+(\lambda)$ the statistical manifold which is obtained by the restriction of the statistical structure from $(\mathbb{R}^4_+, g_0, \sum_{i=1}^4 (x_i)^{-1}(dx_i)^3)$ to the positive quadrant $S^3_+(\lambda)$ of the sphere of radius λ with center at the origin of $\mathbb{R}^4$.

Proof of Theorem 5.4. We put

$$\overline{A} := \max\{4\sqrt{n}, 4\sqrt{n} \cdot A\}.$$

Let U be an open neighborhood of $(\lambda, (2\overline{A})^{-1}, (2\overline{A})^{-1}, (2\overline{A})^{-1}) \in S^3(2/\sqrt{n})$. Here λ is the positive number such that

$$n\lambda^2 + 3n(2\overline{A})^{-2} = 4. \tag{21}$$

We denote by U_+ the intersection $U \cap Cap^3_+(2/\sqrt{n})$. We now choose U so small such that the product $U \times_{n\,times} U \subset S^{4n-1}$ lies in the complement $S^{4n-1} \setminus C$. Since U_+ is a statistical submanifold of $(R^4_+, g_0, \sum_{i=1}^4 x_i^{-1}(dx_i)^3)$, the direct product $U_+ \times_{n\,times} U_+$ is a statistical submanifold of $(\mathbb{R}^{4n}_+, g_0, \sum_{i=1}^{4n} x_i^{-1}(dx_i)^3)$. Hence $U_+ \times_{n\,times} U_+$ is a statistical submanifold of (Cap^{4n}, g^F, T^{A-C}).

We denote by $U(A, r)$ the ball of radius r at the point $(\lambda, (2\overline{A})^{-1}, (2\overline{A})^{-1}, (2\overline{A})^{-1}) \in S^3(2/\sqrt{n})$. First we prove

Lemma 5.2. *For given positive numbers $R > 0$ and $A \geq 0$ there exists a positive number r such that the bounded domain $[0, R] \times_{n\,times} [0, R] \subset (\mathbb{R}^n, g_0, A \cdot T_0)$ can be realized as an immersed statistical submanifold of $U_+(A, r) \times_{n\,times} U_+(A, r) \subset (Cap^{4n}_+, g^F, T^{A-C})$.*

Proof of Lemma 5.2. It suffices to show that there is a statistical immersion $f : ([0, R], dx^2, A \cdot dx^3) \to U_+(A, r)$. On $U_+(A, r)$ we consider the distribution $D(\rho)$ which is defined by

$$D_x(\rho) := \{v \in T_x U_+(A, r) : |v| = 1, T(v, v, v) = \rho\}$$

for any given $\rho > 0$. Clearly the existence of an immersion $f : ([0, \mathbb{R}], dx^2, A \cdot dx^3) \to U_+(A, r)$ is equivalent to the existence of an integral curve with the length R of the distribution $D(A)$. The existence of the desired curve is a consequence of the following Lemma

Lemma 5.3. *There exists an embedded torus T^2 in $U_+(A, r)$ which is provided with a unit vector field V on T^2 such that $T(V, V, V) = A$.*

Proof. Let us denote $x_0 := (\lambda, (2\overline{A})^{-1}, (2\overline{A})^{-1}, (2\overline{A})^{-1}) \in S^3(2/\sqrt{n})$ with λ satisfying (21). We shall need the following

Lemma 5.4 (Sublemma). *Let H be any 2-dimensional subspace in $T_{x_0}U_+(A,r)$. Then there exists a unit vector $w \in H$ such that $T(w,w,w) \geq 2A$.*

Proof of Sublemma 5.4. The subspace H can be defined by two linear equations:

$$< w, \vec{x}_0 >= 0, \tag{22}$$

$$< w, \vec{h} >= 0. \tag{23}$$

Here w is a vector in $H \subset \mathbb{R}^4$ and $\vec{h}$ is a unit vector in $\mathbb{R}^4$ which is not co-linear with $\vec{x}_0$ and which is orthogonal to H. Without loss of generality we can assume that

$$\vec{h} = (0 = h_1, h_2, h_3, h_4) \text{ and } \sum_i h_i^2 = 1.$$

Case 1. Suppose that not all the coordinates h_i of $\vec{h}$ are of the same sign, so we assume that $h_1 = 0, h_2 \leq 0, h_3 > 0$. We put

$$k_2 := \frac{-h_2}{\sqrt{(h_2)^2 + (h_3)^2}}, \; k_3 := \frac{h_3}{\sqrt{(h_2)^2 + (h_3)^2}},$$

$$w := (w_1, w_2 = (1-\varepsilon_2)k_3, w_3 = (1-\varepsilon_2)k_2, 0 = w_4). \tag{24}$$

The equation (23) for w is obviously satisfied. Now we choose w_1, ε_2 from the following equations which are equivalent to (22) and the normalization of w:

$$\lambda \cdot w_1 + (1-\varepsilon_2) \cdot (2\bar{A})^{-1} \cdot (k_2 + k_3) = 0, \tag{25}$$

$$w_1^2 = (2\varepsilon_2 - \varepsilon_2^2). \tag{26}$$

- From (25) we get

$$w_1 = \frac{(1-\varepsilon_2)(k_2+k_3)}{\lambda \cdot 2\bar{A}}. \tag{27}$$

Substituting this into (26), we get

$$\left(\frac{(k_2+k_3)^2}{(\lambda \cdot 2\bar{A})^2} + 1\right)\varepsilon_2^2 - \left(2 + \frac{2(k_2+k_3)^2}{(\lambda \cdot 2\bar{A})^2}\right)\varepsilon_2 + \left(\frac{k_2+k_3}{\lambda \cdot 2\bar{A}}\right)^2 = 0. \tag{28}$$

We shall take one of (2 possible) solutions ε_2 of (28) which is

$$\varepsilon_2 = 1 + \left(\frac{k_2+k_3}{\lambda \cdot 2\bar{A}}\right)^2 - \sqrt{1 + \frac{(k_2+k_3)^2}{(\lambda \cdot 2\bar{A})^2}}. \tag{29}$$

• From (29) we get

$$\varepsilon_2 \leq \frac{5}{16}, y$$

since $0 < k_2 + k_3 \leq 2$ (this follows from (24)), and $\lambda \geq 2n^{-1/2}$ (this follows from (21)), so $\lambda \cdot 2\overline{A} \geq 16$. Since

$$T_{x_0}(w, w, w) = \lambda^{-1} w_1^3 + (2\overline{A})(w_2^3 + w_3^3)$$

we get

$$T(w, w, w) \geq 8\sqrt{n} \cdot A(1 - \frac{5}{16}) \cdot \frac{1}{2\sqrt{2}} \geq 2A.$$

So in this case 1 Sublemma 5.4 holds.

Case 2. Now we shall assume that $h_1 = 0$ and $h_2 \geq h_3 \geq h_4 > 0$. We set

$$w := (w_1, w_2 = -a(1 - \varepsilon_2), w_3 = -a(1 - \varepsilon_2), w_4 = \alpha \cdot a(1 - \varepsilon_2)) \text{ where}$$

$$\alpha := \frac{h_2 + h_3}{h_4} \geq 2, \tag{30}$$

$$a > 0 \text{ such that } a^2(\alpha^2 + 2) = 1. \tag{31}$$

The Eq. (30) ensures that $< w, h >= 0$. In order that w is a unite vector and $w \in H$ the numbers w_1, ε_2 must satisfy the following equations

$$-\lambda w_1 + \frac{(1 - \varepsilon_2)(\alpha - 2)a}{2\overline{A}} = 0, \tag{32}$$

$$w_1^2 = (2\varepsilon_2 - \varepsilon_2^2). \tag{33}$$

• From (32) we get

$$w_1 = \frac{(1 - \varepsilon_2)(\alpha - 2)a}{\lambda 2\overline{A}}. \tag{34}$$

Now substituting (34) into (33) we get a solution ε_2

$$\varepsilon_2 = (1 + 2B^2) - \sqrt{(1 + 2B^2)^2 - B^2(1 + B^2)},$$

where

$$B = \frac{(\alpha - 2)a}{\lambda 2\overline{A}} < \frac{1}{\lambda 2\overline{A}}.$$

Since $0 < (\alpha - 2)a < 1$ we have $\varepsilon_2 < 1/(\lambda 2\overline{A})$ and $\varepsilon_1 < 1/(\lambda 2\overline{A})$. Hence

$$T(w, w, w) = \lambda^{-1} w_1^3 + (2\overline{A}) a^3 (1 - \varepsilon_2)^3 (2 + \alpha^3) > 2A. \qquad \square$$

Completion of the proof of Lemma 5.3. First we choose a small embedded torus T^2 in $U_+(A, r)$ such that for all $x \in T^2$ we have

$$\max_{v \in S^1(T_x T^2)} \{T(v, v, v)\} \geq \frac{3}{2} A. \tag{35}$$

This is possible thank to Sublemma 5.4. Since $T(v, v, v) = -T(v, v, v)$ and $T^2 = \mathbb{R}^2/\mathbb{Z}^2$ is parallelizable, (35) implies that we can find a smooth vector field V on T^2 sastifying the condition in Lemma 5.3.
Completion of the proof of Theorem 5.4. For a given A in Theorem 5.5 we let $A' := A+\varepsilon$ for some small positive ε and we apply Lemma 5.2 to (R, A') which is in fact to apply Lemma 5.3. We can show that the existence of an isostatistical immersion $f' : ([0, R], dx^2, A' \cdot dx^3) \to U_+(A', r)$ implies the existence of an isostatistical embedding $f : ([0, R], dx^2, A \cdot dx^3) to U_+(A', r)$ by using the same argument in the proof of Proposition 4.4.a.
Proof of Main Theorem for the non compact case. We can deal with this case by using the compact decomposition of M^m as Nash did for the isometric embedding for the smooth case [10]. Namely we cover M^m by disk neighborhoods N_i^j, $j = \overline{1, m+1}$ in the following way. For each j let

$$C^j := \cup_i N_i^j. \tag{36}$$

Then we require that the union in (36) is a disjoint union, i.e. $N_i^j \cap N_k^j = \emptyset$, if $i \neq k$. We also require that each N_i^j overlaps only a finite number of other N_l^k. Now we "compactify" N_i^j via an surjective smooth mapping $: \phi_i : N_i^j \to S_i^j$, where S_i^j is a sphere of the same dimension m. The map ϕ_i^j can be extended to the whole M^m, since it maps the boundary of N_i^j into the north point of the sphere. On the other hand, this map ϕ_i^j is injective in a large (enough) sub domain $\overline{N}_i^j \subset N_i^j$. We can furthermore use the unity partition function to define a C^1 statistical structure on each S_i^j such that the (sum of) pull back via ϕ_i^j is the given statistical structure on M. In other words we can consider the C^1 statistical structure on M^m as induced from (infinitely many) spheres S_i^j via the smooth mapping ϕ_i^j.

Now let for each j , $1 \leq j \leq m+1$, we put

$$S^j := \cup_i S_i^j.$$

Using Theorem 5.1 we can find an isostatistical embedding

$$\psi^j : S^j \to (S_+^{4l(m)-1}(\frac{2}{\sqrt{m+1}}), i^*(g_0), i^*(T_S))$$

inductively, since each S_i^j is compact.

Now let us consider the map

$$I_j : M^m \to (S_+^{4l(m)-1}(\frac{2}{\sqrt{m+1}}), i^*(g_0), i^*(T_S)),$$

as a composition of the map $\phi_i^j : M \to S_i^j$ and the map ψ_j.

Finally the product mapping

$$I_0 = I_1 \times \cdots \times I_{m+1}$$

is the desired isostatistical embedding, since the statistical structure on $S^{N(m)-1}$ is induced from the (non-linear) statistical structure on $\mathbb{R}^{N(m)}$. We can see this easily by noticing that both symmetric forms g_0 and T_S are decomposable w.r.t. the embedding $i : \mathbb{R}^n \to \mathbb{R}^{n+l}$ for any $n, l > 0$, i.e. $g_0(\mathbb{R}^n) = i_*(g_0)(\mathbb{R}^{n+l})$, $T_S(R_+^n) = i^*(T_S)(\mathbb{R}_+^{n+l})$.

Acknowledgement

I am thankful to Jürgen Jost and Nihat Ay for their introduction to the field of information geometry and helpful discussions.

References

1. S. Amari and H. Nagaoka, *Methods of Information Geometry*, Trans. of Math. Monograph (2000).
2. S. Amari, *Contemporary Math.* **203**, 81 (1997).
3. N. Ay, *Ann. of Prob.* **30**, 416 (2002).
4. N. N. Chentsov, *Statistical decision rules and optimal inference*, (AMS. 1982), (originally published in Russian, Nauka, Moscow, 1972).
5. M. Gromov, *Partial Differential Relations,* (Springer-Verlag, 1986).
6. J. Jost, *Information geometry,* (lecture at MPI MIS 2005).
7. S. Lauritzen, Statistical manifolds, in *Differential Geometry in Statistical Inference*, IMS Lecture Notes, Monograph Serie 10., (Inst. of Math. Stat. Hayward, California, 1987).
8. T. Matsumoto, *Hiroshima Math. J.* **23**, 327 (1993).
9. E. A. Morozova and N. N. Chentsov, *Markov invariant geometry on manifolds of states* (in Russian), (Itogi Nauki i Techniki 36, 1990) pp. 69-102.
10. J. Nash, *Ann. Math.* **60**, 383 (1954).
11. J. Nash, *Ann. of Math.*, **63**, 20 (1956).
12. E. B. Vinberg, A. L. Onishchik, *Seminars on Lie Groups and Algebraic Groups*, (in Russian) (Nauka, Moscow, 1988).
13. C. R. Rao, *Bull. of the Calcutta Math. Soc.* **37**, 89 (1945).
14. C. R Rao, Differential metrics in probability spaces, in *Differential Geometry in Statistical Inference, IMS Lecture Notes*, Mongraph Serie 10., (Inst. of Math. Stat. Hayward, California, 1987).

Advances in Deterministic and Stochastic Analysis
Eds. N. M. Chuong *et al.* (pp. 255–268)

§13. GRADED ČECH COHOMOLOGY IN NONCOMMUTATIVE GEOMETRY *†

DO NGOC DIEP
Institute of Mathematics,
Vietnam Academy of Science and Technology,
18 Hoang Quoc Viet Road, Cau Giay Dist., 10307 Hanoi, Vietnam
E-mail: dndiep@math.ac.vn

The $\mathbf{Z}_2$-graded Čech cohomology theory is considered in the framework of noncommutative geometry over complex number field and in particular the homotopy invariance and Morita invariance are proven. In some special case we deduce an isomorphism between this noncommutative theory and the classical $\mathbf{Z}_2$-graded Čech cohomology theory.

Keywords: Čech cohomology of a sheaf, cyclic theory
Mathematics Subject Classification 2000: 14F20, 19K35, 46L80, 46M20

1. Introduction

Let us fix in this paper the field of complex numbers as the ground field for algebras, modules, etc.

The general Čech cohomology presheaf $\check{\mathcal{H}}^q$ for an arbitrary presheaf X on a category $\mathcal{C}$, $\check{H}^q(X,G) = \check{\mathcal{H}}^q(G)(X)$ was introduced in Ref. 7.

The idea of Čech cohomology for noncommutative geometry was appeared in Refs.[2] and 6. In this paper we use this idea to define the corresponding (periodic) $\mathbf{Z}_2$-graded Čech theory.

We prove the homotopy invariance and Morita invariance of Čech cohomology in the framework of noncommutative geometry. Our main result is based on a detailed analysis of the structure of C*-algebras. A crucial observation is the fact that for C*-algebras the category of *-representations

*The text is presented at the Second International Conference on Abstract and Applied Analysis, Quinhon, Vietnam, 4-9 June 2005. The author thanks the organizers and especially Professor DSc. Nguyen Minh Chuong for the invitation to give this lecture.
†The work was supported in part by Vietnam National Project of Research in Fundamental Sciences and The Abdus Salam ICTP, UNESCO.

defines exactly the C*-algebra itself, by the well-known Gelfand-Naimark-Segal Theorem. From this we can deduce the Morita invariance and homotopy invariance of the Čech cohomology, the same properties of periodic cyclic homology of the C*-algebra. Since our result is valid not only for C*-algebras, we work in the general context of a noncommutative algebra over complex numbers.

The paper is organized as follows. Taking the Čech cohomology in place of the de Rham theory in the periodic cyclic theory of A. Connes, in Sec. 3 we define the $\mathbf{Z}_2$-graded Čech cohomology theory. Then in Sec. 4 we prove two important properties of the theory as the homotopy invariance and Morita invariance. In the last Sec. 5 we deduce also some kind of Connes-Hochschild-Kostant-Rosenberg theorem. This lets us see a clear relation with the classical case of commutative algebras and ordinary Čech cohomology theory.

Notations: We follow the notations from Refs. 7 and 4

2. Preparation: Differential Systems and the Cyclic Theory

2.1. *Differential Systems, Following Kashiwara*

From M. Kashiwara Ph. D. dissertation:[1] Linear polynomial differential systems

$$\sum_{j=1}^{N} P_{ij}u_j = 0, \qquad \sum_{\mu=1}^{R} Q_{\lambda,\mu}v_\mu = 0$$

correspond to the $\mathcal{D}$-modules

$$\mathcal{M} = \mathcal{D}^N/\mathcal{D}^{N'}P \qquad \mathcal{N} = \mathcal{D}^R/\mathcal{D}^{R'}Q$$

In general if

$$\mathcal{E} \xrightarrow{P} \mathcal{F}$$

is a system of partial differential operators from the vector bundle $\mathcal{E}$ to $\mathcal{F}$ and set

$$\check{\mathcal{E}} = \mathrm{Hom}_{\mathcal{O}}(\mathcal{E}, \mathcal{O}),$$

then

$$\mathcal{D} \otimes \check{\mathcal{E}} \cong \{ \text{ differential operators from } \mathcal{E} \text{ to } \mathcal{O}\}$$

and we get D-module homomorphism

$$\mathcal{D} \otimes \check{\mathcal{E}} \xleftarrow{P} \mathcal{D} \otimes \check{\mathcal{F}}.$$

Let $\mathcal{M}$ be the cokernel, then

$$\mathrm{Hom}_{\mathcal{D}}(\mathcal{D}\otimes\check{\mathcal{E}},\mathcal{O}) = \mathrm{Hom}_{\mathcal{O}}(\mathcal{E}^{\vee},\mathcal{O}) = \mathcal{E}$$

and the $\mathcal{D}$-module $\mathcal{M}$ represents $\mathcal{E} \xrightarrow{P} \mathcal{F}$, e. i. the following short sequence is exact

$$0 \longrightarrow \mathrm{Hom}_{\mathcal{D}}(\mathcal{M},\mathcal{O}) \longrightarrow \mathcal{E} \xrightarrow{P} \mathcal{F}$$

and $\mathrm{Hom}_{\mathcal{D}}(\mathcal{E},\mathcal{O})$ appears as the sheaf of solution of $P:\mathcal{E}\to\mathcal{F}$.

In the same way for a given

$$\mathcal{E} \xrightarrow{P} \mathcal{F} \xrightarrow{Q} \mathcal{G}, \text{ such that } Q\circ P = 0$$

and

$$0 \longleftarrow \mathcal{M} \longleftarrow \mathcal{D}\otimes\check{\mathcal{E}} \xleftarrow{P} \mathcal{D}\otimes\check{\mathcal{F}} \xleftarrow{Q} \mathcal{D}\otimes\check{\mathcal{G}}$$

is exact, then the cohomology of the resolution

$$\mathcal{E} \longrightarrow \mathcal{F} \longrightarrow \mathcal{G}$$

is isomorphic to $\mathrm{Ext}_{\mathcal{D}}(M,\mathcal{O})$. Therefore, $\mathrm{Ext}^1_{\mathcal{D}}(\mathcal{M},\mathcal{O})$ represents the obstruction for an element $f\in\mathcal{F}$ satisfying the compatibility condition $Qf=0$ to be written as $f=Pg$ for some $g\in\mathcal{E}$. Thus the computation of $\mathrm{Ext}^i_{\mathcal{D}}(\mathcal{M},\mathcal{O})$ become an essential problem.

Let us recall the first Spencer sequence. Let $\mathcal{M}$ be a $\mathcal{D}$-module with a filtration $\mathcal{M}=\cup_k\mathcal{M}_k$ and we define the $\mathcal{D}$-homomomorphic derivative

$$\delta:\mathcal{D}\otimes_{\mathcal{O}}\wedge^p\theta\otimes_{\mathcal{O}}\mathcal{M}\to\mathcal{D}\otimes_{\mathcal{O}}\wedge^{p-1}\theta\otimes_{\mathcal{O}}\mathcal{M}$$

by the formula

$$\delta(P\otimes(\mathbf{v}_1\wedge\cdots\wedge\mathbf{v}_p)\otimes\mathbf{u} = \sum_{i=1}^{p}(-1)^{i-1}P\mathbf{v}_i\otimes(\mathbf{v}_1\wedge\cdots\wedge\hat{\mathbf{v}}_i\wedge\cdots\wedge\mathbf{v}_p)\otimes\mathbf{u}+$$

$$+\sum_{i=1}^{p}(-1)^iP\otimes(\mathbf{v}_1\wedge\cdots\wedge\hat{\mathbf{v}}_i\wedge\cdots\wedge\mathbf{v}_p)\otimes\mathbf{v}_i\mathbf{u}+$$

$$+\sum_{1\le i<j\le p}(-1)^{i+j}P\otimes([\mathbf{v}_i,\mathbf{v}_j]\wedge\mathbf{v}_1\wedge\cdots\wedge\hat{\mathbf{v}}_i\wedge\cdots\wedge\hat{\mathbf{v}}_j\wedge\cdots\wedge\mathbf{v}_p)\otimes\mathbf{u}.$$

Theorem 2.1 (D. Quillen[5]). *Assume that the differential system $\mathcal{M}$ is endowed with a good filtration $\mathcal{M}=\cup_k\mathcal{M}_k$. Then locally for sufficiently large k the first Spencer sequence associated to this filtration is exact.*

One of the famous applications of these ideas is the **Cauchy-Kovalevskaya Theorem**: Let us consider an analytic space X, Y an analytic subspace with defining $\mathcal{O}_X$-ideal $\mathcal{I}$. For any $\mathcal{O}_X$-modules $\mathcal{F}$ set

$$H^k_{[Y]}(\mathcal{F}) = \varinjlim_{\nu} \operatorname{Ext}^k_{\mathcal{O}_X}(\mathcal{O}_X/\mathcal{I}^{\nu+1}, \mathcal{F}).$$

There is a canonical homomorphism $H^k_{[Y]}(\mathcal{F}) \to H^k_Y(\mathcal{F})$. In the case where X and Y are complex manifolds, and $f : Y \to X$ an analytic map, define $\mathcal{O}^{(0,n)}_{Y\times X}$ the sheaf of holomorphic functions onf $Y \times X$, we have $\mathcal{D}_f = \mathcal{D}_{Y\to X} = H^n_{[Y]}(\mathcal{O}^{(0,n)}_{Y\times X})$ as the sheaf of differential operators of finite order from Y to X. In local coordinate x_i of X and y_j of Y, the sections of $\mathcal{D}_{Y\to X}$ may be written in form

$$P(y, \partial_x) = \sum_{|\alpha|\leq m} a_\alpha(y)\partial^\alpha_x,$$

where we use multiindex notation. For any $\varphi(x) \in \mathcal{O}_X$, we can define

$$(P\varphi)(y) = \sum_\alpha a_\alpha(y)[\partial^\alpha_x \varphi(x)]_{x=\varphi(y)}.$$

This define the product action

$$\mathcal{D}_{Y\to X} \times f^{-1}\mathcal{O}_X \to \mathcal{O}_Y.$$

It is not hard to prove that

$$\mathcal{D}_{Y\to X} \cong \mathcal{O}_Y \times_{f^{-1}\mathcal{O}_X} f^{-1}\mathcal{D}_X.$$

Define

$$f^*\mathcal{M} = \mathcal{D}_{Y\to X} \otimes_{f^{-1}\mathcal{D}_X} f^{-1}\mathcal{M} = \mathcal{O}_Y \otimes_{f^{-1}\mathcal{O}_X} f^{-1}\mathcal{M},$$

then

$$f^*\mathcal{O}_X = \mathcal{O}_Y, \quad f^*\mathcal{D}_X = \mathcal{D}_{Y\to X}.$$

Theorem 2.2 (Cauchy problem). *Let $f : Y \to X$ be a smooth holomophic map. For any coherent $\mathcal{D}_X$-modules $\mathcal{M}$ and $\mathcal{N}$, the map*

$$\nu : f^{-1}RHom_{\mathcal{D}_X}(\mathcal{M}, \mathcal{N}) \to RHom_{\mathcal{D}_Y}(f^*\mathcal{M}, f^*\mathcal{N})$$

is an isomorphism. Moreover, for any k,

$$f^{-1}\operatorname{Ext}^k_{\mathcal{D}_X}(\mathcal{M}, \mathcal{N}) \to \operatorname{Ext}^k_{\mathcal{D}_Y}(f^*\mathcal{M}, \mathcal{N})$$

is also an isomorphism. In the particular case where Y is an submanifold of the complex manifold X, let T^*X be the cotangent vector bundle of X

and $P^*X = (T^*X \setminus X)/\mathbf{C}^\times$ *the cotangent projective bundle of* X, $T^*_Y X$ *the kernel of the canonical projection*

$$T^*X \times_X Y \to T^*Y, \quad P^*_Y X = (T^*_Y X \setminus Y)/\mathbf{C} \subset P^*X.$$

The $\mathcal{D} = \mathcal{D}_X$*-module is non-characteristic if* $char(\mathcal{M}) \cap P^*_Y X = \emptyset$. *If* $\mathcal{M}$ *is a coherent* $\mathcal{D}_X$*-module and the submanifold* Y *is non-characteristic, then* $\mathcal{M}_Y = \mathcal{O}_Y \otimes_{\mathcal{O}_X} \mathcal{M} = \mathcal{D}_{Y \to X} \otimes_{\mathcal{D}_X} \mathcal{M}$ *is a coherent* $\mathcal{D}_Y$*-module and*

$$\mathrm{Hom}_{\mathcal{D}_X}(M, \mathcal{O}_X)|_Y \to \mathrm{Hom}_{\mathcal{D}_Y}(\mathcal{M}_Y, \mathcal{O}_Y)$$

is an isomorphism.

In the next we construct almost the same construction of homological algebra of C*-algebras and projective finitely generated C*-modules.

2.2. *Fields of C*-algebras and Sheafification*

One of the best results of J. M. G. Fell we need is the following proposition.

Proposition 2.1. *Every C*-algebra of type I can be presented as the C*-algebra* $\Gamma(X, \mathcal{E})$ *of cross-sections on the continuous field of elementary C*-algebras* $\mathcal{E}_\alpha$ *on its dual* $X = \hat{A}$

It is easy to see that using the Fourier-Gelfand transforms

$$a \mapsto \hat{a}; \quad \hat{a}(\pi) = \pi(a),$$

we can define for each open subset $U \subset \hat{A}$ a closed two-sided ideal

$$I(U) := \bigcap_{\pi \in \hat{A} \setminus U} \ker \pi.$$

Proposition 2.2. *Let* A *be a C*-algebra of type I,* $\hat{A}$ *its dual object. There is a bijection between the set of open subsets in* $\hat{A}$ *and the set of closed 2-sided ideals in* A.

Proposition 2.3. *There are naturally localized algebras*

$$\hat{A}(U) = \varinjlim_n A/I^n(U)$$

with respect to the $I(U)$*-adic topology.*

3. $\mathbf{Z}_2$-graded Čech Cohomology

3.1. *Grothendieck Topos*

The main purpose of this section is to formulate and define the functor of (periodic cyclic) $\mathbf{Z}_2$-graded Čech cohomology. The well-known periodic cyclic homology is based on the cyclic homology theory of A. Connes, which is an algebraic framework of the $\mathbf{Z}_2$-graded de Rham cohomology theories. In the algebraic context, it was defined by J. Cuntz and D. Quillen in terms of $\mathcal{X}$-complexes and it has become a new chapter of noncommutative algebraic geometry. The most general Gronthendieck algebraic geometry is purely based in terms of categories. In the generic case this turns out to the algebraic version of the Čech cohomology in place of de Rham cohomology theories. We follows the work of Orlov[4] in particular to formulate the theory. The main references are Refs. 7 and 4.

Many of our results could be obtained in the fields of other characteristics, but we restrict ourselves to the complex case.

Let us denote by $\mathcal{C}$ a fixed category and **Set** the category of sets. Any contravariant functor X from $\mathcal{C}$ to **Set** is called a *presheaf of sets* and the category of all presheaves of sets on $\mathcal{C}$ is denoted by $\hat{\mathcal{C}}$. The category $\mathcal{C}$ can be considered as a subcategory of $\hat{\mathcal{C}}$, consisting of representable functors $h = h_R : \mathcal{C} \to \mathbf{Set}$. If R is an object of $\mathcal{C}$, then there is a natural isomorphism $\mathrm{Hom}_{\hat{\mathcal{C}}}(h_R, X) = X(R)$. For any object $X \in \hat{\mathcal{C}}$ the *category over* X is the category of pairs (R, Φ), where R is an object of $\mathcal{C}$ and $\Phi \in X(R)$, and is denoted by $\mathcal{C}/X$.

Recall that a *sieve* in the category $\mathcal{C}$ is a full subcategory $\mathcal{D} \subseteq \mathcal{C}$ such that any object of $\mathcal{C}$ for which there exists a morphism from it to some object in $\mathcal{D}$ is contained in $Obj(\mathcal{D})$. A sieve on R is nothing more than a *subpresheaf* of R in the category $\hat{\mathcal{C}}$

A *Grothendieck topology* $\mathcal{T}$ on a category $\mathcal{C}$ is defined by giving for each object R in $\mathcal{C}$ a set $J(R)$ of the so called *covering sieves* satisfying the following axioms:

(T1) For any object R the maximal sieve $\mathcal{C}/R$ is in $J(R)$.

(T2) If $T \in J(R)$ and $f : S \to R$ a morphism in $\mathcal{C}$, then the induced sieve

$$f^*(T) := \{U \xrightarrow{\quad\alpha\quad} S | f\alpha \in T\}$$

is in $J(S)$.

(T3) If $T \in J(R)$ is a covering sieve and U is a sieve on R such that $f^*(U) \in J(S)$ for all $f : S \to R$ in T, then $U \in J(R)$.

A *Grothendieck site* $\Phi = (\mathcal{C}, \mathcal{T})$ is a category $\mathcal{C}$ and equipped with a Grothendieck topology $\mathcal{T}$.

It is reasonable to remind the Jacobson topology on the set of all representations of a group or Zariski topology on algebraic varieties.

For the categories with fiber product, a Grothendieck topology can be given by a *Grothendieck pretopology* which is defined by giving for each object R in $\mathcal{C}$ a family $Cov(R)$ of morphism to R such that

(P1) For any family $\{R_\alpha \to R\}_{\alpha\in I}$ in $Cov(R)$ and $S \to R$ a morphism of $\mathcal{C}$, the fiber product family $R_\alpha \times_R S \to S$ is also in $Cov(R)$.
(P2) If $\{R_\alpha \to R\}_{\alpha\in R}$ is in $Cov(R)$ and $\{R_{\beta_\alpha} \to R_\alpha\}_{\beta_\alpha \in J_\alpha}$ is in $Cov(R_\alpha)$ for each $\alpha \in I$, then the total family $\{R_\gamma \to R\}_{\gamma\in\coprod_{\alpha\in I} J_\alpha}$ is in $Cov(R)$.
(P3) The trivial family $\{id_R : R \to R\}$ is in $Cov(R)$.

Any Grothendieck pretopology P on $\mathcal{C}$ generates a Grothendieck topology $\mathcal{T}$ such that a sieve is covering in $\mathcal{T}$ if and only if it contains some covering family in P.

The topos on the category of functors from a category $\mathcal{C}$ to another one $\mathcal{D}$ is defined by the usual rule.

3.2. *The Standard Cosimplicial Complex of a Continuous Functor*

We define in this subsection the $\mathbf{Z}_2$-graded Čech cohomology theory. Let us recall the definition of the Čech cohomology with coefficients in a sheaf M. Let $\mathcal{U} = (U_i \to X)_{i\in I}$ be a covering sieve of X in the category $\mathcal{C}$. Suppose that the cover has the property that all fiber product and pushout diagrams exist, (see Ref. 7, Exp. 4). Denote $\mathcal{A}_\bullet$ the associated standard simplicial complex:

$$\mathcal{A}_\bullet : \quad \cdots \begin{smallmatrix}\longrightarrow\\ \longrightarrow\\ \longrightarrow\\ \longrightarrow\end{smallmatrix} \coprod_{(i,j,k)\in I^3} A(U_i \times U_j \times U_k) \begin{smallmatrix}\longrightarrow\\ \longrightarrow\\ \longrightarrow\end{smallmatrix} \coprod_{(i,j)\in I^2} A(U_i \times U_j) \begin{smallmatrix}\longrightarrow\\ \longrightarrow\end{smallmatrix} \coprod_{i\in I} A(U_i)$$

Let us consider again a ringed cite (category) $(\mathcal{C}, A)$, $\hat{\mathcal{C}}$ the topos of presheaves on $\mathcal{C}$, $\varepsilon : \mathcal{C} \hookrightarrow \hat{\mathcal{C}}$ the canonical functor associating to each object $X \in \mathcal{C}$ the functor h_X, presented by X. Define $H^q(h_X, M) = H^q(X, M)$, (see Ref. 7, Exp IV, 2.3.1), as derived functor of the projective limit functor:

$$H^q(S, M) := R^q \varprojlim_{C/S} M|_S$$

For any A-module M, denote $C^\bullet(\mathcal{U}, M) := \mathrm{Hom}_A(\mathcal{A}_\bullet, M)$:

$$C^\bullet(\mathcal{U}, M): \quad \prod_{i\in I} M(U_i) \overset{\rightarrow}{\underset{\rightarrow}{}} \prod_{(i,j)\in I^2} M(U_i \times U_j) \overset{\longrightarrow}{\underset{\longrightarrow}{\longrightarrow}} \cdots$$

One defines $H^q(\mathcal{U}, M) := H^q(C^\bullet(\mathcal{U}, M))$. If R is a covering sieve generalized by the family $\mathcal{U}$ then

$$H^q(\mathcal{U}, M) \cong H^q(R, M)$$

and the functor $H^q(\mathcal{U}, .)$ commutes with restriction of scalars (see Ref. 7, Exp. IV, Proposition 2.3.4).

One defines, (see Ref. 7, Exp. IV, 2.4)

$$\mathcal{H}^0(M)(X) := H^0(X, M) = M(X),$$

$$\mathcal{H}^q(M)(X) := H^q(X, M).$$

For an arbitrary presheaf G of A-module G, the groups $H^q(R, G)$ are called the *Čech cohomology with respect to the covering sieve R, with coefficients in G.*

For a sheaf M of A-modules over $\mathcal{C}$, the group

$$H^q(\mathcal{U}, M) := H^q(\mathcal{U}, \mathcal{H}^0(M))$$

is defined as the *Čech cohomology group of the sheaf M with respect to he the cover $\mathcal{U}$.*

One has also

$$\check{\mathcal{H}}^0(G)(X) = \varinjlim_{R\hookrightarrow X} G(R)$$

and therefore

$$H^q(G)(X) = \varinjlim_{R\hookrightarrow X} H^q(R, G)$$

which is called *the presheaf of Čech cohomology*

Define

$$\check{H}^q(X, G) := \check{\mathcal{H}}^q(G)(X),$$

one has also

$$\check{H}^q(X, G) = \varinjlim_{\mathcal{U}} H^q(\mathcal{U}, G).$$

For a sheaf M of A-modules, one has

$$\check{H}^q(X, M) = \check{H}^q(X, \mathcal{H}^0(M)), \quad \check{\mathcal{H}}^q(M) = \check{\mathcal{H}}^q(\check{\mathcal{H}}^0(M)).$$

The groups $\check{H}^q(X, M)$ are called the *Čech cohomology groups of the sheaf M.*

3.3. *The Periodic Cyclic Bicomplex*

Lemma 3.1 (The action of $\mathbf{Z}_{k+1}$). *There is a natural action of the cyclic group $\mathbf{Z}_{k+1}$ on the Čech cohomology cochain complex $C^\bullet(\mathcal{U}, M)$ associated with a covering $\mathcal{U}$.*

Proof. The action of the cyclic group $\mathbf{Z}_{k+1}$ is defined a cyclic permutation of indices of $U's$, i.e.

$$(\lambda M)(U_{i_0} \times \cdots \times U_{i_k}) := M(U_{i_k} \times U_{i_0} \times \cdots \times U_{i_{k-1}}).$$

It is not hard also to see that for a covering sieve $\mathcal{U} = \{U_\alpha \to X\}$ there is a natural isomorphism

$$M(U_{i_0} \times \cdots \times U_{i_k}) \cong M(U_{i_0}) \otimes \cdots \otimes M(U_{i_k}).$$

Therefore the Čech cohomology complex becomes the cyclic complex for

$$\varinjlim_{\mathcal{U}} M(\mathcal{U}). \qquad \square$$

Corollary 3.1 (Hochschild differentials and Cyclic operations). *The well-known Hochschild differentials b' and b and Connes cyclic operators λ, $N = 1 + \lambda + \cdots + \lambda^k$, s are well-defined on $\mathbf{Z}_2$-graded Čech cocycles*

Definition 3.1 (Periodic bicomplex). *Let $(\mathcal{C}, A)$ be a ringed U-cite, $\hat{\mathcal{C}}$ the topos of sheaves, M a sheaf of A-modules. Then the bicomplex*

$$\begin{array}{ccccccc}
 & \vdots & & \vdots & & \vdots & \\
 & {\scriptstyle -b'}\uparrow & & {\scriptstyle b}\uparrow & & {\scriptstyle -b'}\uparrow & \\
\cdots \xrightarrow{1-\lambda} & \prod_{(i_1,i_2,i_3)\in I^3} M(U_{i_1}\times U_{i_2}\times U_{i_3}) & \xrightarrow{N} & \prod_{(i_1,i_2,i_3)\in I^3} M(U_{i_1}\times U_{i_2}\times U_{i_3}) & \xrightarrow{1-\lambda} & \cdots & \\
 & {\scriptstyle -b'}\uparrow & & {\scriptstyle b}\uparrow & & {\scriptstyle -b'}\uparrow & \\
\cdots \xrightarrow{1-\lambda} & \prod_{(i,j)\in I^2} M(U_i\times U_j) & \xrightarrow{N} & \prod_{(i,j)\in I^2} M(U_i\times U_j) & \xrightarrow{1-\lambda} & \cdots & \\
 & {\scriptstyle -b'}\uparrow & & {\scriptstyle b}\uparrow & & {\scriptstyle -b'}\uparrow & \\
\cdots \xrightarrow{1-\lambda} & \prod_{i\in I} M(U_i) & \xrightarrow{N} & \prod_{i\in I} M(U_i) & \xrightarrow{1-\lambda} & \cdots &
\end{array}$$

is well-defined and is called the (periodic) Čech bicomplex.

Definition 3.2 (The total complex and $\mathbf{Z}_2$-graded Čech cohomology). *The associated total complex of which is defined as*

$$\mathsf{Tot}\, C(\mathcal{U}, M)^{\pm} := \prod_{i+j=\pm(mod\ 2)} C^{i,j},$$

where $C^{i,j} := \prod_{(i_1,\dots,i_k)\in I^k} M(U_{i_1}\times\ldots\times U_{i_k})$ *and* $\pm = ev$ *(even) or od (odd).*

The cohomology of this total complex is called the $\mathbf{Z}_2$*-graded Čech cohomology of* M *and denoted by* $\mathbf{Z}_2\check{\mathbf{H}}^*(\mathcal{U}, M)$ *and* $\mathbf{Z}_2\check{\mathbf{H}}(X, M) := \varinjlim_{\mathcal{U}} \mathbf{Z}_2\check{\mathbf{H}}(\mathcal{U}, M)$. *It can be also realized as the cohomology of the total complex related with the process of passing through direct limits, i.e. the direct limit bi-complex*

$$\begin{array}{ccccccc}
 & & \vdots & & \vdots & & \vdots \\
 & & \uparrow{\scriptstyle -b'} & & \uparrow{\scriptstyle b} & & \uparrow{\scriptstyle -b'} \\
\dots \xrightarrow{1-\lambda} & & \prod_{(i_1,i_2,i_3)\in I^3} \varinjlim_{\mathcal{U}} M(U_{i_1}\times U_{i_2}\times U_{i_3}) & \xrightarrow{N} & \prod_{(i_1,i_2,i_3)\in I^3} \varinjlim_{\mathcal{U}} M(U_{i_1}\times U_{i_2}\times U_{i_3}) & \xrightarrow{1-\lambda} & \dots \\
 & & \uparrow{\scriptstyle -b'} & & \uparrow{\scriptstyle b} & & \uparrow{\scriptstyle -b'} \\
\dots \xrightarrow{1-\lambda} & & \varinjlim_{\mathcal{U}} \prod_{(i,j)\in I^2} M(U_i\times U_j) & \xrightarrow{N} & \varinjlim_{\mathcal{U}} \prod_{(i,j)\in I^2} M(U_i\times U_j) & \xrightarrow{1-\lambda} & \dots \\
 & & \uparrow{\scriptstyle -b'} & & \uparrow{\scriptstyle b} & & \uparrow{\scriptstyle -b'} \\
\dots \xrightarrow{1-\lambda} & & \varinjlim_{\mathcal{U}} \prod_{i\in I} M(U_i) & \xrightarrow{N} & \varinjlim_{\mathcal{U}} \prod_{i\in I} M(U_i) & \xrightarrow{1-\lambda} & \dots
\end{array}$$

For a C^**-algebra* A*, we define it* $\mathbf{Z}_2$*-graded Čech cohomology* $\mathbf{Z}_2\check{\mathbf{H}}(A) = \mathbf{Z}_2\check{\mathbf{H}}(\mathcal{A})$ *as the* $\mathbf{Z}_2$*-graded Čech cohomology of the category of *-representations of* A.

Remark 3.1. In the first periodic bi-complex without direct limits, all the horizontal lines are acylic, but it is in general not the case for the second periodic bi-complex with direct limits.

4. Homotopy Invariance and Morita Invariance

We prove in this section two main properties of the (periodic cyclic) $\mathbf{Z}_2$-graded Čech cohomology theory: homotopy invariance and Morita invariance, which make the theory easier to compute and being a generalized homology theory.

Definition 4.1 (Chain Homotopy of functors). *Let us consider two functors* $F, G : \mathcal{C} \to \mathcal{D}$. *Denote the corresponding chain functors between complexes by* $\{F_n\}, \{G_n\}$, *where* $F_n, G_n : C_n \to D_n$ *for complexes*

$$\begin{array}{ccccccccc}
0 & \longrightarrow & C_0 & \longrightarrow & C_1 & \longrightarrow & C_2 & \longrightarrow & \dots \\
 & & \downarrow{\scriptstyle F_0,G_0} & & \downarrow{\scriptstyle F_1,G_1} & & \downarrow{\scriptstyle F_2,G_2} & & \\
0 & \longrightarrow & D_0 & \longrightarrow & D_1 & \longrightarrow & D_2 & \longrightarrow & \dots
\end{array}$$

We say that F *and* G *are chain homotopic if there exist augmentation functors* $s_n : C_n \to D_{n-1}$ *such that for all* n

$$F_n - G_n = s_n \circ \partial_{n-1} + \partial_n \circ s_{n+1}.$$

Lemma 4.1. *Two functors* $F, G : \mathcal{A} \to \mathcal{B}$ *are homotopic if and only if for any covering sieve* $\mathcal{U} = (U_i \to X)_{i \in I}$, *there exists a chain homotopy of chain complexes*

$$\coprod_{i \in I} A(U_i) \overset{\longleftarrow}{\underset{\longleftarrow}{}} \coprod_{i,j \in I \times I} A(U_i \times_X U_j) \overset{\longleftarrow}{\underset{\longleftarrow}{\longleftarrow}} \dots$$

and

$$\coprod_{i \in I} B(U_i) \overset{\longleftarrow}{\underset{\longleftarrow}{}} \coprod_{i,j \in I \times I} B(U_i \times_X U_j) \overset{\longleftarrow}{\underset{\longleftarrow}{\longleftarrow}} \dots$$

Remark 4.1. In the case of smooth manifolds the chain complex homotopy is realized by integration of the so called Cartan homotopy formula for the Lie derivative

$$L_\xi = \imath(\xi) \circ d + d \circ \imath(\xi)$$

between de Rham complexes.

Lemma 4.2. *Let* A *and* B *be* C^**-algebras, and let* $\mathcal{A}$ *(resp.* $\mathcal{B}$*) be the category of* $*$*-modules . Then the categories* $\mathcal{A}$ *and* $\mathcal{B}$ *are homotopic one-to-another if and only if the two algebras* A *and* B *are homotopic.*

Proof. Because of the Gelfand-Naimark-Segal theorem, the C*-algebras are exactly defined by the category of *-representations, the category of *-representations of $B \otimes C[0,1]$ is isomorphic to the category of *-representations of B. □

Lemma 4.3. *Let* A *be a* C^**-algebra and* $\mathcal{A}$ *the category of *-representations (i.e.* A*-modules) of* A*, then*

$$\mathbf{Z}_2\check{\mathbf{H}}^*(\mathcal{A}) \cong \mathrm{HP}^*(A).$$

Proof. Let us consider affine covering sieve $U_i \to X = \hat{A} = \operatorname{Spec} A$, the dual object of A. □

Theorem 4.1 (Homotopy Invariance). *Let* $\varphi_t : A \to B, t \in I = [0,1]$ *be a homotopy of algebras, then*

$$\mathbf{Z}_2\check{\mathbf{H}}^*(A) \cong \mathbf{Z}_2\check{\mathbf{H}}^*(B).$$

Proof. Let $\mathcal{A}$ (resp. $\mathcal{B}$) be the category of A-modules (resp., B-modules).

Step 1. Change the homotopy by a piecewise-linear homotopy in the space of functors from the category $\mathcal{A}$ to the category $\mathcal{B}$.

Step 2. A piecewise-linear homotopy gives rise to a chain complex homotopy.

Step 3. Two chain complex homotopical functors induces the same isomorphism of Čech cohomology groups. It is an easy consequence from the results of homological algebra: If F_n and G_n are chain complex homotopic than the induced morphisms satisfies

$$F_n^* - G_n^* = \partial_{n-1}^* \circ s_n^* + s_{n+1}^* \circ \partial_n^*.$$

The second summand is a zero morphism on cohomology and the first summand is a boundary. The sum on the right is therefore a zero morphism. □

Lemma 4.4. *Let us denote by* $\mathrm{Mat}_n(\mathbb{C})$ *the algebra of all square* $n \times n$*-matrices with complex entries. Then we have a natural isomorphism*

$$\mathbf{Z}_2\check{\mathbf{H}}^*(\mathrm{Mat}_n(\mathbb{C})) \cong \mathbf{Z}_2\check{\mathbf{H}}^*(\mathbb{C}).$$

Proof. Every complex matrix can be homotopic to a unitary one. Then, every unitary matrix can be by conjugation reduced to a diagonal matrix of complex numbers of module 1. Every elementary block (in this case, diagonal element)$[e^{i\theta}]$ is homotopic to identity [1] by the classical homotopy $[e^{i\theta t}]_{0\leq t\leq 1}$, i.e.

$$\begin{bmatrix} \cos\theta & -\sin\theta \\ \sin\theta & \cos\theta \end{bmatrix} \sim I = \begin{bmatrix} 1 & 0 \\ 0 & 1 \end{bmatrix}.$$ □

Lemma 4.5 (Adjoint functors). *There is a natural equivalence of functors* Hom *and* $\otimes$*:*

$$\mathrm{Hom}(R \otimes \mathrm{M}_n(\mathbb{C}), M) \cong \mathrm{Hom}(R, \mathrm{Hom}(\mathrm{M}_n(\mathbb{C}), M)).$$

Proof. This isomorphism of functors is a particular case of the general adjointness between Hom and $\otimes$ in homological algebra. □

Lemma 4.6. *There is a natural isomorphism of derived functors*

$$R^q\,\mathrm{Hom}(\mathrm{M}_n(\mathbb{C}), M) \cong R^q\,\mathrm{Hom}(\mathbb{C}, M).$$

Proof. It is an easy exercise from homological algebra. □

Theorem 4.2 (Morita Invariance).

$$\mathbf{Z}_2\check{\mathbf{H}}^*(A \otimes \mathrm{Mat}_n(\mathbf{C})) \cong \mathbf{Z}_2\check{\mathbf{H}}^*(A).$$

Proof. Let us remark that $A \to A \otimes \mathrm{Mat}_n(\mathbf{C})$ is a fiber bundle. Now apply the Grothendieck's Leray-Serre spectral sequence for this fibration. Following the previous lemmas 4.5 and 4.6, there is a natural isomorphism of functors

$$R^p \operatorname{Hom}(R \otimes \mathrm{M}_n(\mathbb{C}), R^q(M) \cong R^p \operatorname{Hom}(R, R^q \operatorname{Hom}(\mathrm{M}_n(\mathbb{C}), M)),$$

which are the E^2 term of a Leray-Serre spectral sequence converging to the Čech cohomology. □

Corollary 4.1. *The $\mathbf{Z}_2$-graded Čech cohomology theory is a generalized cohomology theory.*

5. Comparison with the Classical Čech Cohomology Theory

In this section we show that a generalization of the Connes-Hochschild-Kostant-Rosenberg Theorem can be obtained easily.

Theorem 5.1. *Let A be a stable continuous C^*-algebra with spectrum a smooth compact manifold X, in fact $A = C(X, \mathcal{K}(P))$ is the algebra of continuous sections of a smooth, locally trivial bundle $\mathcal{K}(P) := P \times_{PU} \mathcal{K}$ on X with fibre the algebra $\mathcal{K}$ of compact operators on a separable Hilbert space associated to a principal PU bundle P on X via the adjoint action of PU on $\mathcal{K}$. Let $\delta(P) \in H^3(X; \mathbf{Z})$ be the Dixmier-Douady invariant, that classifies such algebras A and $c(P)$ some closed 3-form on X, that presents the class $2\pi i\delta(P)$ in the real cohomology. Let $\mathcal{A}$ be the category of all *-representations of the C^*-algebra A. Then the $\mathbf{Z}_2$-graded Čech cohomology $\mathbf{Z}_2\check{\mathbf{H}}^*(\mathcal{A})$ is isomorphic to the de Rham cohomology $H^*(X; c(P))$ which is isomorphic to the classical $\mathbf{Z}_2$-graded Čech cohomology $\mathbf{Z}_2\check{\mathbf{H}}^*(X; c(P))$.*

Proof. In this situation, the $\mathbf{Z}_2$-graded Čech cohomology of the category $\mathcal{A}$ is isomorphic to the Connes periodic cyclic homology $HP_*(C^\infty(X, \mathcal{L}^1(P)))$, where $C^\infty(X, \mathcal{L}^1(P))$ is consisting of all smooth section of the sub-bundle $\mathcal{L}^1(P) = P \times_{PU} \mathcal{L}^1$ of $\mathcal{K}(P)$ with fibre the algebra $\mathcal{L}^1$ of trace class operators on the Hilbert space with the same structure groups PU, see Ref. 3 for a more detailed proof in the language of periodic cyclic homology. □

Acknowledgments

The main part of this work was done while the author was visiting The Abdus Salam ICTP in Trieste, Italy. The author is grateful to ICTP and in particular would like to express his sincere thanks to Professor Dr. Le Dung Trang for invitation and support.

The text of this paper is presented at the Second International Conference on Abstract and Applied Analysis, Quinhon, Vietnam, 4-9 June 2005. The author thanks the organizers and especially Professor Dr. Sc. Nguyen Minh Chuong for the invitation to give this lecture.

References

1. M. Kashiwara, *Algebraic study of systems of partial differential equations*, Mémoire No. 63, Suplément Bull. SMF, Tome 123, No. 4, 1995.
2. A. Rosenberg and M. Kontsevich, *Noncommutative smooth spaces*, The Gelfand mathematical Seminar, 1996-1999, pp. 85–107.
3. V. Mathai and D. Stevenson, *On a generalized Connes-Hochschild-Kostant-Rosenberg theorem,* arXiv: math.KT/0404329 v1, 19 April 2004.
4. D. Orlov, *Quasicoherent sheaves in commutative and noncommutative geometry,* MPI-1999-31 Preprint, 1999.
5. D. Quillen, *Formal properties of overdetermined systems of linear differential equations*, Thesis (1964).
6. A. L. Rosenberg,Compositio Math. **112**, 93 (1998).
7. M. Artin, A. Grothendieck, and J. L. Verdier, *Théorie des topos et cohomologie des schémas*, SGA 4, tome 2, Springer Lecture notes in Mathematics, No. 270, 1972.

Advances in Deterministic and Stochastic Analysis
Eds. N. M. Chuong *et al.* (pp. 269–278)

§14. SOBOLEV SPACES WITH WEIGHT ON RIEMANNIAN MANIFOLDS

NGUYEN MINH CHUONG† and LE DUC THINH

Institute of Mathematics
18 Hoang Quoc Viet, Cau Giay, 10307 Hanoi, Vietnam
†*E-mail: nmchuong@math.ac.vn*

Sobolev spaces with weight are defined on Riemannian manifolds. The class of such spaces, even in $\mathbb{R}^n$, is very wide and very interesting (see Ref. 10, p. 353). It includes the usual Sobolev spaces of nonnegative integer orders as special cases. Properties of such spaces are studied.

1. Introduction

Many famous geometers, mathematicians have contributed to solve the well known Yamabe's problem.[3,5] To solve this geometrical problem one must use various tools, among which mathematical analysis and Sobolev spaces are very powerful. From this geometrical problem arises a very celebrated elliptic problem with critical Sobolev exponents.[4,6]

So far, the Sobolev spaces used on Riemannian manifolds are of nonegative interger orders.

The aim of this paper is to introduce Sobolev spaces with weighted norms on Riemannian manifolds including usual Sobolev spaces mentioned above as a special case.

2. Sobolev Spaces with Weighted Norm on Riemannian Manifold

Let (M_n, g) be an n-dimensional smooth Riemannian manifold with the metric tensor g. In local coordinates, g has the form:

$$g = \sum_{i,j=1}^{n} g_{ij} dx^i \otimes dx^j.$$

Let D be the Levi-Civita connection on M_n. D is extended uniquely to covariant derivative of tensor fields, denoted also by D. For $f \in C^k(M_n)(k \geq$

0, an integer), let us denote by $\triangle^k f$ the covariant derivative of f of order k:

$$\triangle^0 f = f, \triangle^1 f = Df, \triangle^{r+1} f = D(\triangle^r f), \quad (0 \le r \le k-1).$$

Thus $\triangle^k f$ is a tensor field of kth covariance. In local chart (U, φ) of M_n, we denote by $\triangle_{\alpha_1 \ldots \alpha_k}$ f the components of $\triangle^k f$:

$$\triangle^k f = \sum_{\alpha_1, \ldots, \alpha_k = 1}^{n} \triangle_{\alpha_1 \ldots \alpha_k} f \; dx^{\alpha_1} \otimes \ldots \otimes dx^{\alpha_k}$$

The metric on the tangent bundle of the M_n generates the metric on all tensor bundles over M_n. So, we can define the norm of $\triangle^k f$, denoted by $|\triangle^k f|$:

$$|\triangle^k f|^2 = \sum g^{\alpha_1 \beta_1} \ldots g^{\alpha_k \beta_1} (\triangle_{\alpha_1 \ldots \alpha_k f})(\triangle_{\beta_1 \ldots \beta_k} f),$$

where (g^{ij}) is the inverse matrix of the matrix (g_{ij}). For $k = 0, |\triangle^0 f| = |f|$.

Definition 2.1. For a general Riemannian manifold M_n , we call a weight function a positive continuous function on M_n. In the case the Riemannian manifold is $\mathbb{R}^n$, a positive function ω is called a weight function if there exist positive constants C and N such that:

$$\omega(\xi + \eta) \le (1 + C|\xi|)^N \omega(\eta); \quad \forall \xi, \; \eta \in \mathbb{R}^n.$$

(These weight functions are introduced by L. Hörmander in Ref. 8).

We denote by $\mathcal{K}(M_n)$ the set of all weight functions on M_n.

It is well known that $\omega_s(\xi) = (1 + |\xi|^2)^{\frac{s}{2}}$, $s \in \mathbb{R}$, belongs to $\mathcal{K}(\mathbb{R}^n)$.

If $\omega \in \mathcal{K}(\mathbb{R}^n)$ then the functions:

$$M_\omega(\xi) = \sup_\eta \frac{\omega(\xi + \eta)}{\omega(\eta)}; \quad \breve{\omega}(\xi) = \omega(-\xi)$$

belong to $\mathcal{K}(\mathbb{R}^n)$ (see Ref. 8).

Definition 2.2. Let $\omega \in \mathcal{K}(M_n)$, p is a real number, $p \ge 1$ and k is a nonnegative integer. Let $\mathcal{F}_{p,k,\omega}(M_n)$ be the vector space of all functions $f \in C^\infty(M_n)$ such that $\omega|\nabla^\ell f| \in L_p(M_n)$, for all integers ℓ with $0 \le \ell \le k$. The Sobolev space with the weight function ω, denoted by $H_{p,k,\omega}(M_n)$, is the completion of $\mathcal{F}_{p,k,\omega}(M_n)$ with respect to the norm:

$$||f||_{p,k,\omega} = \sum_{\ell=0}^{k} ||\omega|\nabla^\ell f|||_p.$$

Let us denote by $C_\circ^\infty(M_n)$ the space of smooth function of M_n with compact support, $H^\circ_{p,k,\omega}(M_n)$ the closure of $C_\circ^\infty(M_n)$ with respect to the above mentioned norm.

For $k = 0$, $H_{p,0,\omega}(M_n) = L_{p,\omega}(M_n)$ is a Banach space equipped with the norm:

$$||f||_{p,\omega} = ||\omega f||_p.$$

As in Ref. 2 we can get Theorems 2.1 and 2.2 below:

Theorem 2.1. *$C_\circ^\infty(\mathbb{R}^n)$ is dense in $H_{p,k,\omega}(\mathbb{R}^n)$, that is $H^\circ_{p,k,\omega}(\mathbb{R}^n) = H_{p,k,\omega}(\mathbb{R}^n)$.*

Theorem 2.2. *Let M_n be a complete Riemannian manifold. Then $C_\circ^\infty(M_n)$ is dense in $H_{p,1,\omega}(M_n)$.*

Theorem 2.3. *If $\omega_1, \omega_2 \in \mathcal{K}(M_n)$ and there exists a constant $C > 0$ such that $\omega_2(x) \leq C\omega_1(x), \forall x \in M_n$, then $H_{p,k,\omega_1}(M_n) \subset H_{p,k,\omega_2}(M_n)$, and the inclusion operator is continuous.*

Theorem 2.4. *If $\omega_1, \omega_2 \in \mathcal{K}(\mathbb{R}^n)$ and $H_{p,k,\omega_1}(\mathbb{R}^n) \subset H_{q,\ell,\omega_2}(\mathbb{R}^n)$ for some p, q, k, ℓ, then there exists a constant $C > 0$ such that:*

$$\omega_2(x) \leq C\omega_1(x), \forall x \in \mathbb{R}^n.$$

Proof. Consider the inclusion operator:

$$I : H_{p,k,\omega_1}(\mathbb{R}^n) \hookrightarrow H_{q,\ell,\omega_2}(\mathbb{R}^n).$$

The operator I has closed graph. Indeed, if $u_n \to u$ in $H_{p,k,\omega_1}(\mathbb{R}^n)$ and $Iu_n \to v$ in $H_{q,\ell,\omega_2}(\mathbb{R}^n)$ then $u_n \to u$ in $L_{p,\omega_1}(\mathbb{R}^n)$ and $Iu_n \to v$ in $L_{q,\omega_2}(\mathbb{R}^n)$. Consequently there exists a subsequence denoted also by $\{u_n\}$ such that $u_n \to u$ almost everywhere, and $u_n = Iu_n \to v$ almost everywhere. So $u(x) = v(x)$ almost everywhere in $\mathbb{R}^n$, and $Iu = v$. By the closed graph Theorem, the operator I is continuous. Therefore there exists a constant $A > 0$ such that:

$$||f||_{q,\ell,\omega_2} \leq A||f||_{p,k,\omega_1}, \quad \forall f \in H_{p,k,\omega_1}(\mathbb{R}^n) \tag{1}$$

Fix a function $f \in C_\circ^\infty(\mathbb{R}^n)$, $f \not\equiv 0$. For each $\eta \in \mathbb{R}^n$, consider a function f_η, defined by:

$$f_\eta(\xi) = f(\xi - \eta).$$

We have:

$$D^\alpha f_\eta(\xi) = (D^\alpha f)(\xi - \eta), \forall \alpha.$$

So:

$$|\triangle^\ell f_\eta|(\xi) = |\triangle^\ell f|(\xi - \eta), \forall \ell \geq 0.$$

Since $f \in C_\circ^\infty(\mathbb{R}^n)$, we get $f_\eta \in C_\circ^\infty(\mathbb{R}^n)$, thus $f_\eta \in H_{p,k,\omega_1}(\mathbb{R}^n)$.

On the other hand:

$$\omega_2(\xi) \geq \frac{\omega_2(\eta)}{M_{\omega_2}(\eta - \xi)} = \frac{\omega_2(\eta)}{\check{M}_{\omega_2}(\xi - \eta)}.$$
$$\omega_1(\xi) \leq \omega_1(\eta) M_{\omega_1}(\xi - \eta).$$

Hence:

$$||f_\eta||_{q,l,\omega_2} \geq \omega_2(\eta).||f||_{q,l,1/\check{M}_{\omega_2}}.$$
$$||f_\eta||_{p,k,\omega_2} \leq \omega_1(\eta).||f||_{p,k,M_{\omega_1}}.$$

By applying (1) to f_η we obtain:

$$\omega_2(\eta) \leq C\omega_1(\eta),$$

where $C = A\|f\|_{p,k,M_{\omega_1}} / \|f\|_{q,l,1/\check{M}_{\omega_2}}$. □

Theorem 2.5. *Assume that* $\omega_1, \omega_2 \in \mathcal{K}(\mathbb{R}^n)$ *and* $p > n$. *If* $\dfrac{\omega_2(\xi)}{\omega_1(\xi)} \to 0$ *as* $\xi \to \infty$ *then the inclusion operator* $C_\circ^\infty(\mathbb{R}^n) \cap H_{p,2,\omega_1}(\mathbb{R}^n) \hookrightarrow L_{p,\omega_2}(\mathbb{R}^n)$ *is completely continuous.*

Proof. Let $\{f_i\}_{i \geq 1} \subset C_\circ^\infty(\mathbb{R}^n) \cap H_{p,2,\omega_1}(\mathbb{R}^n)$ be an arbitrary sequence such that $||f_i||_{p,2,\omega_1} \leq 1, \forall i$. We shall prove that there exists a subsequence converging uniformly on every compact subset of $\mathbb{R}^n$.

Let K be an arbitrary compact set. Then there exists an open ball S such that $K \subset S$. The closure $\bar{S}$ of S is a n-dimensional compact Riemannian manifold with smooth boundary in $\mathbb{R}^n$. Since $p > n$, applying Sobolev imbedding Theorems to compact manifold with boundary (see Ref. 2), we see that $H_2^p(S)$ is imbedded into $C^1(\bar{S})$. Consequently there exists a positive constant C such that:

$$\sup_{\overline{S}} |f_i| \leq C\|f_i\|_{p,2,S}, \quad \forall\, i \geq 1,$$
$$\sup_{\overline{S}} |\triangle f_i| \leq C\|f_i\|_{p,2,S}, \quad \forall i \geq 1.$$

Note that:

$$\|f\|_{C^1(\bar{S})} = \max\{\sup_{\overline{S}}|f|,\ \sup_{\bar{S}}|\triangle f|\},$$
$$\|f\|_{p,2,S} = \sum_{\ell=0}^{2}\Big(\int_S |\triangle^\ell f|^p\Big)^{\frac{1}{p}}.$$

By the continuity of ω_1, we have:

$$B = \inf_{\xi\in\overline{S}}\omega_1(\xi)\ >0 \quad \text{and} \quad ||f_i||_{p,2,S} \leq \frac{1}{B}||f_i||_{p,2,\omega_1} \leq \frac{1}{B},\ \forall\, i\geq 1.$$

Hence the sequence $\{f_i\}$ and $\{|\triangle f_i|\}$ are uniformly bounded on K, so the sequence $\{f_i\}$ is uniformly bounded and is equicontinuous on K. Since $\mathbb{R}^n$ is separable and countable at infinity, the sequence $\{f_i\}$ possesses a subsequence converging uniformly on every compact subset of $\mathbb{R}^n$. This subsequence, may be denoted by $\{f_i\}$. Given any $\epsilon > 0$, by $\omega_2(\xi)/\omega_1(\xi) \to 0$ as $\xi \to \infty$, there exists a large enough closed ball S such that:

$$\frac{\omega_2(\xi)}{\omega_1(\xi)} \leq \epsilon, \forall\xi \in \mathbb{R}^n\backslash S.$$

By Minkowski inequality, we obtain:

$$\begin{aligned}||\omega_2(f_i - f_j)||_p &\leq \Big(\int_{\mathbb{R}^n\backslash S}(\omega_2|f_i - f_j|)^p\Big)^{\frac{1}{p}} + \Big(\int_S(\omega_2|f_i - f_j|)^p\Big)^{\frac{1}{p}}\\ &\leq \epsilon||f_i - f_j||_{p,2,\omega_1} + C\Big(\int_S |f_i - f_j|^p\Big)^{\frac{1}{p}}.\end{aligned}$$

where $C = \sup_{\overline{S}}\omega_2 > 0$. Since $||f_i - f_j||_{p,2,\omega_1} \leq 2$ and the sequence $\{f_i\}$ uniformly converges on S we get:

$$\|\omega_2(f_i - f_j)\|_p \to 0 \ as\ i,j \to \infty.$$

So the sequence $\{f_i\}$ is Cauchy in $L_{p,\omega_2}(\mathbb{R}^n)$ and because $L_{p,\omega_2}(\mathbb{R}^n)$ is complete, the sequence $\{f_i\}$ converges in $L_{p,\omega_2}(\mathbb{R}^n)$. □

Remark 2.1.

(1) With the assumptions of the Theorem 2.5, it is obvious that the inclusion operator $C_\circ^\infty(\mathbb{R}^n) \cap H_{p,k,\omega_1}(\mathbb{R}^n) \hookrightarrow L_{p,\omega_2}(\mathbb{R}^n)$ is completely continuous, for all integers k ≥ 2.

(2) When M_n is a Riemannian manifold, it is also a metric space that is separable and countable at infinity. But a ball S in M_n is not necessary to be a Riemannian submanifold with boundary, so we cannot apply the Sobolev embedding theorem as above to S. However, the above theorem holds true for the class of Cartan-Hadamard manifold, that is, a simply connected complete Riemannian manifold with non-positive sectional curvature. By the exponential map at any point such a manifold is diffeomorphic to an Euclidean space of the same dimension. Consequently a closed ball in a Cartan-Hadamard manifold is a Riemannian submanifold. So we get:

Theorem 2.6. *Assume that M_n is a Cartan-Hadamard manifold and $\omega_1, \omega_2 \in \mathcal{K}(M_n)$. If $p > n$ and $\dfrac{\omega_2(Q)}{\omega_1(Q)} \to 0$ as $d(P,Q) \to \infty$ (P is a fixed point of M_n), then the inclusion operator from $C_\circ^\infty(M_n) \bigcap H_{p,2,\omega_1}(M_n)$ to $L_{p,\omega_2}(M_n)$ is compact.*

Theorem 2.7. *Assume that $\omega_1, \omega_2 \in \mathcal{K}(\mathbb{R}^n)$ and the embedding operator from $C_\circ^\infty(\mathbb{R}^n)$ to $L_{q,\omega_2}(\mathbb{R}^n)$ is completely continuous for reals $p, q \geq 1$ and for some non-negative k. Then $\dfrac{\omega_2(\xi)}{\omega_1(\xi)} \to 0$ as $\xi \to \infty$.*

Proof. It is sufficient to prove that if $\{\xi_\nu\}_{\nu \geq 1}$ is an arbitrary sequence such that $\xi_\nu \to \infty$ as $\nu \to \infty$ then $\dfrac{\omega_2(\xi_\nu)}{\omega_1(\xi_\nu)} \to 0$ as $\nu \to \infty$.

Let us fix a function $f \in C_\circ^\infty(\mathbb{R}^n)$, $f \not\equiv 0$. Consider sequence $\{f_\nu\}$ defined by:

$$f_\nu(x) = \frac{f(x - \xi_\nu)}{\omega_1(\xi_\nu)}.$$

From the proof of Theorem 4 we have:

$$\|f_\nu\|_{p,k,\omega_1} \leq \|f\|_{p,k,M_{\omega_1}},$$

$$\|f_\nu\|_{q,\omega_2} \geq \frac{\omega_2(\xi_\nu)}{\omega_1(\xi_\nu)} \|f\|_{q,\frac{1}{M_{\omega_2}}}.$$

From the first inequality it follows that the sequence $\{f_\nu\}$ is bounded in $H_{p,k,\omega_1}(\mathbb{R}^n) \bigcap C_\circ^\infty(\mathbb{R}^n)$. The assumptions of the theorem give us the relative compactness of the sequence $\{f_\nu\}$ in $L_{q,\omega_2}(\mathbb{R}^n)$. So there exists a compact set $G \subset \mathbb{R}^n$ such that:

$$\Big(\int_{\mathbb{R}^n \setminus G} (\omega_2 |f_\nu|)^q \Big)^{\frac{1}{q}} < \epsilon, \forall \nu \geq 1,$$

where ϵ is any positive number (see Ref. 1, pages 31-33). By the Minkowski inequality, we obtain:

$$\|f_\nu\|_{q,\omega_2} \leq \Big[\int_{\mathbb{R}^n \setminus G} (\omega_2 |f_\nu|)^q \Big]^{\frac{1}{q}} + \frac{1}{\omega_1(\xi_\nu)} \Big[\int_G \omega_2^q(x) |f(x - \xi_\nu)|^q dx \Big]^{\frac{1}{q}}$$

Since G and $\operatorname{supp} f$ are compact sets, for ν large enough, we have $x - \xi_\nu \notin \operatorname{supp} f$, $\forall\, x \in G$. So:

$$\frac{1}{\omega_1(\xi_\nu)} \Big(\int_G \omega_2^q(x) |f(x - \xi_\nu)|^q dx \Big)^{\frac{1}{q}} \to 0 \ as \ \nu \to \infty$$

that implies $\|f_\nu\|_{q,\omega_2} \to 0$ as $\nu \to \infty$, and by the second above inequality we obtain:

$$\frac{\omega_2(\xi_\nu)}{\omega_1(\xi_\nu)} \to 0 \ as \ \nu \to \infty. \qquad \square$$

Similarly to the proof of Proposition 2.11 from [2] (pp. 36-37), we get:

Proposition 2.1. *If $H_{q_0,1,\omega}(M_n)$ is imbedded into $L_{p_0}, \omega(M_n)$ for $\frac{1}{p_0} = \frac{1}{q_0} - \frac{1}{n}(1 \leq q_0 < n)$, then $H_{q,k,\omega}(M_n)$ is imbedded into $H_{p_\ell,\ell,\omega}(M_n)$ for $\frac{1}{p_\ell} = \frac{1}{q} - \frac{k-\ell}{n} > 0 (k > \ell \geq 0)$.*

In [2], Lemma 2.22, T. Aubin proved that:

$$\sup |\varphi| \leq C(q) ||\varphi||_{q,1}, \forall \varphi \in C_\circ^\infty(M_n),$$

here $q > n$ and $C(q)$ is a constant, M_n is a complete Riemannian manifold having positive injectivity radius and upper bounded sectional curvature $(K \leq b^2)$.

If M_n is a compact Riemannian, then all these above conditions are satisfied, so we get the desired inequality.

However, in the case of a compact manifold M_n, another proof will be shown.

Theorem 2.8. *Let M_n be a compact Riemannian manifold. Then there exists a positive constant C such that* $\sup |f| \leq C ||f||_{q,1}, \forall f \in C_\circ^\infty(M_n)$.

Proof. Since M_n is compact, there exists $\rho > 0$ such that $\forall\ P \in M_n$, the polar coordinates exist on the closed ball:

$$\bar{B}_P(\rho) = \{Q \in M_n | d(P, Q) \leq \rho\}$$

(see Ref. 9, p. 22).

Because of the compactness of M_n, there exists also the points $P_i \in M_n (i = 1, ..., N)$ such that $\{B_{P_i}\left(\frac{\rho}{4}\right)\}_{i=1}^N$ is an open covery of M_n ($B_{P_i}\left(\frac{\rho}{4}\right)$ is an open ball of radius $\frac{\rho}{4}$ with center P_i).

Let $\{\alpha_i\}_{i=1}^N$ be the unity partition corresponding to the above open covery. It will be shown that for each i, there exists $C_i > 0$ such that:

$$sup|\alpha_i f| \leq C_i ||\triangle(\alpha_i f)||_q, \forall f \in C_\circ^\infty(M_n).$$

Set $f_i = \alpha_i f, K_i = supp\alpha_i$. Obviously there exists $Q_i \in K_i$ such that $sup|\alpha_i f| = |f_i(Q_i)|$.

Consider a closed ball $\tilde{B}_\rho$ of radius ρ with center O in $\mathbb{R}^n$. The exponential map exp_{P_i} at P_i is a diffeomorphism from $\tilde{B}_\rho$ onto $\bar{B}_{P_i}(\rho)$. The function $h(x) = f_i(exp_{P_i}(x))$ is defined on $\tilde{B}_\rho$. Set $x_0 = exp_{P_i}^{-1}(Q_i)$. Let (r, θ) be polar coordinate in $\mathbb{R}^n$ with center at x_0. For every $x \in \mathbb{R}^n$, satisfying $d(x_0, x) = \frac{2\rho}{3}$, we have $d(0, x) > \frac{\rho}{4}$ and $d(0, x) < \rho$, so $x \in \tilde{B}_\rho$ and $h(x) = 0$. Consequently:

$$h(x_0) = -\int_0^{\frac{2\rho}{3}} \partial_r h(r, \theta) dr.$$

Hence:

$$|f_i(Q_i)| = |h(x_0)| \leq \int_0^{\frac{2\rho}{3}} |\partial_r h(r, \theta)| dr.$$

Integrating of the two sites of this inequality over S^{n-1} with respect to θ yields:

$$\omega_{n-1} |f_i(Q_i)| \leq \int_{\tilde{B}} |\partial_r h(r, \theta)| dr d\theta,$$

where $\tilde{B}$ is a closed ball of radius $\frac{2\rho}{3}$, with center at x_0 in $\mathbb{R}^n$ and ω_{n-1} is the volume of S^{n-1}. By using Hölder inequality we obtain:

$$\int_{\tilde{B}} |\partial_r h(r,\theta)| dr d\theta \leq \Big(\int_{\tilde{B}} |\partial_r h(r,\theta)|^q r^{n-1} dr d\theta \Big)^{\frac{1}{q}} . \Big(\int_{\tilde{B}} r^{\frac{(1-n)q'}{q}} dr d\theta \Big)^{\frac{1}{q'}}$$

$$\leq \Big[\int_{\tilde{B}} |\triangle h(x)|^q dE \Big]^{\frac{1}{q}} . \Big[\omega_{n-1} \int_0^{\frac{2\rho}{3}} r^{\frac{(1-n)q'}{q}} dr \Big]^{\frac{1}{q'}},$$

where dE is the canonical volume element in $\mathbb{R}^n$ and $\dfrac{1}{q} + \dfrac{1}{q'} = 1$. Since K_i is compact there exists constants λ, μ with $0 < \lambda \leq \mu < \infty$, such that:

$$\lambda |\xi|^2 \leq \sum_{i,j} g_{ij}(P) \xi_i \xi_j \leq \mu |\xi|^2, \forall \xi = (\xi_1, ..., \xi_n) \in \mathbb{R}^n, \forall P \in K_i.$$

Consequently:

$$dE \leq \lambda^{-\frac{n}{2}} dV, |\triangle h| \leq \mu^{\frac{1}{2}} |\triangle f_i| \circ exp_{P_i}$$

on K_i, where dV is the Riemannian volume element on M_n. Hence:

$$\omega_{n-1} |f_i(Q_i)| \leq \lambda^{\frac{-n}{2q}} \mu^{\frac{1}{2}} \|\triangle f_i\|_q \Big[\omega_{n-1} \frac{q-1}{q-n} \Big(\frac{2\rho}{3} \Big)^{\frac{q-n}{q-1}} \Big]^{\frac{q-1}{q}}.$$

We can choose:

$$C_i = (\omega_{n-1})^{\frac{-1}{q}} \lambda^{\frac{-n}{2q}} \mu^{\frac{1}{2}} \Big\{ \frac{q-1}{q-n} \Big(\frac{2\rho}{3} \Big)^{\frac{q-n}{q-1}} \Big\}^{\frac{q-1}{q}}.$$

Therefore, $\forall f \in C^\infty_\circ(M_n)$ we get:

$$\sup |f| \leq \sum_{i=1}^N \sup |\alpha_i f| \leq \sum_i C_i \|\triangle(\alpha_i f)\|_q$$

$$\leq \sum_{i=1}^N C_i (\sup |\triangle \alpha_i| . ||f||_q + ||\triangle f||_q).$$

□

References

1. R. A. Adams, *Sobolev spaces,* Academic Press, (New York, 1975).
2. T. Aubin, *Nonlinear Analysis on manifolds*, Monge-Ampere Equations, (Springer-Verlag, New York, 1982).
3. T. Aubin, *A Course in Differential Geometry*, AMS, Providence, Rhode Island, (2000).
4. T. Aubin, *Bull. Sci. Math.* 2e Série, **100**, 149 (1976).
5. H. Brezis and L. Nirenberg, *Comm. Pure Appl. Math.* **36**, 437 (1983).

6. N. M. Chuong and T. D. Ke, *Existence and nonexistence of solutions for a semelinear degenerate elliptic system*, J. Abstract and Applied Analysis (submitted).
7. S. Gallot, D. Hulin, and J. LaFontaine, *Riemannian geometry*, (Spinger-Verlag, 1993).
8. L. Hörmander, *Linear partial differential operators*, (Springer-Verlag, Berlin, 1976).
9. Jürgen Jost, *Riemannian goemetry and geometric analysis*, (Springer, 1998).
10. H. Triebel, *Theory of functions spaces*, (in Russian) Akdemische Verlagsgesellschaft Geest & Portig K. G, (Leipzig Birkhauser Verlag, Basel_Boston_Stuttgart, 1986).

Chapter IV

STOCHASTIC AND INFINITE-DIMENTIONAL ANALYSIS

Advances in Deterministic and Stochastic Analysis
Eds. N. M. Chuong *et al.* (pp. 281–296)

§15. STOCHASTIC POPULATION CONTROL AND RSDE WITH JUMPS

SITU RONG

Department of Mathematics, Zhongshan University,
Guangzhou 510275, China
E-mail: mcsstr@zsu.edu.cn

We obtain and improve results on the existence, convergence, stability of solutions to reflecting stochastic differential equation (RSDE) with jumps and apply them to discuss the properties of stochastic population solutions and finally solve the optimal stochastic population control problem.

1. A Deterministic Population Dynamical System

Consider a population dynamics system. Let us denote by x_t^i the size of the population with age between $[i, i+1)$, and r_m – the possible largest age of the people. Suppose that the year variable is discrete as $t = 1, 2, \cdots$. Then as time t evolves, the size $x_t^i, i = 1, 2, \cdots, r_m$ also changes. Then intuitively, the increment of $\triangle x_t^i$ caused by the time increment $\triangle t = 1$ can be expressed as

$$\begin{cases} \triangle x_t^1 = [-(1+\eta_t^1)x_t^1 + \sum_{i=r_1}^{r_2} b_t^i x_t^i \beta_t] \triangle t, \\ \triangle x_t^i = [-(1+\eta^i)x_t^i + x_t^{i-1}] \triangle t, \\ i = 2, \cdots, r_m; \end{cases} \tag{1}$$

where β_t is the specific fertility rate of females, $b_t^i = (1-\mu_t^0)k_t^i h_t^i \neq 0$, $i = r_1, \cdots, r_2; 1 < r_1 < r_2 < r_m$; μ_t^0 is the death rate of babies, η_t^i is the forward death rate by ages, k_t^i is the sex rate, h_t^i – the fertility model, and $[r_1, r_2]$ is the age interval that women can give babies. The absolute value $(1+\eta_t^1)x_t^1$ of the first term in the right hand side of the first equation in (1) is the amount of persons who will leave the age interval $[1, 2)$ in the next year $t+1$, and the value $\sum_{i=r_1}^{r_2} b_t^i x_t^i \beta_t$ of the second term is the amount of persons who will arrive at age interval $[1, 2)$ in the next year $t+1$. Similar physical meaning can be made for $\triangle x_t^i, 2 \leq i \leq r_m$. If one considers that the time variable changes continuously, and writes the corresponding ordinary differential equation in a matrix form, then he will get the following deterministic population control system:[5]

$$\begin{cases} dx_t = (A_t x_t + B_t x_t \beta_t)dt \\ x_0 = x, \ t \in [0,T], \end{cases} \tag{2}$$

where $x_t = (x_t^1, \cdots, x_t^{r_m})$, and

$$A_t = \begin{bmatrix} -(1+\eta_t^1) & 0 & \cdots & 0 \\ 1 & -(1+\eta_t^2)\ 0 \cdots & & 0 \\ \cdots & \cdots & \cdots & \cdots \\ 0 & \cdots & 0\ 1 & -(1+\eta_t^{r_m}) \end{bmatrix}, \tag{3}$$

$$B_t = \begin{bmatrix} 0 \cdots 0\ b_t^{r_1} \cdots b_t^{r_2}\ 0 \cdots 0 \\ 0 \cdots \cdots \cdots \cdots \cdots \cdots 0 \end{bmatrix}. \tag{4}$$

Notice that the size of the population should be non-negative, so one should require that

$$x_t^i \geq 0, 1 \leq i \leq r_m. \tag{5}$$

Under appropriate conditons (5) is valid.

2. RSDE for Stochastic Population Dynamical System

In case that the size of the population is disturbed by some stochastic perturbation the situation is quite different. For example, in the simplest case even it is disturbed by a standard Brownian Motion process (BM), caused by the strange properties of the BM (5) still cannot be guaranteed. So, we need to consider the reflecting stochastic differential equations (RSDEs) as the stochastic population dynamic system model. Let us first introduce a more general $d-$dimensional RSDE with jumps as follows: (Here we write "RCLL" for "right continuous with left limit".)

$$\begin{cases} dx_t = b(t,x_t,\omega)dt + \sigma(t,x_t,\omega)dw_t + \int_Z c(t,x_{t-},z,\omega)\widetilde{N}_k(dt,dz) + d\phi_t, \\ x_0 = x, \ x_t \in \overline{R}_+^d, \text{for all } t \geq 0, \\ \text{where } R_+^d = \left\{x = (x^1,\cdots,x^d) \in R^d : x^i > 0, 1 \leq i \leq d\right\}, \\ \phi_t \text{ is a } R^d - \text{valued } \mathfrak{F}_t - \text{adapted RCLL process with finite variation} \\ |\phi|_t \text{ on each finite interval } [0,t] \text{ such that } \phi_0 = 0, \text{and} \\ |\phi|_t = \int_0^t I_{\partial R_+^d}(x_s)d|\phi|_s, \\ \phi_t = \int_0^t n(s)d|\phi|_s, \\ n(t) \in \mathcal{N}_{x_t}, \text{ as } x_t \in \partial R_+^d, \end{cases} \tag{6}$$

where w_t is a $d-$dimensional BM, and $\widetilde{N}_k(dt,dz)$ is a Poisson martingale measure generated by a $\mathfrak{F}_t-$Poisson point process $k(\cdot)$ with a compensator $\pi(dz)dt$, $\pi(.)$ is a $\sigma-$finite measure on a measurable space $(Z,\Re(Z))$ such that

$\widetilde{N}_k(dt,dz) = N_k(dt,dz) - \pi(dz)dt,$

and

$\mathcal{N}_x = \cup_{r\neq 0}\mathcal{N}_{x,r}, \mathcal{N}_{x,r} = \left\{n \in R^d : |n| = 1, B(x-nr,r)\cap\Theta = \emptyset\right\},$
$B(x_0,\varepsilon) = \left\{y \in R^d : |y-x_0| < \varepsilon\right\}.$

Condition $|\phi|_t = \int_0^t I_{\partial R^d_+}(x_s)d\,|\phi|_s$ in (6) obviously means that $|\phi|_s$ increases only when $x_s \in \partial R^d_+$, and the condition $\phi_t = \int_0^t n(s)d\,|\phi|_s$ means that the reflection $d\phi_t$ happens along the direction of the inner normal vector $n(t)$, so we may call such a reflection a normal reflection.

The geometrical meaning of $\mathcal{N}_x$ is that it is the set of all inner normal vectors at x, when $x \in \partial R^d_+$. Obviously, when $x \in R^d_+$, by definition $\mathcal{N}_x = \Phi$ (the empty set). Actually, in the case that $x_t \in R^d_+$ no refelction is needed, so we do not need an inner normal vector.

More precisely, for (6), by definition as $x \in R^d_+$,

$\mathcal{N}_x = \Phi$ (empty set).

Notice that

$\partial R^d_+ = \cup_{k=1}^d \cup^d_{i_1<i_2<\cdots<i_k, i_1,i_2,\cdots,i_k=1} A_k,$

where

$A_k = \left\{(x_1,x_2,\cdots,x_d) : x_{i_p} = 0, x_j > 0, \forall j \neq i_p, 1 \leq p \leq k\right\}$

So, if $x \in A_k$, then $n = (n_1,n_2,\cdots,n_d) \in \mathcal{N}_x$, iff, for $k \geq 3$,

$n_{i_1} = \sin\theta\cos\varphi_1\cos\varphi_2\cdots\cdots\cdots\cos\varphi_{k-2},$
$n_{i_2} = \sin\theta\cos\varphi_1\cos\varphi_2\cdots\cos\varphi_{k-3}\sin\varphi_{k-2},$
$\cdots\cdots\cdots$
$n_{i_{k-2}} = \sin\theta\cos\varphi_1,$
$n_{i_{k-1}} = \sin\theta\sin\varphi_1,$
$n_{i_k} = \cos\theta,\ \theta \in [0,\frac{\pi}{2}], \varphi_1,\varphi_2\cdots,,\varphi_k \in [0,\frac{\pi}{2}],$
$n_j = 0,$ as $j \neq i_1,i_2,\cdots,i_k;$

and for $k = 2$,

$n_{i_1} = \cos\theta, n_{i_2} = \sin\theta,\ \theta \in [0,\frac{\pi}{2}],$
$n_k = 0, \forall k \neq i_1,i_2;$

and for $k = 1$,

$n_{i_1} = 1,\ n_j = 0, \forall j \neq i_1.$

From this one sees that the inner normal vector at a boundary point of the domain R^d_+ is not necessary unique. However, from (6) one knows that there exists a unique $n(t)$ such that as $x_t \in \partial R^d_+$, $n(t) \in \mathcal{N}_{x_t}$ and the reflection $d\phi_t$ happens along the direction of the inner normal vector $n(t)$.

Since all $n_i \geq 0, i = 1, 2, \cdots, d$. So we have the following Claims.

Claim 2.1. *all* $\phi_t^i, i = 1, 2, \cdots, d$, *in (6) are increasing, where* $\phi_t = (\phi_t^1, \phi_t^2, \cdots, \phi_t^d)$.

Claim 2.2. *If* (x_t, ϕ_t) *satisfies (6), then* $t \geq 0$, $\int_0^t x_s^i d\phi_s^i = 0,\ i = 1, 2, \cdots, d$.

Claim 2.2 is easily obtained by (6) and by the property of $\mathcal{N}_x$.

From now on we always make the following Hypothesis on the jump term:

(H_1): $x + c(t, x, z, \omega) \in \overline{R_+^d}$, for all $t \geq 0, z \in Z, \omega \in \Omega$ and $x \in \overline{R_+^d}$.

(H_2) $\pi(dz) = dz/|z|^{d+1}$, $(Z = R^d - \{0\})$.

The meaning of (H_1) is that if the state x_{t-} before the jump lives in $\overline{R_+^d}$, then after the jump, the state $x_t = x_{t-} + c(t, x, z, \omega)$ still stays in $\overline{R_+^d}$. The Hypothesis (H_2) gives a more concrete measure for the compensator of the jump Poisson point process. Notice that under assumption (H_1) if (x_t, ϕ_t) satisfies (6), then ϕ_t should be continuous[1, Lemma 329]. So under assumption (H_1) in (6) we may write that

$$\phi_t \text{ is a continuous function.} \tag{7}$$

Now for the solution of RSDE (6) with (7) we introduce the following definition.

Definition 2.1.

1) We say that (x_t, ϕ_t) is a strong solution of (6) with (7), if x_t is RCLL and ϕ_t is continuous and they are $\mathfrak{F}_t^{w,\widetilde{N}_k}$–adapted, where $\mathfrak{F}_t^{w,\widetilde{N}_k} = \sigma(w_s, \widetilde{N}_k((0, s\}, U); s \leq t, U \in \mathfrak{B}(Z))$, and (x_t, ϕ_t) satisfies (6) with (7).

2) We say that the pathwise uniqueness for solutions of (6) with (7) holds, if for any two solutions $(x_t^i, \phi_t^i), i = 1, 2$, of (6) with (7), which are defined on the same probability space $(\Omega, \mathfrak{F}, (\mathfrak{F}_t), P)$ with the same BM w_t and Poisson martingale measure $\widetilde{N}_k(dt, dz)$,

$P(\sup_{t\geq 0} |x_t^1 - x_t^2| = 0) = 1$, $P(\sup_{t\geq 0} |\phi_t^1 - \phi_t^2| = 0) = 1$.

3. Existence of Solutions to RSDE with Jumps

We can have the following general existence theorems of solutions for RSDE with jumps which are some generalization of Theorem 335 and 336 for $\Theta = R_+^d$ in Ref. 1.

Theorem 3.1. *Assume that for* $t \geq 0, x \in \overline{R^d_+}, z \in Z$
1° $b(t,x), \sigma(t,x)$*, and* $c(t,x,z)$ *are non-random, jointly measurable; moreover,* $b(t,x), \sigma(t,x)$ *are jointly continuous, and as* $|x-y| \to 0, |t-s| \to 0$

$\int_Z |c(t,x,z) - c(s,y,z)|^2 \pi(dz) \to 0,$

furthermore, $b(t,x), \sigma(t,x)$ *and* $\int_Z |c(t,x,z)|^2 \pi(dz)$ *are locally bounded for* x*; that is, for each* $r > 0$*, as* $|x| \leq r$*,*

$|b(t,x)|^2 + \|\sigma(t,x)\|^2 + \int_Z |c(t,x,z)|^2 \pi(dz) \leq k_r,$

where $k_r \geq 0$ *is a constant depending only on* r*;*
2° there exists a real function $V(t,x) \in C^{1,2}([0,\infty) \times R^d)$*, i.e. it has a first continuous derivative with respect to* t *and up to second continuous derivatives with respect to* x*, such that one of the following conditions is satisfied:*

(i) $\mathcal{L}V(t,x) = \frac{\partial}{\partial t}V(t,x) + \sum_{i=1}^d b_i(t,x)\frac{\partial}{\partial x_i}V(t,x)$
$+\frac{1}{2}\sum_{i,j=1}^d \sum_{k=1}^d \sigma_{ik}\sigma_{jk}(t,x)\frac{\partial^2}{\partial x_i \partial x_j}V(t,x)$
$+\int_Z [V(t,x+c(t,x,z)) - V(t,x) - \sum_{i=1}^d c_i(t,x,z)\frac{\partial}{\partial x_i}V(t,x)]\pi(dz)$
$\leq c_1(t),$

where $c_1(t) \geq 0$*,* $\int_0^T c_1(t)dt < \infty, \forall T < \infty$*;*

(ii) $\mathcal{L}V(t,x) \leq c_1(t)V(t,x)$*,*

where $c_1(t)$ *has the same property as in the above (i);*
3° there exists a $a_1(t)$ *with* $a_1(|t|) \geq 0$ *such that* $a_1(t) \uparrow \infty$*, as* $t \uparrow \infty$*; and as* $|x|$ *is large enough,*

$V(t,x) \geq a_1(|x|),$

where $V(t,x)$ *is the same function occured in 2°; and as* $x_i \geq 0$*,*

$\frac{\partial}{\partial x_i}V(t,x) \leq k_0 x_i, \ i = 1, 2, \cdots, d;$

4° $2(x-y)\cdot(b(t,x) - b(t,y)) + \|\sigma(t,x) - \sigma(t,y)\|^2$
$+\int_Z |c(t,x,z) - c(t,y,z)|^2 \pi(dz) \leq k_{N,T}(t)\rho_{N,T}(|x-y|^2),$
as $|x|, |y| \leq N, t \in [0,T]$*,*

where for each $T < \infty$*,* $0 \leq k_{N,T}(t)$*,* $\int_0^T k_{N,T}(t)dt < \infty$ *and* $\rho_{N,T}(u)$ *is concave and strictly increasing in* $u \geq 0$ *such that* $\rho_{N,T}(0) = 0$*, and*

$\int_{0+} du/\ \rho_{N,T}(u) = \infty$*, for each* $N = 1, 2, \ldots.$

Then (6) with (7) has a pathwise unique strong solution.

Theorem 3.1 has the following obvious applications.

Corollary 3.1. *Under assumptions 1° and 4° in Theorem 3.1 if, furthermore, there exists a constant* $k_0 \geq 0$ *such that*

$2\langle x, b(t,x)\rangle + \|\sigma(t,x)\|^2 + \int_Z |c(t,x,z)|^2 \pi(dz) \leq k_0(1+|x|^2),$

then (6) with (7) has a pathwise unique strong solution.

Corollary 3.1 obviously includes Theorem 335 for $\Theta = R^d_+$ in Ref. 1. To show this Corollary by Theorem 3.1 one only needs to take $V(t,x) = 1 + |x|^2$, $a_1(t) = t^2$, $c_1(t) = k_0$. Then one easily sees that under the assumption of Corollary 3.1

$$\mathcal{L}V(t,x) \leq c_1(t)V(t,x),$$

and Theorem 3.1 applies.

Corollary 3.2. *Under assumptions 1° and 4° in Theorem 3.1 if, in addition,*

$$\langle x, b(t,x)\rangle \leq c_1(t)(1 + |x|^2 \textstyle\prod_{k=1}^m g_k(x)),$$
$$\|\sigma(t,x)\|^2 \leq c_1(t)(1 + |x|^2 \textstyle\prod_{k=1}^m g_k(x)),$$
$$\textstyle\int_Z |c(t,x,z)|^2 \pi(dz) \leq c_1(t),$$

where

$$g_k(x) = 1 + \underbrace{\ln(1 + \ln(1 + \cdots \ln}_{k-times}(1 + |x|^{2n_0}))),$$

(m and n_0 are some natural numbers), and $c_1(t) \geq 0$ is non-random such that for each $T < \infty$, $\int_0^T c_1(t)dt < \infty$;

then (6) with (7) has a pathwise unique strong solution.

Corollary 3.2 obviously includes the statement of the second part of Theorem 336 for $\Theta = R^d_+$ in Ref. 1. To show this Corollary by Theorem 3.1 one only needs to take $V(t,x) = g_{m+1}(x)$, $a_1(t) = g_{m+1}(t)$, $c_1(t) = c_1(t)$. Then one easily sees that under the assumption of Corollary 3.2

$$\mathcal{L}V(t,x) \leq k_0 c_1(t),$$

where $k_0 \geq 0$ is a constant. So Theorem 3.1 applies.

Now let us prove Theorem 3.1.

Proof. Let

$$f^N(t,x) = \begin{cases} f(t,x), & \text{as } |x| \leq N, \\ f(t, N\frac{x}{|x|}), & \text{as } |x| > N, \end{cases}$$

where $f = b, \sigma$; and

$$c^N(t,x,z) = \begin{cases} c(t,x,z), & \text{as } |x| \leq N, \\ c(t, N\frac{x}{|x|}, z), & \text{as } |x| > N. \end{cases}$$

Then by Theorem 334 in Ref. 1 there exists a pathwise unique strong solution (x_t^N, ϕ_t^N) satisfying RSDE (6) with (7) but with coefficients b^N, σ^N and c^N. Let

$$\tau^{N,k} = \inf\left\{t \geq 0 : \left|x_t^N\right| > k\right\}.$$

Then by Ito's formula one easily sees that under the assumption (ii) of 2° in Theorem 3.1 for any given $T < \infty$ as $0 \leq t \leq T$

$EV(t\wedge\tau^{N,N},x^N_{t\wedge\tau^{N,N}})=EV(0,x_0)+E\int_0^{t\wedge\tau^{N,N}}\mathcal{L}V(s,x^N_s)ds.$
$\leq k_0'+\int_0^t c_1(s)EV(s\wedge\tau^{N,N},x^N_{s\wedge\tau^{N,N}})ds.$
Hence as $0\leq t\leq T$
$EV(t\wedge\tau^{N,N},x^N_{t\wedge\tau^{N,N}})\leq k_0'e^{\int_0^T c_1(t)dt},$
and
$a(N)P(\tau^{N,N}<T)\leq EV(\tau^{N,N}\wedge T,x^N_{\tau^{N,N}\wedge T})I_{\tau^{N,N}<T}$
$\leq EV(\tau^{N,N}\wedge T,x^N_{\tau^{N,N}\wedge T})\leq k_0'e^{\int_0^T c_1(t)dt}.$
Therefore, as $N\to\infty$,
$P(\tau^{N,N}<T)\to 0.$

For the case that assumption (i) of 2° in Theorem 3.1 holds, the same result is similarly derived and the derivation is even simpler. From this result one easily sees that $P-a.s.\ \lim_{N\to\infty}\tau^{N,N}=\infty$. Now set
$x_t=x^N_t,\phi_t=\phi^N_t$, as $0\leq t<\tau^{N,N}$.
It is easily seen that (x_t,ϕ_t) is the pathwise unique strong solution of RSDE (6) with (7). □

4. Stochastic Population Solutions and Their Properties

Now let

$$b(t,x)=A_tx+B_tx\beta_t(x), \tag{8}$$

where A_t and B_t are defined by (3) and (4), and let $d=r_m$, then (6) with (7) and (8) becomes the stochastic population dynamic system. By Corollary 3.1 we immediately obtain the following result.

Theorem 4.1. *Assume that coefficients σ and c satisfy conditions 1° and 4° in Theorem 3.1, and assume that A_t and B_t are bounded and $\beta_t(x)$ is bounded and jointly continuous such that*
$|\beta_t(x)-\beta_t(y)|\leq k_{N,T}(t)\rho_{N,T}(|x-y|^2),$
as $|x|,|y|\leq N, t\in[0,T],\forall T<\infty;$
where $k_{N,T}(t)$ and $\rho_{N,T}(u)$ have the same properties as those in the condition 4° of Theorem 3.1. Then the population SDE, i.e. (6) with (7) and with the coefficient (8), has a pathwise unique strong solution.

A simple case for σ and c to satisfy the above conditions in Theorem 4.1 is given as follows:

Corollary 4.1. *Let*

$$\begin{cases} \sigma(t,x) = C_0 + (C_1^{(1)}x, \cdots, C_1^{(d)}x) \\ c(t,x,z) = (\overline{C}_0 + \overline{C}_1 x) f(z), \end{cases} \tag{9}$$

where $C_0, C_1^{(i)}, i = 1, \cdots, d$ and $\overline{C}_1$ are constant $d \times d$ matrices, and $\overline{C}_0$ is a constant R^d-vector, (i.e. all elements in the matrices or vectors are constants), and , in addition, all elements in $\overline{C}_0$ and $\overline{C}_1$ are non-negative, i.e. $\overline{c}_0^i, \overline{c}_1^{ij} \geq 0, i,j = 1, 2, \cdots, d$; where $\overline{C}_0 = (\overline{c}_0^1, \cdots, \overline{c}_0^d)^T, \overline{C}_1 = (\overline{c}_1^{ij})_{i,j=1}^d$; and $f(z) \geq 0$ is such that $\int_{R^d - \{0\}} \frac{f(z)^2}{|z|^{d+1}} dz < \infty$. Then one easily checks that assumption (H_1) and all conditions for σ and c in the above theorem hold. So Theorem 4.1 applies.

Let us discuss the properties of the stochastic population solutions

1) Convergence Property. First we can consider the more general RSDEs with jumps. If (x_t^n, ϕ_t^n) satisfies (6) with (7) and with coefficients $b^n(t,x)$, $\sigma^n(t,x)$, and $c^n(t,x,z)$ and with the initial condition x_0^n as well, then we have the following convergence theorem.

Theorem 4.2. *Assume that*

1° $|b^n(t,x)|^2 + \|\sigma^n(t,x)\|^2 + \int_Z |c^n(t,x,z)|^2 \pi(dz) \leq k_0(1 + |x|^2)$, $n = 0, 1, 2, ..,$ *where $k_0 \geq 0$ is a constant;*

2° *there exists a real function $0 \leq V(t,x) \in C^{1,2}([0,\infty) \times R^d)$ such that*

$$\begin{aligned}(\mathcal{L}_1 V)(t, x-y) &= \frac{\partial}{\partial t} V(t,x-y) + \sum_{i=1}^d (b_i^0(t,x) - b_i^0(t,y)) \frac{\partial}{\partial x_i} V(t,x-y) \\ &+ \frac{1}{2} \sum_{i,j=1}^d \sum_{k=1}^d (\sigma_{ik}^0 \sigma_{jk}^0(t,x) - \sigma_{ik}^0 \sigma_{jk}^0(t,y)) \frac{\partial^2}{\partial x_i \partial x_j} V(t,x-y) \\ &+ \int_Z [V(t, x-y+c^0(t,x,z) - c^0(t,y,z)) - V(t,x-y) \\ &- \sum_{i=1}^d (c_i^0(t,x,z) - c_i^0(t,y,z)) \frac{\partial}{\partial x_i} V(t,x-y)] \pi(dz) \leq c_1(t) V(t,x-y),\end{aligned}$$

and there exists a $a_1(t)$ with $a_1(|t|) \geq 0$ such that $a_1(t), t \geq 0$ is strictly increasing, and

$$V(t,x) \geq a_1(|x|),$$

moreover,

$$\frac{\partial}{\partial x_i} V(t,y) = c_1(t) g(y) y_i, \quad i = 1, 2, \cdots, d;$$

$$\left|\frac{\partial^2}{\partial x_i \partial x_j} V(t,y)\right| \leq c_1(t), \quad i,j = 1,2,\cdots,d,$$

where $|g(y)| \leq k_0$, *and* $c_1(t) \geq 0$, $\int_0^T c_1(t)dt < \infty, \forall T < \infty$;

3°
$lim_{n\to\infty} \int_0^t [\sup_{x\in R^d} |b^n(s,x) - b^0(s,x)|^2 + \sup_{x\in R^d}(\|\sigma^n(s,x) - \sigma^0(s,x)\|^2$
$+ \int_Z |c^n(s,x,z) - c^0(s,x,z)|^2 \pi(dz))]ds = 0$, *for all* $t > 0$;

4° $x_0^n \to x_0^0$, *as* $n \to \infty$,

where the initail values $x_0^n, n = 0,1,2,\cdots$ *are all non-random constants.*

Then for all $t \geq 0$
$\lim_{n\to\infty} E\sup_{s\leq t} V(t, x_s^n - x_s^0) = 0.$

In particular, for any $\varepsilon > 0$ *and* $t \geq 0$

$$\lim_{n\to\infty} \sup_{s\leq t} P(|x_s^n - x_s^0| > \varepsilon) = 0.$$

In addition, assume that

5° $lim_{|y-x|\to 0} \sup_{s\leq t} |g^0(s,x) - g^0(s,y)| = 0$,

where $g = b, \sigma$; *and*

$$\lim_{|y-x|\to 0} \sup_{s\leq t} \int_Z |c^0(s,x,z) - c^0(s,y,z)|^2 \pi(dz) = 0.$$

Then for all $t \geq 0$, *as* $n \to \infty$

$$P(|\phi_t^n - \phi_t^0| > \varepsilon) \to 0.$$

Proof. By condition 1° and using Ito's formula one easily obtains that[1]

$$E\sup_{t\leq T} |x_t^n| \leq k_T, \quad \forall n = 0,1,2,\cdots,$$

where $k_T \geq 0$ is a constant depending on $T < \infty$ only. Now applying Ito's formula to $V(t, x_t^n - x_t^0)$, one has that as $t \leq T$

$$E\sup_{s\leq t} V(s, x_s^n - x_s^0) \leq V(0, x_0^n - x_0^0) + f(t) + \int_0^t c_1(s) EV(s, x_s^n - x_s^0) ds,$$

where

$$\begin{aligned} f(t) = \overline{k}_T\{(&\int_0^t \sup_{x\in R^d} [|b^n(s,x) - b^0(s,x)|^2 ds)^{1/2} \\ &+ \int_0^t \sup_{x\in R^d} [\|\sigma^n(s,x) - \sigma^0(s,x)\|^2 \\ &+ \int_Z |c^n(s,x,z) - c^0(s,x,z)|^2 \pi(dz)]ds\}. \end{aligned}$$

So, by Gronwall's inequality[1, Lemma 118]

$$a_1(\varepsilon)\sup_{s\le t} P(|x_s^n - x_s^0| > \varepsilon) \le E\sup_{s\le t} V(t, x_s^n - x_s^0)$$

$$\le V(0, x_0^n - x_0^0) + f(t) + \int_0^t e^{\int_0^t c_1(s)ds} c_1(s)[f(s) + V(0, x_0^n - x_0^0)]ds.$$

Hence the 1st and 2nd conclusions are easily derived.
Notice that

$$\begin{aligned} d(x_t^n - x_t^0) &= (b^n(t, x_t^n, \omega) - b^0(t, x_t^0, \omega))dt \\ &\quad + (\sigma^n(t, x_t^n, \omega) - \sigma^0(t, x_t^0, \omega))dw_t \\ &\quad + \int_Z (c^n(t, x_{t-}^n, z, \omega) - c^0(t, x_{t-}^n, z, \omega))\widetilde{N}_k(dt, dz) \\ &\quad + d(\phi_t^n - \phi_t^0). \end{aligned}$$

From this the last conclusion is also easily obtained. □

Corollary 4.2. *Under assumptions 1°, 3° and 4° in Theorem 4.2 if*

6° $2(x-y)\cdot(b^0(t,x) - b^0(t,y)) + \|\sigma^0(t,x) - \sigma^0(t,y)\|^2$
$+ \int_Z |c^0(t,x,z) - c^0(t,y,z)|^2 \pi(dz) \le c_1(t)|x-y|^2,$

where $c_1(t) \ge 0$, $\int_0^T c_1(t)dt < \infty, \forall T < \infty$; *then for all* $t \ge 0$

$\lim_{n\to\infty} E(\sup_{s\le t} |x_s^n - x_s^0|^2) = 0.$

In addition, if 5° in Theorem 4.2 is also satisfied, then for all $t, \varepsilon \ge 0$,

$P(\sup_{s\le t} |\phi_s^n - \phi_s^0| > \varepsilon) \to 0$, *as* $n \to \infty$.

To see that the first conclusion in Corollary 4.4 is true, one only needs to take $V(t,x) = |x|^2, a_1(t) = t^2$ and applies Theorem 4.4. The second conclusion can be proved directly.

Corollary 4.3. *Under assumptions 1°, 3° and 4° in Theorem 4.2 if*

$(x-y)\cdot(b^0(t,x) - b^0(t,y)) \le c_1(t)|x-y|^2 \prod_{k=1}^m g_k(x-y),$

$\|\sigma^0(t,x) - \sigma^0(t,y)\|^2 + \int_Z |c^0(t,x,z) - c^0(t,y,z)|^2 \pi(dz)$

$\le c_1(t)|x-y|^2 \prod_{k=1}^m g_k(x-y),$

where $c_1(t)$ *has the same property as that in Corollary 4.4, and* $g_k(x)$ *is defined in Corollary 3.2; then for all* $t \ge 0$

$\lim_{n\to\infty} E\sup_{s\le t} g_m(x_s^n - x_s^0) = 0,$

and for all $t, \varepsilon \ge 0$,

$\sup_{s\le t} P(|x_s^n - x_s^0| > \varepsilon) \to 0$, *as* $n \to \infty$.

In addition, if 5° in Theorem 4.2 is also satisfied, then for all $t, \varepsilon \ge 0$,

$P(|\phi_s^n - \phi_s^0| > \varepsilon) \to 0$, *as* $n \to \infty$.

To show Corollary 4.4, one only needs to take $V(t,x) = g_m(x)$, $a_1(t) = g_m(t)$ and applies Theorem 4.4.

Applying Corollary 4.2 we immediately have the following result for the convergence of solutions of population SDEs.

Theorem 4.3. *Assume that*
1° there exists a constant $k_0 \geq 0$, *for all* $n = 0, 1, 2, ...$

$|A^n(t)| + |B^n(t)| + |\beta_t^n| \leq k_0,$

where $A^n(t)$, $B^n(t)$, *and* β_t^n *are non-random such that they do not depend on* x, *and* $\sigma^n(t,x)$ *and* $c^n(t,x,z)$ *satisfy the condition 1° in Theorem 4.2;*
2° $\sigma^0(t,x)$ *and* $c^0(t,x,z)$ *satisfy the condition 6° in Corollary 4.2;*
3° $\lim_{n\to\infty} \int_0^T [|A^n(t) - A^0(t)| + |B^n(t) - B^0(t)| + |\beta_t^n - \beta_t^0|]dt = 0,$
and the condition 3° in Theorem 4.2 for $\sigma^n(t,x), \sigma^0(t,x)$ *and* $c^n(t,x,z), c^0(t,x,z)$ *holds;*
4° the condition 4° in Theorem 4.2 holds.
Then for all $t \geq 0$

$\lim_{n\to\infty} E(\sup_{s\leq t} |x_s^n - x_s^0|^2) = 0.$

In addition, if 5° in Theorem 4.2 is also satisfied, then for all $t, \varepsilon \geq 0$, *as* $n \to \infty$

$$P(\sup_{s\leq t} |\phi_s^n - \phi_s^0| > \varepsilon) \to 0.$$

This indicates that if coefficients in the population SDE can be approximatly calculated, then solutions of the population SDE also can be approximatly (in probability) obtained. For more convergence results on the solutions of population SDE one can refer to Ref. 1.

2) Stability Property.

For the stability of solutions to RSDE (6) with (7) we have the following theorem.

Theorem 4.4. *Assume that conditions 1°, 3°, 4° in Theorem 3.1 hold, and assume that condition 2° in Theorem 3.1 is changed to be*

2°′ : there exists a real function $0 \leq V(t,x) \in C^{1,2}([0,\infty) \times R^d)$ *such that*

$\mathcal{L}V(t,x) \leq -c_1(t)V(t,x),$

where $c_1(t) \geq 0$, $\int_0^\infty c_1(t)dt = \infty$, *and as* $x_i \geq 0$,

$\frac{\partial}{\partial x_i} V(t,x) \leq k_0 x_i$, $i = 1, 2, \cdots, d.$

Then (6) with (7) has a pathwise unique strong solution, which satisfies that

$EV(t,x_t) \leq e^{-\int_0^t c_1(s)ds} EV(0,x_0).$

Furthermore, if $b(t,0) = 0, \sigma(t,0) = 0, c(t,0,z) = 0$, *and there exists a real function* $a_1(t) \geq 0, t \geq 0$, *which is increasing and* $a_1(t) > 0$, *as* $t > 0$, *such that*

$V(t,x) \geq a_1(|x|)$,

then $(0,0)$ *is the pathwise unique strong solution of (6) with (7) and with the initial condition* $x_0 = 0$; *and for any given* $\varepsilon > 0$,

$\lim_{t\to\infty} P(|x_t| > \varepsilon) = 0$,

i.e. the strong solution of (6) with (7) is stable in probability.

Proof. First, by Theorem 3.1 (6) with (7) has a pathwise unique strong solution for any initial condition x_0. Now applying Ito's formula to $V(t,x_t)$, one has that as $t \leq T$

$EV(t, x_{t\wedge\tau_N}) \leq V(0,x_0) - \int_0^t c_1(s)EV(s, x_{s\wedge\tau_N})ds$,

where $\tau_N = \inf\{t \geq 0 : |x_t| > N\}$. By using Gronwall's inequality and Fatou's lemma the first conclusion is obtained. Secondly, under additional condition one has that for any given $\varepsilon > 0$

$a_1(\varepsilon)P(|x_t| > \varepsilon) \leq EV(t,x_t) \leq e^{-\int_0^t c_1(s)ds}EV(0,x_0)$.

So the second conclusion is also obtained. □

An immediate corollary is as follows:

Corollary 4.4. *Assume that 1°, 4° in Theorem 3.1 hold, and*

$2\langle x, b(t,x)\rangle + \|\sigma(t,x)\|^2 + \int_Z |c(t,x,z)|^2 \pi(dz) \leq -c_1(t)|x|^2$,

where $c_1(t) \geq 0$, $\int_0^\infty c_1(t)dt = \infty$.

Then (6) with (7) has a pathwise unique strong solution, which satisfies that

$$E|x_t|^2. \leq e^{-\int_0^t c_1(s)ds}E|x_0|^2. \tag{10}$$

Corollary 4.4 obviously follows from Theorem 4.4, if we take $V(t,x) = |x|^2$, and $a_1(t) = t^2$.

Now let us apply the above Corollaries to give the stability results on the stochastic population dynamics (6) with (7) and with $b(t,x) = A_t x + B_t x\beta_t(x)$. Suppose that $\sigma(t,0) = 0$ and $c(t,0,z) = 0$ then under conditions of the existence Theorem 4.1 (6) with (7) and with the coefficient (8), has a pathwise unique strong solution $(0,0)$ with $x_0 = 0$. Furthermore, if $2\langle A_t x, x\rangle + 2\langle B_t\beta_t(x)x, x\rangle + \|\sigma(t,x)\|^2 + \int_Z |c(t,x,z)|^2 \pi(dz) \leq -c_1(t)|x|^2$, where $c_1(t) \geq 0$, $\int_0^\infty c_1(t)dt = \infty$; then (10) holds, i.e. the solutions of population SDE is exponentially stable in the mean square. More precisely, if $\sigma(t,x) = \left(C_1^{(1)}, \cdots, C_1^{(r_m)}\right)x$, $c(t,x,z) = \overline{C}_1 x I_U(z)$, where $\pi(U) = 1$, and $C_1^{(i)}, i = 1, \cdots, d$; $\overline{C}_1$ are defined in Corollary 4.1, and $0 \leq \beta_t(x) \leq \beta_0$. Let

$\delta_0 = \min_{t\geq 0}(\frac{1}{2} + \eta_1(t), \eta_2(t), \cdots, \eta_{r_m-1}(t), \frac{1}{2} + \eta_{r_m}(t)),$
$b_M = \max_{t\geq 0}(b_{r_1}(t), \cdots, b_{r_2}(t)).$
Then as
$\delta_1 = 2\delta_0 - \sum_{i=1}^{r_m} \left\| C_1^{(i)} \right\|^2 - \left|\overline{C}_1\right|^2 - \beta_0 b_M (r_2 - r_1 + 1) > 0,$
$E\left|x_t\right|^2 \leq E\left|x_0\right|^2 e^{-\delta_1 t}$, for all $t \geq 0$.

That is, the population dynamics of (6) with (7) and with the coefficient (8) is exponentially stable in the mean square under the above assumptions. In particular, for a given constant $a > 0$ we easily derive from the above inequality that
$t \geq \frac{1}{\delta_1} \ln \frac{E|x_0|^2}{a} \implies E\left|x_t\right|^2 \leq a.$

This indicates that if the stochastic perturbation is not too large (that is, $\sum_{i=1}^{d} \left\| C_1^{(i)} \right\|^2 + \left|\overline{C}_1\right|^2$ is very small), and the forward death rate is greater than a positive constant (that is, $\delta_0 > 0$), then the stochastic population dynamics (6) with (7) and with the coefficient (8) can be exponentially stable "in the mean square" if the fertility rate of females is small enough. Furthermore, given $a > 0$, we can find out when the population size vector "in the mean square" can be less than this target a.

3) Comparison property.

Theorem 4.5. *Assume that*
1° $\|\sigma(t,x)\|^2 + \int_Z |c(t,x,z)|^2 \pi(dz) \leq k_0(1 + |x|^2)$;
where $k_0 \geq 0$ *is a constant;*

2° $\sigma(t,x) = (\sigma_{ik}(t,x))_{i,k=1}^d$ *satisfies the condition that*
$|\sigma_{ik}(t,x) - \sigma_{ik}(t,y)|^2 \leq k_{N,T}(t)\rho_{N,T}(|x_i - y_i|^2),$
as $|x|, |y| \leq N, t \in [0,T], \forall i, k = 1, 2, \cdots, d$;
where $0 \leq k_{N,T}(t)$, $\int_0^T k_{N,T}(t)dt < \infty$, *for each* $T < \infty$,
and $\rho_{N,T}(u) > 0$, *as* $u > 0$; $\rho_{N,T}(0) = 0$; *and* $\rho_{N,T}(u)$ *is strictly increasing in* u *and such that*
$\int_{0+} du/\rho_{N,T}(u) = \infty$, *for* $N = 1, 2, ...$ *and* $T < \infty$;
3° $c(t,x,z) = (c_i(t,x,z))_{i=1}^d$ *satisfies conditions that*
$x + c(t,x,z) \in \overline{R}_+^d, \forall t \geq 0, x \in \overline{R}_+^d, z \in Z;$
$x_i \geq y_i \implies x_i + c_i(t,x,z) \geq y_i + c_i(t,y,z),$
$\forall t \geq 0,\ x, y \in \overline{R}_+^d, z \in Z,$
where $x = (x_1, \cdots, x_d), y = (y_1, \cdots, y_d)$;
4° $I_{x^i > y^i}(\beta_t(x) - \beta_t(y)) \leq k_{N,T}(t)\rho_{N,T}(\sum_{i=r_1}^{r_2}(x^i - y^i)^+), \forall x, y \in \overline{R}_+^d,$
as $|x|, |y| \leq N, t \in [0,T]$,
where $k_{N,T}(t)$ *and* $\rho_{N,T}(u)$ *have the same properties as those in 2°;*
5° $0 \leq \eta_t^i, b_t^j, \beta_t(x) \leq k_0, \forall i = 1, 2, \cdots, d; j = r_1, \cdots, r_2.$

If (x_t, ϕ_t) *and* $(\overline{x}_t, \overline{\phi}_t)$ *are solutions of the stochastic population dynamics (6) with (7) and with the coefficient (8) and with the initial value* x_0*, the fertility rate of females* $\beta_t(x)$ *and* $\overline{x}_0, \overline{\beta}_t(x)$*, respectively; then*
$x_0^i \geq \overline{x}_0^i, \forall i = 1, \cdots, d$*; and*
$\beta_t(x) \geq \overline{\beta}_t(x), \forall t \geq 0, \forall x \in \overline{R}_+^d,$
implies that $P - a.s.$
$x_t^i \geq \overline{x}_t^i, \forall t \geq 0, \forall i = 1, \cdots, d.$

This indicates that larger initial size and lager fertility rate of female always result larger population size.

Theorem 4.5 can be proved by Tanaka's formula as in Ref. 1. (See the proof of Theorem 346 in Ref. 1. However, now here $\beta_t(x)$ depends on x).

5. The Optimal Stochastic Population Control

Now let us discuss the optimal stochastic population control problem for the stochastic population dynamics (6) with (7) and with the coefficient (8). Suppose that the admissible control set is

$$\widetilde{U} = \left\{ \begin{array}{r} \beta : \beta = \beta_t(x) \text{ is jointly continuous,} \\ \overline{\beta}_0 \leq \beta \leq \beta_0, \end{array} \right\},$$

where $0 \leq \overline{\beta}_0$ and β_0 are constants. Let us minimize the following functional

$$J(\beta) = E[\int_0^T F(t, x_t^\beta)dt + G(x_T^\beta)],$$

among all $\beta \in \widetilde{U}$, where $0 \leq T < \infty$, and $(x_t^\beta, \phi_t^\beta)$ is the pathwise unique strong solution of (6) with (7) and with the coefficient (8) corresponding to the given $\beta \in \widetilde{U}$; where $F(t, x)$ and $G(x)$ are Borel measurable.

Theorem 5.1. *Assume that* $\sigma(t, x)$ *and* $c(t, x, z)$ *satisfy all conditions in Theorem 4.1 and Theorem 4.5;* $F(t, x)$ *and* $G(x)$ *are jointly Borel measurable functions defined on* $t \in [0, T]$ *and* $x \in \overline{R}_+^d$ *such that as* $x_i \leq y_i$*,* $1 \leq i \leq d$*;* $x, y \in \overline{R}_+^d, \forall t \in [0, T]$*, where we denote* $d = r_m$*,*

$$G(x) \leq G(y), F(t, x) \leq F(t, y).$$

Then the smallest constant $\overline{\beta}_0$ *is an admissible optimal control for the minimization of the functional* $J(\beta)$*; that is,*
$\overline{\beta}_0 \in \widetilde{U}$*, and* $J(\overline{\beta}_0) = \inf_{\beta \in \widetilde{U}} J(\beta)$*.*
Furthermore, assume that

$$\sigma_{ik}(t, x) = c_1^{ik} x_i, c_i(t, x, z) = \overline{c}_1^i x_i I_U(z),$$

where $c_1^{ik}, \overline{c}_1^i, i, k = 1, \cdots, d$, *are all constants,* $\overline{c}_1^i \geq 0$, *and* $\pi(U) = 1$, *let*

$\delta_0 = \min_{t \geq 0}(\frac{1}{2} + \eta_1(t), \eta_2(t), \cdots, \eta_{r_m - 1}(t), \frac{1}{2} + \eta_{r_m}(t))$,

$b_M = \max_{t \geq 0}(b_{r_1}(t), \cdots, b_{r_2}(t))$.

Then one has that

$$E\left|x_t^{\overline{\beta}_0}\right|^2 \leq E|x_0|^2 e^{-\delta_1 t}, \text{ for all } t \geq 0;$$

as $\delta_1 = 2\delta_0 - c_1^M - \overline{c}_1^M - \overline{\beta}_0 b_M(r_2 - r_1 + 1) > 0$, *where*

$c_1^M = \max(\sum_{k=1}^{r_m}(c_1^{ik})^2, 1 \leq i \leq r_m)$, $\overline{c}_1^M = \max((\overline{c}_1^{ik})^2, 1 \leq i \leq r_m)$,

and $x_t^{\overline{\beta}_0}$ *is the so-called optimal trajectory corresponding to the optimal control* $\overline{\beta}_0$. *Moreover, for a given constant* $a > 0$ *we have that*

$$t \geq \frac{1}{\delta_1} \ln \frac{E\left|x_t^{\overline{\beta}_0}\right|^2}{a} \implies E\left|x_t^{\overline{\beta}_0}\right|^2 \leq a.$$

This theorem actually tells us the following facts:

1) If a target functional monotonically depends on the population size, for example, the energy, the consumption, and so on, spent by the population, then to minimize this target functional we should control the fertility rate of females to take a value as small as possible.

2) The optimal population size can be exponentially stable in mean square, if we take the fertility rate of females small enough and when the stochastic perturbation is not too large. Furthermore, given $a > 0$, we can find out when the optimal population size vector "in the mean square" can be less than this target a.

Theorem 5.1 can be derived by applying Theorem 4.1, Theorem 4.5 and Corollary 4.4.

To summarize the discussion we have the following conclusion.

6. Conclusion

1) RSDE (6) with (7) and with the coefficient (8) is an appropriate model for the stochastic population dynamic system.

2) The stochastic population size from this model continuously depends on all parameters (i.e. the coefficients in the drift term). If one can "approximately" evaluate the parameters, then the population solution size can also be "approximately" obtained.

3) The stochastic population size monotonely depends on the initail size and the fertility rate of females.

4) The stochastic population size can be "exponentially stable", if we can take the fertility rate of females small enough, and the stochastic perturbation is not too large. Furthermore, we even can find out when the size vector can be smaller than a given level in the mean square.

5) If some target functional, e.g. energy consumption etc., monotonically depends on the population size, then to minimize this functional one only needs to take the fertility rate of females as small as possible.

References

1. Situ Rong (Rong Situ), *Theory of Stochastic Differential Equations with Jumps and Applications.* (Springer, 2005).
2. Situ Rong, *Reflecting Differential Equations with Jumps and Applications*, Research Notes in Mathematics 408, Chapman & Hall/CRC, (2000).
3. Situ Rong, Reflecting differential equations with jumps and stochastic population control, in *Control Theory, Stochastic Analysis and Applications*, eds. S. Chen and J. Yong, etc., (World Scientific, Singapore, 1991), pp. 193-202.
4. Situ Rong & W. L. Chen, *Stochastic Analysis and Applications* **10**, 45 (1992).
5. J. Y. Yu, B. Z. Guo, and G. T. Zhu, *J. Syst. & Math. Scien.* **7**, 214 (1987) (In Chinese).

Advances in Deterministic and Stochastic Analysis
Eds. N. M. Chuong *et al.* (pp. 297–320)

§16. NONCAUSAL STOCHASTIC CALCULUS REVISITED – AROUND THE SO-CALLED OGAWA INTEGRAL

SHIGEYOSHI OGAWA

Dept. of Mathematical Sciences, Ritsumeikan University,
Kusatsu, Shiga, 525-8577 Japan
E-mail: ogawa-s@se.ritsumei.ac.jp

1. Introduction

The causal theory of stochastic calculus originated by K.Itô in 1942 is founded on the Hypothesis of Causality;

"In order that the stochastic integral $\int f(t,\omega)dZ_t$ is well defined the function $f(t,\omega)$, $(t \geq 0, \omega \in \Omega)$ should *be adapted to the increasing family of σ-fields*, generated by the underlying basic process $Z_t(\omega)$, which is a square integrable semi-martingale like the Brownian motion."

The hypothesis seems well fit to the principle of causality in physics, where the variable "t" appears as time parameter. Moreover it endows the theory a remarkable feature of being in natural concordance with the notion of martingale which plays indeed an essential role in Itô's Calculus.
However it is also true that the Hypothesis of Causality has imposed a significant restriction on the applicability of the causal theory of stochastic calculus. Let us take for example the case of such SDEs driven by general stochastic processes, like the fractional Brownian motion, that do not have the martingale property,[8] or the case of usual SDEs but under noncausal situations, that is the SDE with noncausal coefficients and/or noncausal initial values. Let us take moreover the case where the parameter "t" stands for the *space* parameter, or the case where "t" is multi-dimensional (ie., "a stochastic calculus" for the random field[14]). The notion of Causality looses its sound meaning in such cases because of the lack of natural sense of *time direction*. Even in the case of physical problems where "t" appears as *time parameter*, we can find various situations of noncausal nature, such as the Cauchy problem in the theory of Brownian particle equations,[13] noncausal

version of the SDEs, the White noise analysis[23] etc.
Guided by these motivations the author had presented the noncausal theory of stochastic calculus in 1979,[23] by introducing the noncausal stochastic integral which is now refereed by author's name. In this note we will give a unified sketch of that noncausal stochastic calculus. We will show recent development of the theory (Refs. 10 and 13) and we will also refer to some typical applications of the theory to mathematical sciences including mathematical finance.

2. Noncausal Problems in Stochastic Analysis

To motivate our study we like to show in this paragraph some typical examples of noncausal nature in stochastic analysis. Hereafter we fix once for all a probability space $(\Omega, \mathcal{F}, P)$ on which all random quantities are defined. Among these we understand by $W_t(\omega), t \geq 0$ the standard Brownian motion and denote by $\{\mathcal{F}_t\}$ the natural filtration generated by the $W.$, namely the family of σ fields $\mathcal{F}_t = \sigma\{W_s; 0 \leq s \leq t\} \subset \mathcal{F}$.

By the random functions we understand those real valued functions $f(t, \omega)$ which are measurable in (t, ω) with respect to the field $B_{[0,T]} \times \mathcal{F}$ and satisfy the condition,

$$P\{\int_0^T f^2(t, x, \omega) dt < \infty\} = 1$$

We will denote by $\mathbf{H}$ the set of all such random functions. A random function $f(t, \omega) \in \mathbf{H}$ is called *causal* (or non-anticipative) if, for each $t \in [0, T]$ it is adapted to the σ field $\mathcal{F}_t$. We will denote by $\mathbf{M}$ the totality of causal random functions. For the random function f of this class, we have the Itô integral which we will denote by $\int f \, dW_t$ throughout the paper, while for the B-differentiable function $f \in \mathbf{M}$ the integral of symmetric type (or noncausal type) is denoted by the symbol, $\int f \, d_* W_t$ (see, Paragraph 3).

Example 2.1 (Stochastic conservation law).

Suppose given two stochastic processes, X_t, Y_t $(t \geq 0)$ which obey a conservation law as follows;

$$E(X_t, Y_t, t, \alpha, \beta, \gamma, ...) = Const,$$

where α, β, γ etc are causal constants, that is the deterministic or at least the random variables which are independent of the W. Now suppose that *the process X_t is unknown but another one is known to be generated by the following mechanism,*

$$dY_t = a(Y_t)dt + b(Y_t)dW_t, \quad Y_0 = y_0 \in R^1.$$

Given these we are to derive the equation of dynamics for the process X_t. We are already familiar with such problem especially in mathematical physics, but also in the mathematical finance we face to this. For example some authors like M.Schweizer, E.Platen this idea for the modelling of the price process in some specific situation, and we, S. Ogawa and M. Mancino,[11] followed the same way to derive the SDE model for price process in such market that admits a feedback of the future information about the delta hedging strategy.

For the derivation of the desired SDE, we are only to apply the Itô formula to the conservation equation. But an essential problem can arise when the parameters α, β, γ etc. are allowed to arbitrary random variables dependent of the $W_\cdot$, for in such case we can no longer suppose that the process X_t is causal and we can not apply the Itô theory.

Let us take an example again from the mathematical finance.

Example 2.2 (Black-Sholes model in noncausal situation). *The SDE in Black-Sholes model is as follows;*

$$dS_t = rS_t dt + \sigma^2 S_t dW_t, \tag{1}$$

which can be interpreted in the following form of the SDE with the symmetric stochastic integrals,

$$dS_t = (r - \frac{\sigma^2}{2})S_t dt + \sigma S_t d_* W_t, \tag{2}$$

where the term $\int d_* W_t$ represents the symmetric integral,[18] that is; for the B-differentiable causal function $f(t,\omega)$,

$$\int f(t,\omega) d_* W_t = \int f dW_t + \frac{1}{2} \int \frac{\partial}{\partial W_t} f dt,$$

$\frac{\partial}{\partial W_t} f$ is the B-derivative of the f. The symmetric integral is a special case of the noncausal integral that we are to study in the next paragraph.

To this simple model there correspond the following two variations of noncausal nature;

(1) Case 1: The case where the constants r, σ in the B-S equation are replaced by arbitrary random variables.

(2) *Case 2: The case where the Brownian motion W in the equation is replaced by such random process which does not have a martingale property, for example the Fractional Brownian motion.*

The first modification can arise when we need to study the B-S model in such situation of admitting the insider trading, and the latter case has been already discussed as the fractional B-S model by many authors, like A. Shiryaev,[6] P. Cheridito[1] for example, but without giving a justification of such noncausal type "SDEs" in terms of the noncausal calculus.

For the study of these modified models, first of all we need the introduction of the noncausal stochastic calculus and then we are to give precise meaning to such noncausal SDEs as follows;

$$\begin{aligned} dX_t &= a(t, X_t, \eta(\omega))dt + b(t, X_t, \zeta(\omega))dW_t, \\ X_0(\omega) &= \xi(\omega). \end{aligned} \tag{3}$$

In view of the application to the SDEs of the cases 1 and 2, we are concerned with the study of the noncausal SDEs, with a special interest on the validity of a noncausal Itô formula. We will study these subjects in the paragraphs 3,4 and 5 in a much more general situation.

Example 2.3 (SIE of Fredholm type).
Let us consider the boundary value problem for the second order SDE as follows;

$$\begin{aligned} &\{\frac{d}{dt}p(t,\omega)\frac{d}{dt} + q(t,\omega)\}X(t) = X(t)\frac{dZ}{dt}(t,\omega) + h(t,\omega) \\ &X(0) = \xi_0(\omega),\ X(1) = \xi_1(\omega), \end{aligned} \tag{4}$$

where ξ_0 and ξ_1 are arbitrary random variables.
where Z_t, $t \in [0,1]$ is an arbitrary stochastic process defined on the $(\Omega, \mathcal{F}, P)$, with square integrable sample paths.
Then as we do for the boundary value problem of ordinary differential equations, by using the Green's function $K(t,s,\omega)$ corresponding to the above situation,we may get in a very formal way the following integral equation of Fredholm type;

$$X(t,\omega) = f(t,\omega) + \int_0^1 L(t,s,\omega)X(s)ds + \int_0^1 K(t,s,\omega)X(s)d_\varphi Z(s), \tag{5}$$

where $L(t,s,\omega) = h(s)K(t,s,\omega)$ and $\int d_\varphi Z(s)$ represents the stochastic integral of noncausal type with respect to an orthonormal basis $\{\varphi_n\}$ in

$L^2(0,1)$.

This is a very typical subject in the noncausal stochastic calculus, since in such situation we can no more suppose that the solution X_t is still causal (it i.e. adapted) to the natural filtration generated by the underlying fundamental process Z_t, hence the stochastic integral term $\int X_t dZ_t$ loses its meaning in the framework of the Itô's theory.
In the article of 1986[16] *the author studied this subject in connection with the boundary value problems of stochastic differential equations and he showed the existence and the uniqueness of solutions (see Theorem 1 in Ref. 16), under some reasonable assumptions on the choice of the fundamental pair $(Z, \{\varphi_n\})$ and on the sample regularity of the kernels K and L.*

As a natural extension of such SIE (stochastic integral equation) to the case where the Z_t, X_t are the random fields, namely the stochastic processes with multi-dimensional parameters $t \in J = [0,1]^d$, we can think of the SIE for the random fields. Imagine for example the case where the driving force process Z_t is the Brownian sheet.

$$X(t,\omega) = f(t,\omega) + \int_J L(t,s,\omega)X(s)ds + \int_J K(t,s,\omega)X(s)d_\varphi Z(s), \quad (6)$$

For the review of this subject we would refer to the coming article,[9] *which will appear in another monograph of lecture notes of the same conference. So far the examples of noncausal problems are given where the noncausality has entered into the situation through the noncausal quantities, such as the noncausal initial values, the noncausal coefficients in SDE, or the driving stochastic process that does not have the martingale or semi- martingale property. But the noncausal problem can arise even in the ordinary situation where all the random quantities are supposed to be causal, that is adapted to the natural filtration $\{\mathcal{F}_t\}$. Here is another example which is typical in this sense.*

Example 2.4 (The SPDE called BPE).
Given the real Brownian motion W_t and smooth coefficients

$$a(t,x), b(t,x), A(t,x), B(t,x), C(t,x), u_0(x), \ (t,x) \in [0,T] \times R^1,$$

we consider the Cauchy problem of a stochastic partial differential equation as follows;

$$\frac{\partial}{\partial t}u + \{a(t,x) + b(t,x)\frac{dW_t}{dt}\}\frac{\partial}{\partial x}u = A(t,x)u(t,x,\omega)\frac{dW_t}{dt} + Bu + C(t,x)$$
$$u(0,x,\omega) = u_0(x) \tag{7}$$

By the solution of this problem, we understand the random function $u(t,x,\omega)$ *measurable in* (t,x,ω) *with respect to the field* $B_{[0,T]} \times B_{R^1} \times \mathcal{F}$, *especially causal in* (t,ω) *for each fixed* $x \in R^1$, *and satisfies the following weak solution equality with probability one for any test function* $\varphi(t,x)$ *(cf.*[24]*);*

$$\begin{aligned} &\int\int_G [\{\varphi_t + (a\varphi)_x + A\varphi\}u(t,x,\omega) + C\varphi]dtdx \\ &+\int_{R^1} dx \int_0^T (b\varphi)_x u(t,x,\omega)d_*W_t + \int_{R^1} u_0(x)\varphi(0,x)dx = 0, \end{aligned} \tag{8}$$

where $G = [0,T] \times \mathbf{R}^1$, $\varphi_t = \frac{\partial}{\partial t}\varphi$, $\varphi_x = \frac{\partial}{\partial x}\varphi$ *and the term* $\int d_*W_t$ *stands for the symmetric stochastic integral(see the Paragraph 3 below) as it is remarked in the introduction.*

The equation of this type, called the "Brownian particle equation", was introduced by the author ([24]*) in the study of wave propagation in random media as a mathematical model for such stochastic propagation phenomenon carried by the Brownian or diffusive particles, namely the propagation along the stochastic trajectory* X_t *governed by the SDE,*

$$dX_t = a(t,X_t)dt + b(t,X_t)d_*W_t.$$

Now the noncausal situation arises in this problem when we try to construct the solution $u(t,x,\omega)$ *by means of the method of stochastic characteristics which was first studied by the author; By the formal application of the well known method of characteristics in the theory of the partial differential equation of the first order, or of hyperbolic type, we will get in this case the following system of (symmetric) stochastic integral equations.*

$$X^{(t,x)}(s) = x + \int_s^t a(r, X^{(t,x)}(r))dr + \int_s^t b(r, X^{(t,x)}(r))d_*W_r, \quad (0 \le s \le t \le T)$$

$$\begin{aligned} u(t,x,\omega) = u_0(X^{(t,x)}(0)) &+ \int_0^t \{Bu(s, X^{(t,x)}(s),\omega) + C(s, X^{(t,x)}(s))\}ds \\ &+ \int_0^t Au(s, X^{(t,x)}(s),\omega)d_*W_s. \end{aligned} \tag{9}$$

Here the first equation gives the characteristic curve $X^{(t,x)}(s)$ $(s \leq t)$ passing through the fixed point (t, x), along which the phenomenon propagates, and the second equation is derived by integrating the PDE along that stochastic calculus. Now remembre that the solution $u(t, x, \omega)$ is adopted to the σ-field $\mathcal{F}_t$, while the characteristic curve $X^{(t,x)}(s)$, $0 \leq s \leq t \leq T$ is measurable with respect to the σ-field $\mathcal{F}_t^s := \sigma\{W(t) - W(r); s \leq r \leq t\}$. The composed function $u(s, X^{(t,x)}(s), \omega)$ appeared in the second equation above is no more causal with respect to the Brownian motion and hence the term of stochastic integral loses its meaning.

Such are the examples of noncausal problems, for the treatment of which we need another theory of stochastic calculus that is free from the restriction of the *Causality*. The calculus introduced by the author in 1979 is one of such theories and has many advantageous properties compared to other calculus from the viewpoint of the tool for the modelling and analysis of noncausal random problems. We are going to give in what follows a resume of that noncausal theory of stochastic calculus, brief but sufficient for us to see that these problems can be treated in very natural way.

3. Review of the Noncausal Stochastic Calculus

For the rigorous study of those noncausal problems listed in the previous paragraph, we need to introduce a noncausal stochastic calculus that is free from the restriction of causality and one of such calculus was introduced by the author in 1979 (,[23],[18],[2220] and,[15] etc.). We are going to give in this paragraph a rapid review of some fundamental results in the theory of noncausal stochastic calculus, mainly following the recent article.[13] For its special importance of the Brownian motion in the stochastic theory and also for the concreteness of the discussion, we will show mainly the case of the noncausal calculus with respect to the Brownian motion W_t, but as we will see (cf. Remark 3.2) easily the formalism can work even for the ase of more general stochastic processes instead of the W_t.

3.1. *Causal functions and the B-differentiability*

Following the,[25] we will say that an **H**-class random function $g(t, \omega)$ is differentiable with respect to the Brownian motion W_t (or B-differentiable) provided that there exists an **M**-class random function say $\frac{\partial}{\partial W_t} g(t, \omega)$ such that, for small enough $h > 0$,

$$\sup_{t,s,|t-s|<h} E|g(t, \omega) - g(s, \omega) - \int_s^t \frac{\partial}{\partial W_r} g(r, \omega) dW_r|^2 = o(h)$$

where the integral $\int dW$ stands for the Itô's stochastic integral. The function $\frac{\partial}{\partial W_t} g$ is called the B-derivative of the g. It is not difficult to see that if the function $g(t,\omega)$ is B-differentiable then its B-derivative is uniquely determined (see[25]). The B-differentiability of the random function with respect to the multi-dimensional Brownian motion is defined in a similar way.

Remark 3.1. Let $g(t,\omega)$ be a functional of the multi-dimensional Brownian motion, $\mathbf{W}_t = (W_t^1, W_t^2, \cdots, W_t^n)$ where the W^i, $(1 \le i \le n)$ are independent copies of the 1-dim. Brownian motion W_t. Then the B-derivative of such function, say $\nabla_w g$, can be defined in the following way: the $\nabla_w g = (\frac{\partial}{\partial W_t^1} g, \frac{\partial}{\partial W_t^2} g, \cdots, \frac{\partial}{\partial W_t^n} g)^t$ is a causal random vector such that,

$$\sup_{t,s|t-s|<h} E|g(t,\omega) - g(s,\omega) - \sum_{k=1}^{n} \int_s^t \frac{\partial}{\partial W_r^k} g(r,\omega) dW_r^k|^2 = o(h)$$

Here we notice that the Itô integral is defined for the causal random functions $f(t,\omega) \in \mathbf{M}$ and roughly speaking the symmetric integrals (i.e. $\mathcal{I}_{1/2}$ of Ogawa[25] and Stratonovich-Fisk integral) are defined for the causal and B-differentiable functions. That is, the symmetric integral $\mathcal{I}_{1/2}(f)$ of a B-differentiable function was introduced as the limit (in probability) $\lim_{|\Delta| \to 0} \mathcal{I}_\Delta(f)$ of the sequence $\{\mathcal{I}_\Delta(f)\}$ of Riemannian sums,

$$\mathcal{I}_\Delta(f) = \sum_{t_i \in \Delta} f(\frac{t_i + t_{i+1}}{2})\{W(t_{i+1}) - W(t_i)\}$$

where, $\Delta = \{0 \le t_1 < \cdots < t_n \le 1\}$ is a partition of the interval $[0,1]$ and $|\Delta| = \max_i (t_{i+1} - t_i)$.

The following result was established by the author in 1970,

Theorem 3.1 (Ref. 25). *The limit (in probability)* $\mathcal{I}_{1/2}(f) = \lim_{|\Delta| \to 0} \mathcal{I}_\Delta(f)$ *exists and is represented in the following form:*

$$\mathcal{I}_{1/2}(f) = \frac{1}{2} \int_0^1 \frac{\partial}{\partial W_t} f \, dt + \int_0^1 f(t,\omega) \, dW_t$$

3.2. *Noncausal stochastic integral*

Given a random function $f(t,\omega) \in \mathbf{H}$ and an arbitrary complete orthonormal system $\{\varphi_n\}$ in $L^2([0,1])$, we consider the formal random series

$$\sum_n^{\infty} \int_0^1 f(t,\omega)\varphi_n(t)dt \int_0^1 \varphi_n(t)dW_t.$$

The stochastic integral of noncausal type introduced by the author in 1979 ([23]), is given in the following way,

Definition 3.1. A random function $f(t,\omega) \in \mathbf{H}$ is said to be integrable with respect to the basis $\{\varphi_n\}$ (or φ-integrable) when the random series above converges in probability and the sum, denoted by $\int_0^1 f(t,\omega)d_{\varphi}W_t$, is called the stochastic integral of noncausal type with respect to the basis $\{\varphi_n\}$.

Remark 3.2. The validity of the above definition is not limited to the case of Brownian motion or to other square integrable semi-martingales. Indeed it can apply even to the case of general square integrable processes say Z_t that do not posses the property of semi-martingale, as long as the quantities

$$\int_0^1 \varphi_n(t)dZ_t$$

are well defined. The simplest example is when we employ the system of Haar functions $\{H_{0,0}(t), H_{n,i}(t), 0 \le i \le 2^{n-1}-1, n \in \mathbf{N}\}$ as orthonormal basis. Let us remembre that the $H_{n,i}(t)$ are as follows;

$$H_{0,0}(t) = 1,$$

$$H_{n,i}(t) = 2^{(n-1)/2}\{\mathbf{1}_{[2^{-n}2i,2^{-n}(2i+1))}(t) - \mathbf{1}_{[2^{-n}(2i+1),2^{-n}2(i+1))}(t)\} \tag{10}$$

$$0 \le i \le 2^{n-1}-1,\ n \ge 1.$$

In this case the quantity above is defined in the natural way as follows;

$$\int_0^1 H_{0,0}(t)dZ_t = Z(1) - Z(0), \tag{11}$$

$$\int_0^1 H_{n,i}(t)dZ_t = 2^{n/2}[\{Z(\frac{2i+1}{2^n}) - Z(\frac{2i}{2^n})\} - \{Z(\frac{2i+2}{2^n}) - Z(\frac{2i+1}{2^n})\}].$$

The noncausal integral with respect to the fractional Brownian motion can be introduced by this way.[8]

In general case, the way of convergence of the random series being conditional, the integrability and the sum should depend on the basis, even on the order of the same complete system of orthonormal functions. On the relation between the noncausal integrals with respect to different bases, very few is known except the following,

Theorem 3.2 (1984, Ref. 21). *If the random function $f(t,\omega) \in \boldsymbol{H}$ is integrable in the L^1- sense (ie. convergent in $L^1(\Omega, P)$ sense) with respect to the system of trigonometric functions,*

$$\{1, \sqrt{2}\cos 2n\pi x, \sqrt{2}\sin 2n\pi x; n \in \mathbf{N}\}.$$

Then the f is integrable with respect to the system of Haar functions and the value of two integrals coincide.

If the function is integrable with respect to any basis $\{\varphi_n\}$ and the sum does not depend on the choice of the basis, we will say that the function is *universally* integrable (or shortly u-integrable).

3.3. *Equivalent expressions and variants*

Here are some equivalent expressions and a possible variation of the above definition, which are worth to be remarked so that we can have a better understanding of the nature of our noncausal integral.

(a) As a limit of the sequence of random Stieltjes integrals:
Given the pair $(W_t, \{\varphi_n\})$ we introduce the sequence of approximation processes $W_n^{\varphi}(t)$ in the following way.

$$W_n^{\varphi}(t) = \sum_{k=1}^{n} \int_0^t \varphi_k(s)ds \int_0^1 \varphi_k(s)dW_s. \tag{12}$$

It is immediate to see that this gives a pathwise smooth approximation of the Brownian motion $W(t,\omega)$. Moreover, by virtue of the famous theorem due to K.Itô and M.Nishio,[2] we know that for any choice of the basis $\{\varphi_n\}$ the sequence $\{W_n^{\varphi}(t)\}$ converges uniformly in $t \in [0,1]$ as $n \longrightarrow \infty$ with probability one. Now we notice that our noncausal integral can be expressed as the limit (in probability) of the sequence of random Stieltjes integrals;

Proposition 3.1. *It holds that,*

$$\int_0^1 f \, d_{\varphi}W_t := \lim \int_0^1 f \, dW_n^{\varphi}(t) \quad \textit{(in probability)}.$$

(b) Riemannian definition:
Let us take the Haar functions $\{H_{n,i}(t)\}$ for basis $\{\varphi_n\}$. This is a case of special interest because we have the following,

Lemma 3.1 (1984, Ref. 21). *Let us define the approximation process $W_n^H(t)$ for this case by the following formula,*

$$W_n^H(t) = \sum_{k \le n} \sum_{i=0}^{2^k-1} \int_0^1 H_{k,i}(s)dW_s \int_0^t H_{k,i}(s)ds.$$

Then each $W_n^H(t)$ is the Cauchy polygonal approximation of the process W_t taken over the set of dyadic points $\{k/2^n; 0 \le k \le 2^n\}$, that is,

$$W_n^H(t) = W(\frac{k}{2^n}) + 2^n\{W(\frac{k+1}{2^n}) - W(\frac{k}{2^n})\}(t - \frac{k}{2^n}), \quad \text{for } t \in [\frac{k}{2^n}, \frac{k+1}{2^n}). \tag{13}$$

To check this, we introduce the indicator function, $\chi_{n,i}(t) = 2^{n/2}\mathbf{1}_{[2^{-n}i, 2^{-n}(i+1))}(t)$. It is immediate to see that

$$(\chi_{n,i}, H_{m,k}) = 0 \quad \text{for } {}^\forall(m,k) \text{ with } m \ge n+1,$$

here the symbol $(\cdot, \cdot)$ denotes the inner product in $L^2(0,1)$.
Therefore each $\chi_{n,i}$ should be represented as linear combination of the membres $\{H_{m,k}, m \le n\}$, say;

$$\chi_{n,i}(t) = C(n,j;0,0)H_{0,0} + \sum_{1 \le m \le n} \sum_{k=0}^{2^{(m-1)}-1} C(n,i;m,k)H_{m,k}(t).$$
$$C(n,i;m,k) = (\chi_{n,i}, H_{m,k}).$$

It is also easy to see that we have the following relation;

$$\sum_{i=0}^{2^n-1} C(n,i;m,k)C(n,i;l,j) = \delta_{m,l}\delta_{k,j}.$$

Based on this relation we can get the following equality,

$$W_n^H(t) = \sum_{k=0}^{2^n-1} \Delta_k^n W \chi_{n,k}(t), \quad \Delta_k^n W = W(\frac{k+1}{2^n}) - W(\frac{k}{2^n}),$$

and this is what we want to see.

Now applying this result to the expression given in the Proposition 3.1 of (a), we see that in this case the defining formula of the noncausal integral is given as the Riemannian sum,

$$\int_0^1 f d_H W_t = \lim_{n\to\infty} \sum_{i=0}^{2^n-1} 2^n \int_{2^{-n}i}^{2^{-n}(i+1)} f(s)ds \cdot \{W(2^{-n}(i+1)) - W(2^{-n}i)\}. \tag{14}$$

This type of definition can be found in recent publications of some authors. But as we have seen in here ([21]), this is merely a special case of our integral.

(c) Let $D_n(t,s)$ be the kernel given by,

$$D_n(t,s) = \sum_{k=1}^{n} \varphi_k(t)\varphi_k(s), \quad (t,s \in [0,1]).$$

Then we have the following representation for the noncausal integral,

$\int_0^1 f d_\varphi W(t) = \lim_{n\to\infty} \int_0^1 dt \int_0^1 f(t,\omega) D_n(t,s) dW_s$ (limit in probability).

For the case of trigonometric functions, the kernel $D_n(t,s)$ is the Dirichlet kernel appearing in the theory of Fourier series.

(d) A generalization of the above view: Replace the kernels $\{D_n(t,s)\}$ in the above interpretation by any δ- sequence say $\{K_n(t,s)\}$, then we will get a generalized formula for the noncausal integral:

$$\int_0^1 f \, d_K W := \lim_{n\to\infty} \int_0^1 dt \int_0^1 f(t,\omega) K_n(t,s) dW_s$$

3.4. *Condition for the integrability – in the framework of the Homogeneous Chaos theory*

Let $\mathbf{H}_0$ be the totality of all random functions $f(t,\omega) \in \mathbf{H}$ such that, $E \int_0^1 |f(t,\omega)|^2 dt < \infty$. By Wiener-Itô's theory of Homogeneous Chaos, we know that such function $f \in \mathbf{H}_0$ can be decomposed into the sum of multiple Wiener integrals, that is:

There exists a set of kernels, say $\{k_n^f(t; t_1, \cdots, t_n)\}_{n=0}^{\infty}$, such that

$$k_n^f \in L^2([0,1]^{n+1}) \text{ with } \sum_n n! ||k_n^f||_{n+1}^2 < \infty,$$

symmetric in n-parameters $(t_1, \cdot, t_n) \in [0,1]^n$ and that,

$$f(t,\omega) = \sum_{n=0}^{\infty} I_n(k_n^f(t;\cdot)),$$

$$I_n(k_n^f(t;\cdot)) = \int\int\cdots\int k_n^f(t;t_1,\cdots,t_n)dW_{t_1}dW_{t_2}\cdots dW_{t_n}$$

where $\|\cdot\|_n$ stands for the norm in $L^2([0,1]^n)$ space.

We will denote by $\mathbf{H}_1$ the totality of all $\mathbf{H}_0$- functions $f(t,\omega)$ such that, $\sum_{n=1}^{\infty} nn!||k_n^f||_{n+1}^2 < \infty$. Given a function $f \in \mathbf{H}_1$ we introduce its stochastic derivative $\mathcal{D}f$ by the following formula,

$$\mathcal{D}f(t,s) = \sum_{n=1}^{\infty} nI_{n-1}(k_n^f(t;s,\cdot)).$$

Since $E\int_0^1\int_0^1 (\mathcal{D}f(t,s))^2 dtds = \sum nn!||k_n^f||_{n+1}^2$, we notice that the stochastic derivative $\mathcal{D}f(t,s)$ is well defined for the $f \in \mathbf{H}_1$. Now we can state the condition for the φ-integrability of the $\mathbf{H}_1$-class functions in the following theorem that was obtained by the author in 1984.

Theorem 3.3 (Ref. 22). *Let $f \in \mathbf{H}_1$ and let $\{\varphi_n\}$ be an arbitrary orthonormal basis. Then the necessary and sufficient condition for the random function f to be φ-integrable is that the* $\lim_{n\to\infty}\int_0^1\int_0^1 \mathcal{D}f(t,s)D_n(t,s)dtds$ *exists in probability.*

3.5. *Relation between symmetric and noncausal integrals*

We call a random function $f(t,\omega)$ semi martingale when it admits the decomposition, $f(t,\omega) = a(t,\omega) + \int^t \hat{f}dW_s$ where $\hat{f} \in \mathbf{M}$ and $a(t)$ is such that almost every sample path is of bounded variation in t over $[0,1]$. Notice that if $\sup_{t,s|t-s|<h} E|a(t)-a(s)|^2 = o(h)$ then f is B-differentiable.

The followings are the basic results concerning the relation between the symmetric integrals with the noncausal integral.

Theorem 3.4 (Ref. 18). *Every causal B- differentiable function is integrable in noncausal sense with respect to the system of Haar functions and the sum coincides with that of the symmetric integrals:*

$$\int_0^1 f d_H W = \int_0^1 f dW + \frac{1}{2}\int_0^1 \hat{f} dt$$

We say that an orthonoral basis $\{\varphi_n\}$ is *regular* provided that it satisfies the next condition:

$$\sup_n ||u_n||_2 < \infty, \quad \text{where} \quad u_n(t) = \sum_{k \le n} \varphi_k(t) \int_0^t \varphi_k(s) ds \tag{15}$$

Remark 3.3. Notice that this condition (15) is equivalent to the fact that,

$$w - \lim_{n \to \infty} u_n = \frac{1}{2} \quad (\text{in } L^2)$$

namely to the fact that, for any $f(t) \in L^2(0,1)$ it holds the following,

$$\lim_{n \to \infty} \int_0^1 u_n(t) f(t) dt = \frac{1}{2} \int_0^1 f(t) dt$$

Theorem 3.5 (Ref. 18). *Every semi martingale (causal or not) becomes φ-integrable, iff the basis $\{\varphi_n\}$ is regular. In this case the noncausal integral coincides with the symmetric integrals.*

Related to this result is a natural and interesting question asking whether there can or can not be a basis $\{\varphi_n\}$ which is not regular. This question is affirmatively answered by P.Mayer and M.Mancino.[4] We can go on further. The next result shows that *a smoothness in* W_t of the integrand ensures the integrability with respect to any orthonormal basis.

Theorem 3.6 (Ref. 18). *Every semi martingale that is twice B-differentiable, namely the B-derivative $\hat{f}$ is again a semi martingale, is u-integrable.*

In their articles,[7],[5] of the authors M.Zakai and D.Nualart, the noncausal integral was referred as the *intrinsic* Ogawa integral. We like to remark at this stage that, precisely saying what they referred there was our noncausal integral for the case of u-integrable functions,

Now let us finish this paragraph by giving another result on the way of convergence of noncausal intgerals for semi-martingales, for its usefulness in appications.

Proposition 3.2 (1985, Ref. 18).
If the B-derivative $\hat{f}$ of the semi-martingale f, $df = \hat{f} dW_t + g(t,\omega)dt$, satisfies the following condition,

$$P\{\int_0^1 \hat{f}^4(t)dt < \infty\} = 1.$$

Then the noncausal integral $\int_0^t f d_* W_t$ with respect to the regular basis converges uniformely in $t \in [0,1]$.

4. Applications to the Noncausal SDEs

Now we like to show basic results concerning the Cauchy problem of the noncausal SDEs and the validity of the noncausal Itô formula, which will give us at the same time the appropriate answers to those noncausal problems listed in the Paragraph 2.

4.1. *Noncausal Cauchy problem*

First notice that the SDE in (3) becomes meaningful in the framework of the noncausal stochastic calculus, that is:

$$\begin{aligned} dX_t &= a(t, X_t, \eta(\omega))dt + b(t, X_t, \eta(\omega))d_\varphi W_t, \ t \in (0,T], \\ X_0(\omega) &= \xi(\omega) \end{aligned} \tag{16}$$

here the $\{\varphi_n\}$ is a regular basis in $L^2(0,1)$, which we will fix throughout the discussion.

We notice at this stage that when the parameters $\xi(\omega), \eta(\omega)$ are not random and the solution X_t can be supposed to be causal, then by virtue of the Theorem 3.5 the SDE in (16) is reduced to the usual SDE with symmetric integration,

$$\begin{aligned} dX_t &= a(t, X_t, \eta)dt + b(t, X_t, \eta)dW_t, \ t \in (0,T], \\ X_0(\omega) &= \xi. \end{aligned} \tag{17}$$

The Cauchy problem for the noncausal SDE was first studied by the author[19] for such simple case where the parameter η is not random or does not appear in $a(t,x)$, $b(t,x)$ and only the initial data $\xi(\omega)$ arises as a noncausal factor.

$$\begin{aligned} dX_t &= a(t, X_t)dt + b(t, X_t)d_\varphi W_t, \; t \in (0, T], \\ X_0(\omega) &= \xi(\omega). \end{aligned} \tag{18}$$

For this case, the results on the existence and a kind of uniqueness properties of the solution are proved under a milder assumption on the regularity of the coefficients $a(\cdot), b(\cdot)$ as follows:

Assumption 4.1. *The coefficients $a(t,x)$, $b(t,x)$ are sufficiently regular in such sense that,*

(1) $a(t,x), \frac{\partial^2}{\partial x^2} b(t,x)$ are of C^1-class,
(2) $a(t,x), b(t,x)$ are sufficiently regular in the sense that the causal Cauchy problem 17 admits the unique strong solution $X(t,\omega;\xi)$ and that the $X(t,\omega;\xi)$ is continuous in (t,ξ) with probability one.

Notice that under such conditions the composite

$$\tilde{X}(t,\omega) = X(t,\omega;\xi(\omega))$$

of the strong solution $X(t,\xi,\omega)$ of the (17) and the random variable $\xi(\omega)$ is well defined, which we expect to be a solution of the noncausal Cauchy problem (18). In fact we have the following,

Theorem 4.1 (1985, Ref. 19). *The composite $\tilde{X}(t,\omega)$ is a solution of the noncausal Cauchy problem (18).*

We have also found that this solution $\tilde{X}(t,\omega)$ verifies the Itô formula of noncausal type, that is:

Proposition 4.1 (1985, Ref. 17). *For any function $F(x) \in C^4$ it holds the equality,*

$$dF(\tilde{X}_t) = F'(\tilde{X}_t)\{a(t,\tilde{X}_t)dt + b(t,\tilde{X}_t)d_\varphi W_t\} \quad 0 \le t \le 1$$

As an application of this we can show the following result that concerns the uniqueness of the solution for the noncausal problem (18),

Corollary 4.1 (Ref. 17). *When the $b(x) \neq 0$, the composed function $\tilde{X}(t,\omega) = X(t,\omega;\xi(\omega))$ is the unique solution among all random functions verifying the Itô formula 4.1 of noncausal type.*

The proof of this together with that of the previously presented Proposition 4.1 will be given in the next paragraph for a more general case.

4.2. *Discussions for the more general cases*

We are going to give in this paragraph the results on the Cauchy problem for the more general case (16).

Assumption 4.2. *We suppose that the coefficients $a(t,x;\eta), b(t,x;\eta)$ are sufficiently regular in such sense that, for an arbitrary couple of parameters (ξ,η) the causal Cauchy problem 17 admits the unique strong solution $X(t,\omega;\xi,\eta)$ and that the $X(t,\omega;\xi,\eta)$ is continuous in (t,ξ,η) with probability one.*

Remark 4.1. The assumption is satisfied when, for example, the $a(t,x;\eta), b(t,x;\eta)$ are of the C^4-class in x, of C^1-class in η and all derivatives are bounded on $[0,1]\times R^1$.

We also notice that under the Assumption 4.2 the composite

$$\tilde{X}(t,\omega) = X(t,\omega;\xi(\omega),\eta(\omega))$$

of the strong solution $X(t,\omega;\xi,\eta)$ of the (17) and the random variables $\xi(\omega),\eta(\omega)$ is well defined, and as in the previous case we expect this composite $\tilde{X}(t,\omega)$ to be a solution of the noncausal Cauchy problem (16). In fact we have the following,

Theorem 4.2. *The $\tilde{X}$ gives a noncausal solution of the noncausal Cauchy problem 16.*

For the verification of this, we need some preparations.

Proposition 4.2. *Let $f(t,\omega;\xi,\eta)$ $(\xi,\eta\in[-A,A])$ be a semi-martingale such that for each fixed (ξ,η),*

$$df(t,\omega;\xi,\eta) = g(t,\omega;\xi,\eta)dt + h(t,\omega;\xi,\eta)dW_t \tag{19}$$

where $g(\cdot), h(\cdot)$ are causal random functions satisfying the following condition,

$$P[\int_{-A}^{A} d\xi \int_{-A}^{A} d\eta \int_0^1 \{g^2(t,\omega;\xi,\eta) + h^2(t,\omega;\xi,\eta)\}dt < \infty] = 1$$

(i) Then for any regular basis $\{\varphi_n\}$ in $L^2(0,1)$, the following equality holds,

$$\lim_{n\to\infty} \int_{-A}^{A} d\xi \int_{-A}^{A} d\eta [\int_0^1 f(t,\omega;\xi,\eta)\{d_\varphi W(t) - dW_n^\varphi(t)\}]^2 = 0 \tag{20}$$

(in probability)

(ii) Moreover if the coefficient $h(t,\omega;\xi,\eta)$ in the decomposition (19) again becomes a semi-martingale satisfying the same condition as the $f(\cdot)$, then the equality (20) still holds true for any basis $\{\varphi_n\}$.

Proof. Put $f = f_1 + f_2$ where,

$$f_1(t,\omega;\xi,\eta) = f(0,\omega;\xi,\eta) + \int_0^t g(s,\omega;\xi,\eta)ds$$

and

$$f_2(t,\omega;\xi,\eta) = \int_0^t h(s,\omega;\xi,\eta)dW_s.$$

Then we have for f_1, the equality:

$$\begin{aligned}&\int_0^1 f_1(s,\omega;\xi,\eta)\{d_\varphi W(s) - dW_n^\varphi(s)\}\\ &= f_1(1,\omega;\xi,\eta)\{W(1) - W_n^\varphi(1)\} - \int_0^1 \{W(s) - W_n^\varphi(s)\}g(s,\omega;\xi,\eta)ds\end{aligned}$$

Hence with the help of the Theorem of Nishio-Itô we confirm that,

$$\lim_{n\to\infty}\int_{-A}^{A} d\xi \int_{-A}^{A} d\eta [\int_0^1 f_1(t,\omega;\xi,\eta)\{d_\varphi W(t) - dW_n^\varphi(t)\}]^2 = 0$$

(in probability)

For the term f_2 we have the decomposition,

$$\int_0^1 f_2(t,\omega;\xi,\eta)\{d_\varphi W(t) - dW_n^\varphi(t)\} = \sum_{i=1}^4 I_{i,n}(\xi,\eta)$$

where,

$$\begin{aligned}I_{1,n} &= \sum_{k=n+1}^\infty f_2(1,\omega;\xi,\eta)\tilde{\varphi}_k(1)Z_k,\ (\tilde{\varphi}_k(t) = \int_0^t \varphi_k(s)ds)\\ I_{2,n} &= \sum_{k=n+1}^\infty \int_0^1 \tilde{\varphi}_k(t)\varphi_k(t)h(t,\omega;\xi,\eta)dt\\ I_{3,n} &= \sum_{k=n+1}^\infty \int_0^1 \varphi_k(t)dW(t)\int_0^t \tilde{\varphi}_k(s)h(s,\omega;\xi,\eta)dW(s)\\ I_{4,n} &= \sum_{k=n+1}^\infty \int_0^1 \tilde{\varphi}_k(t)h(t,\omega;\xi,\eta)dW(t)\int_0^t \varphi_k(s)dW(s)\end{aligned}$$

and, $Z_n = \int_0^1 \varphi_n(t)dW(t)$.
We are to show that:

$$\lim_{n\to\infty} \int_{-A}^{A} d\xi \int_{-A}^{A} I_{i,n}^2(\xi,\eta) = 0 \quad \text{(in probability)}, \quad (1 \le i \le 4).$$

Since for the quantities $I_{i,n}$ $(i = 1, 3, 4)$ this could be easily done by a usual routine, it would suffice to show the result only for the term $I_{2,n}$.
By taking the Remark3.3 into account, we see that for each fixed (ξ, η) we have,

$$\lim_{n\to\infty} I_{2,n}(\xi,\eta) = \lim_{n\to\infty} \int_0^1 h(t,\omega;\xi,\eta)\{\frac{1}{2} - u_n(t)\}dt = 0$$

On the other hand we have,

$$I_{2,n}^2 \le (\frac{1}{2} + 2U^2) \int_0^1 h^2(t,\omega;\xi,\eta)dt \quad \text{where, } U = \sup_n ||u_n||_{L^2} < \infty.$$

Hence we confirm the result, $\lim_{n\to\infty} \int_{-A}^{A} \int_{-A}^{A} I_{2,n}^2(\xi,\eta)d\xi d\eta = 0$ □

Now given the unique solution $X(t,\omega;\xi,\eta)$ of the causal problem 17, we introduce the sequence of random functions in the following way,

$$X_n^{\varphi}(t,\omega;\xi,\eta) = \xi + \int_0^t a(s, X(s,\omega;\xi,\eta);\eta)ds + \int_0^t b(s, X(s,\omega;\xi,\eta);\eta)dW_n^{\varphi}(s) \tag{21}$$

where W_n^{φ} is the approximate process of the Brownian motion introduced in the previous paragraph.

We easily see by the Theorem 3.5 that for each fixed $t, (\xi, \eta)$, we have $\lim_{n\to\infty} X_n^{\varphi}(t,\omega;\xi,\eta) = X(t,\omega;\xi,\eta)$, (in probablity). Moreover we can see that this convergence is uniform in (ξ,η) on every finite set $C_A = [-A, A] \times [-A, A]$.

Proposition 4.3. *For an arbitrarily large $A > 0$ it holds the following relation at each fixed $t \in [0,1]$,*

$$\lim_{n\to\infty} \sup_{(\xi,\eta)\in C_A} |X_n^{\varphi}(t,\omega;\xi,\eta) - X(t,\omega;\xi,\eta)| = 0 \quad (\textit{in probability})$$

Proof. Put

$$\Delta_n(t,\omega;\xi,\eta) = X_n^{\varphi}(t,\omega;\xi,\eta) - X(t,\omega;\xi,\eta)$$

From equations (17),(21) we obtain the following:

$$\Delta_n(t,\omega;\xi,\eta) = \int_0^t b(X(s;\xi,\eta);\eta)\{dW_n^{\varphi}(s) - d_{\varphi}W(s)\} \tag{22}$$

On the other hand we have the following expression,

$$\begin{aligned}\Delta_n(t,\omega;\xi,\eta) &= \int_{-A}^{\xi} d\xi_1 \int_{-A}^{\eta} \frac{\partial^2}{\partial\xi\partial\eta}\Delta_n(t,\omega;\xi_1,\eta_1)d\eta_1 \\ &+ \int_{-A}^{\xi} \frac{\partial}{\partial\xi}\Delta_n(t,\omega;\xi_1,-A)d\xi_1 \\ &+ \int_{-A}^{\eta} \frac{\partial}{\partial\eta}\Delta_n(t,\omega;-A,\eta_1)d\eta_1 + \Delta_n(t,\omega;-A,-A)\end{aligned}$$

which implies that,

$$\sup_{(\xi,\eta)\in C_A} |\Delta_n(t,\omega;\xi,\eta)| \le J_1(n) + J_2(n) + J_3(n)$$

where

$$J_1(n) = 4A^2\int_{-A}^{A} d\xi_1 \int_{-A}^{A} d\eta_1 |\frac{\partial^2}{\partial\xi\partial\eta}\Delta_n(t,\omega;\xi_1,\eta_1)|^2 + |\Delta_n(t,\omega;-A,-A)|$$

$$J_2(n) = 2A\int_{-A}^{A} |\frac{\partial}{\partial\xi}\Delta_n(t,\omega;\xi_1,-A)|^2 d\xi_1,$$

$$J_3(n) = 2A\int_{-A}^{A} |\frac{\partial}{\partial\eta}\Delta_n(t,\omega;-A,\eta_1)|^2 d\eta_1$$

We are to show that for each fixed t these $J_1(n), J_2(n), J_3(n)$ tend to zero in probability as $n \longrightarrow \infty$. Since at this stage the parameters ξ,η remain as deterministic constants, we notice that the $X(t,\omega;\xi,\eta)$ is causal and derivable in ξ,η. In fact under the assumption (4.1) on the regularity of the coefficients $a(\cdot), b(\cdot)$ it is easy to verify that the derivatives,

$$X_1(t) = \frac{\partial}{\partial\xi}X(t,\omega;\xi,\eta),\ \ X_2(t) = \frac{\partial}{\partial\eta}X(t,\omega;\xi,\eta),\ \ X_3(t) = \frac{\partial^2}{\partial\xi\partial\eta}X(t,\omega;\xi,\eta),$$

are given as the solutions of the following symmetric type SDEs, which can be solved explicitly:

$$\begin{cases} dX_1(t) &= a_x(t,X,\eta)X_1(t)dt + b_x(t,X,\eta)X_1(t)dW_t, \\ X_1(0) &= 1 \end{cases}$$

$$\begin{cases} dX_2(t) &= \{a_\eta(t,X,\eta)dt + b_\eta(t,X,\eta)dW_t\} \\ & \qquad +\{a_x(t,X,\eta)X_2(t)dt + b_x(t,X,\eta)X_2(t)dW_t\}, \\ X_2(0) &= 0, \end{cases}$$

$$\begin{cases} dX_3(t) = \{a_{xx}(t,X,\eta)X_2(t) + a_{x,\eta}(t,X,\eta)\}X_1(t)dt \\ \qquad +\{b_{xx}(t,X,\eta)X_2(t) + b_{x,\eta}(t,X,\eta)\}X_1(t)dW_t \\ \qquad +a_x(t,X,\eta)X_3(t)dt + b_x(t,X,\eta)X_3(t)dW_t, \\ X_3(0) \ = 0 \end{cases}$$

This combined with the expression (22) would imply that the quantity $\Delta_n(t,\omega;\xi,\eta)$ is derivable in ξ, η and that the order of the derivation in ξ, η and the integration is exchangeable. For example,

$$\frac{\partial^2}{\partial\xi\partial\eta}\Delta_n = \int_0^t \frac{\partial^2}{\partial\xi\partial\eta}b(s,X(s;\xi,\eta))\{dW_n^{\varphi}(s) - dW_{\varphi}(s)\}$$

Hence by virtue of the Proposition 4.2 we only need to show that the following quantities,

$$\frac{\partial^2}{\partial\xi\partial\eta}b(X(t;\xi,\eta);\eta), \quad \frac{\partial}{\partial\xi}b(X(t;\xi,-A);-A), \quad \frac{\partial}{\partial\eta}b(X(t;-A,\eta);\eta)$$

are semi-martingales satisfying the condition in that Proposition. Since this can be verified by a simple routine work, we see that we are done. □

Now we are going to give the proof for our Theorem,

Proof. Fix a positive A in an arbitrary way and put,

$$\xi_A(\omega) = \xi(\omega)1_{C_A}(\xi(\omega)) - A\mathbf{1}_{(-\infty,-A]}(\xi(\omega)) + A\mathbf{1}_{[A,\infty)}(\xi(\omega))$$

$$\eta_A(\omega) = \eta(\omega)1_{C_A}(\eta(\omega)) - A\mathbf{1}_{(-\infty,-A]}(\eta(\omega)) + A\mathbf{1}_{[A,\infty)}(\eta(\omega))$$

For an arbitrary positive ϵ we have,

$$P\{|X_n^{\varphi}(t,\omega;\xi(\omega),\eta(\omega)) - X(t,\omega;\xi(\omega),\eta(\omega))| > \epsilon\}$$

$$\leq P\{|X_n^{\varphi}(t,\omega;\xi_A(\omega),\eta_A(\omega)) - X(t,\omega;\xi_A(\omega),\eta_A(\omega,ega))| > \epsilon\}$$

$$+P(|\xi(\omega)| > A) + P(|\eta(\omega)| > A)$$

Since $|\xi_A(\omega)|, |\eta_A(\omega)| \leq A$, we see that

$$\lim_{n\to\infty} P\{|X_n^{\varphi}(t,\omega;\xi_A(\omega),\eta_A(\omega)) - X(t,\omega;\xi_A(\omega),\eta_A(\omega))| > \epsilon\} = 0$$

by virtue of the Proposition 4.3. The A being arbitrary this implies that,

$$\lim_{n\to\infty} P\{|X_n^{\varphi}(t,\omega;\xi(\omega),\eta(\omega)) - X(t,\omega;\xi(\omega),\eta(\omega))| > \epsilon\} = 0 \qquad \square$$

5. Question of Uniqueness – Noncausal Itô Formula

The noncausal solution of the problem (16), $\tilde{X}(t,\omega) = X(t,\omega;\xi(\omega),\eta(\omega))$ constructed in the Theorem 4.2, has a remarkable property as stated in the next,

Theorem 5.1 (Noncausal Itô formula). *For any random variable $\zeta(\omega)$ and any function $F(x,y)$, which is differentiable in (x,y) and of C^4-class in x with bounded derivatives, it holds the following equality:*

$$dF(\tilde{X}_t,\zeta(\omega)) = (\partial_x F)(\tilde{X}_t,\zeta(\omega))\{a(\tilde{X}_t;\eta(\omega))dt + b(\tilde{X}_t;\eta(\omega))d_\varphi W_t\} \quad (23)$$

Proof. Let $X(t,\omega;\xi,\eta)$ be the unique solution of the causal SDE (17) with deterministic parameters (ξ,η). Then by the usual Itô formula for causal functions, we have for each fixed deterministic parameters $(\xi,\eta,\ \zeta)$, the following relation:

$$\begin{aligned} &F(X(t;\xi,\eta),\zeta) &(24)\\ &= F(\xi,\zeta) + \int_0^t (\partial_x F)(X(s,\omega;\xi,\eta),\zeta)\{a(X_s;\eta)ds + b(X_s;\eta)dW_s\} \end{aligned}$$

Here the stochastic integral $\int dW_t$ stands for the causal symmetric integral. Given this we introduce the approximation sequence as follows,

$$\begin{aligned} &F^n(t,\omega;\xi,\eta,\zeta) &(25)\\ &= F(\xi,\zeta) + \int_0^t (\partial_x F)(X(s,\omega;\xi,\eta),\zeta)\{a(X_s;\eta)ds + b(X_s;\eta)dW_n^\varphi(s)\} \end{aligned}$$

Following the same argument as in the proof of Proposition 4.2, we would easily verify that for each fixed $t \in [0,1]$ the sequence $F^n(t,\omega;\xi,\eta,\zeta)$ converges to $F(X(t,\omega;\xi,\eta),\zeta)$ in probability as $n \to \infty$, uniformly in $(\xi,\eta,\zeta) \in C'_A$ on any finite set $C'_A = [-A,A]^3$. Hence we confirm, again following the same argument as in the proof of the Theorem 4.2, that for each fixed t the sequence $F^n(t,\omega;\xi(\omega),\eta(\omega),\zeta(\omega))$ converges in probability to the $F(X(t,\omega;\xi(\omega),\eta(\omega),\zeta(\omega)) = F(\tilde{X}(t,\omega),\zeta(\omega))$. Now from the equation (25) we see that the following limit,

$$\lim_{n\to\infty} \int_0^t (\partial_x F)(X(s,\omega;\xi(\omega),\eta(\omega)),\zeta(\omega))b(X(s,\omega;\xi(\omega),\eta(\omega));\eta(\omega))dW_n^\varphi(s)$$

should converge in probability to the limit,

$$\int_0^t (\partial_x F)(\tilde{X}(s,\omega),\zeta(\omega))b(\tilde{X}(s,\omega);\eta(\omega))dW_\varphi(s)$$

by definition of the $\tilde{X}(t,\omega)$ and by definition of the noncausal integral with respect to the basis $\{\varphi_n\}$.
Thus from this fact we get the desired equality (23), by letting $n \to \infty$ on both sides of the equality (25). □

As we have mentioned in the previous paragraph, this fact that the solution $\tilde{X}(t,\omega) = X(t,\omega;\xi(\omega),\eta(\omega))$ of the noncausal problem (16) satisfies the Itô formula of noncausal type (23) would give us a partial answer to the question of uniqueness of the solution of our noncausal problem. In fact we have the following result that is valid for the case of 1-dimensional SDE.

Corollary 5.1. *If the $b(t,x;\eta)$ does not depend on the t and $b(x;\eta) > 0$ (or < 0) for all (t,x,η), then the solution $\tilde{X}(t,\omega)$ is unique among the all random functions that verify the noncausal Itô formula (23). We will call such solution the regular solution of the Cauchy problem.*

Proof. Without loss of generality we suppose that $b(\cdot) > 0$. Put $Y(t) = F(\tilde{X}(t))$ where $F(x)$ is as follows,

$$F(x) = \int_0^x \frac{dy}{b(y,\eta)}.$$

Then we have, $\tilde{X}(t,\omega) = F^{-1}(Y(t,\omega))$. By applying the noncausal Itô formula to the function $Y(t)$ we get,

$$Y(t) = F(\xi(\omega)) + \int_0^t (\frac{a}{b})(F^{-1}(Y(s));\eta(\omega))ds + W(t)$$

Since this is merely a family of ordinary integral equations parametrized by the ω, we see the uniqueness of its solution $Y(t)$ for each fixed and hence the uniqueness of the $\tilde{X}(t,\omega)$. This completes the proof. □

References

1. P. Cheridito, *Arbitrage in fractional Brownian motion models*, (2002).
2. K. Itô and M. Nishio, *Osaka J. Math.* **5**, 35 (1968).
3. M. Mancino & L. Pratelli, *Rendiconti Acad. Nat. delle Scienze detta dei XL, Memorie di Matematica*, 112 Vol. **xviii**, fasc 1, (1994) pp. 89-101
4. P. Mayer & M. Mancino, *Séminaire de Probabilités* **xxxi**, 198 (1995).
5. D. Nualart and M. Zakai *Ann. Probab.* **17**, 1536 (1989).
6. A. N. Shiryaev, On arbitrage and replication for fractional models, *Research Report* **20**, MaPhySto, (Dept.of Math.Sci., Univ. of Aarhus, 1998)
7. M.Zakai & D. Nualart, *Proba. and Math. Statist.* **3**, 37 (1987).

8. S. Ogawa & T. Miura, The fractional B–S model interpreted by the Ogawa integral, *presented at* the "poster session" of the Congres on RISK Control Theory, Firenze (2005)
9. S. Ogawa, Stochastic integral equations of Fredholm type, *(to appear in) Harmonic, Wavelet and P-adic Analysis*, (2005)
10. S. Ogawa, On noncausal Cauchy problem for the noncausal SDEs, In: S.Watanabe et al. (Eds.): *Stochastic Processes and Applications to Mathematical Finance*, (World Scientific, Singapore, pp. 289–304, 2004).
11. S. Ogawa & M. E. Mancino, Nonlinear feedback effects by hedging strategies, In: S.Watanabe et al. (Eds.): *Stochastic Processes and Applications to Mathematical Finance*, (World Scientific, Singapore, pp. 255–270, 2004)
12. S. Ogawa & M. Mancino, On a noncausal insider trading equilibrium model of asset pricing (*in preparation*)
13. S. Ogawa, *Rendiconti Acad. Nazionale delle Scienze detta dei XL,119*, Vol. **XXV** (2001)
14. S. Ogawa, On a stochastic integral equation for the random fields, *Séminaire de Proba.* **vol.xxv**, (Springer Verlag, Berlin, pp. 324–339. 1991)
15. S. Ogawa, Topics in the theory of noncausal stochastic integral equations, *Diffusion Processes and Related Problems in Analysis*, edt by M.Pinsky, (Birkhäuser Boston, 1990).
16. S. Ogawa, On the stochastic integral equation of Fredholm type, *Waves and Patterns* (monograph) (Kinokuniya and North-Holland, pp. 597-605, 1986).
17. S. Ogawa & T. Sekiguchi, On the Itô's formula of noncausal type, *Proc.Japan Acad. Sci*, (1985).
18. S. Ogawa, The stochastic integral of noncausal type as an extension of the symmetric integrals, *Japan J.Applied Math.*, (1985).
19. S. Ogawa, Sur la question d'existence de solution de l'equation stochastique differentielle du type noncausal, *Kyoto J. of Math.*, (Kyoto Univ., 1985).
20. S. Ogawa, Quelques propriétés de l'integrale stochastique du type noncausal, *Japan J.Applied Math.*, (1984)
21. S. Ogawa, *Tôhoku Math. J.*, **36**, 41 (1984).
22. S. Ogawa, Noncausal Problems in Stochastic Calculus (in japanese) in *Proc. of the Workshop on Noncausal Calculus and The Related Problems* held at RIMS Kyoto Univ. in February 1984, (Suuri Kagaku Kokyuuroku (RIMS Report) 527, 1984).
23. S. Ogawa, *C.R. Acad. Sci., Paris*, t.288, Série A, pp.359-362, (1979).
24. S. Ogawa, *Z. W.verw. Geb.*, **28**, 53 (1973).
25. S. Ogawa, *Proc. Japan Acad. Sci.*, 70 (1970).

Advances in Deterministic and Stochastic Analysis
Eds. N. M. Chuong *et al.* (pp. 321–339)

§17. INFINITE-DIMENSIONAL STOCHASTIC ANALYSIS AND FOUNDATIONS OF QUANTUM MECHANICS

ANDREI KHRENNIKOV

International Center for Mathematical Modeling in Physics and Cognitive Sciences, MSI, University of Växjö, S-35195, Sweden
E-mail: Andrei.Khrennikov@msi.vxu.se

We show that by using an asymptotic expansion of Gaussian integrals of analytic functions on infinite-dimensional spaces it is possible to represent quantum mechanics as a projection of classical statistical mechanics on infinite-dimensional space.

1. Introduction

The problem of reduction of quantum mechanics to classical statistical mechanics has been discussed from the first days of quantum mechanics, see, e.g., Refs. 1–4. Now days this problem is known as the problem of hidden variables or completeness of quantum mechanics, see, e.g., Refs. 5–8 for debates. There is a rather common opinion that quantum mechanics is complete and that it is impossible to introduce "hidden variables" providing more detailed description than quantum mechanics. In Ref. 9 it was demonstrated that in the opposition to this opinion it is possible to represent quantum mechanics as projection of classical statistical mechanics on *infinite-dimensional space.* In this paper we present this approach (which was called in Ref. 9 *Prequantum Classical Statistical Field Theory* – PCSFT) on the mathematical level of rigorousness; in particular, some functional spaces introduced in Ref. 9 should be modified to obtain the correct results. In the present paper we also find connection of PCSFT with background Gaussian random field on Hilbert space. Finally, we extend the approach of Ref. 9 by considering unbounded operators, see Sec. 6. There are also differences in interpretations of a small parameter of our asymptotic procedure of *dequantization.* In [9] this parameter was identified with the Planck constant h. In this paper we introduce a new parameter α giving the dispersion of prequantum fluctuations, see Ref. 10 for more details on physical interpretation. To simplify considerations, in this paper we consider quantum formalism over the field of real numbers, see Ref. 10 for dequantization of complex quantum mechanics. To exclude possible misunderstanding, we emphasize from the very beginning that our paper *is not about deformation quantization for systems with the infinite number of degrees of freedom,* see, e.g., Ref. 11, but about dequantization of conventional quantum mechanics for systems with a finite number of degrees of freedom by using analysis on infinite-dimensional space.

2. Infinite-dimensional analysis

Let H be a real Hilbert space and let $A : H \to H$ be a continuous self-adjoint linear operator. The basic mathematical formula which will be used in this paper is the formula for a Gaussian integral of a quadratic form $f(\psi) \equiv f_A(\psi) = (A\psi, \psi)$.

Let $d\rho(\psi)$ be a σ-additive Gaussian measure on the σ-field F of Borel subsets of H, see Ref. 12. This measure is determined by its covariation operator $B : H \to H$ and mean value $m \equiv m_\rho \in H$. For example, B and m determines the Fourier transform of ρ : $\tilde{\rho}(y) = \int_H e^{i(y,\psi)} d\rho(\psi) = e^{\frac{1}{2}(By,y)+i(m,y)}, y \in H$. In what follows we restrict our considerations to Gaussian measures with zero mean value $m = 0$, where $(m, y) = \int_H (y, \psi) d\rho(\psi) = 0$ for any $y \in H$. Sometimes there will be used the symbol ρ_B to denote the Gaussian measure with the covariation operator B and $m = 0$. We recall that the covariation operator $B \equiv \text{cov}\ \rho$ is defined by $(By_1, y_2) = \int (y_1, \psi)(y_2, \psi) d\rho(\psi), y_1, y_2 \in H$, and has the following properties: a). $B \geq 0$, i.e., $(By, y) \geq 0, y \in H$; b). B is a self-adjoint operator, $B \in \mathcal{L}_s(H)$; c). B is a trace-class operator and $\text{Tr}\ B = \int_H ||\psi||^2 d\rho(\psi)$. This is *dispersion* $\sigma^2(\rho)$ of the probability ρ. Thus $\sigma^2(\rho) = \text{Tr}\ B$.

We pay attention that the list of properties of the covariation operator of a Gaussian measure differs from the list of properties of a von Neumann density operator [4] only by one condition: Tr D = 1, for a density operator D.

We can easily find the Gaussian integral of the quadratic form $f_A(\psi)$:

$$\int_H f_A(\psi) d\rho(\psi) = \text{Tr}\ BA \tag{1}$$

The differential calculus for maps $f : H \to \mathbf{R}$ does not differ so much from the differential calculus in the finite dimensional case, $f : \mathbf{R}^n \to \mathbf{R}$. Instead of the norm on $\mathbf{R}^n$, one should use the norm on H. We consider so called Frechet differentiability. Here a function f is differentiable if it can be represented as $f(\psi_0 + \Delta\psi) = f(\psi_0) + f'(\psi_0)(\Delta\psi) + o(\Delta\psi)$, where $\lim_{\|\Delta\psi\|\to 0} \frac{\|o(\Delta\psi)\|}{\|\Delta\psi\|} = 0$. Here at each point ψ the derivative $f'(\psi)$ is a continuous linear functional on H; so it can be identified with the element $f'(\psi) \in H$. Then we can define the second derivative as the derivative of the map $\psi \to f'(\psi)$ and so on. A map f is differentiable n-times iff: $f(\psi_0 + \Delta\psi) = f(\psi_0) + f'(\psi_0)(\Delta\psi) + \frac{1}{2}f''(\psi_0)(\Delta\psi, \Delta\psi) + ... + \frac{1}{n!}f^{(n)}(\psi_0)(\Delta\psi, ..., \Delta\psi) + o_n(\Delta\psi)$, where $f^{(n)}(\psi_0)$ is a symmetric continuous n-linear form on H and $\lim_{\|\Delta\psi\|\to 0} \frac{\|o_n(\Delta\psi)\|}{\|\Delta\psi\|^n} = 0$. For us it is important that $f''(\psi_0)$ can be represented by a symmetric operator $f''(\psi_0)(u, v) = (f''(\psi_0)u, v), u, v \in H$ (this fact is well know in the finite dimensional case: the matrice representing the second derivative of any two times differentiable function $f : \mathbf{R}^n \to \mathbf{R}$ is symmetric). We remark that $f(\psi) = f(0) + f'(0)(\psi) + \frac{1}{2}f''(0)(\psi, \psi) + ... + \frac{1}{n!}f^{(n)}(0)(\psi, ..., \psi) + o_n(\psi)$.

For a real Hilbert space H, denote by the symbol $H^{\mathbf{C}}$ its complexification: $H^{\mathbf{C}} = H \oplus iH$. We recall that a function $f : H^{\mathbf{C}} \to \mathbf{C}$ is analytic if it can be expanded into the Taylor series:

$$f(\psi) = f(0) + f'(0)(\psi) + \frac{1}{2}f''(0)(\psi, \psi) + ... + \frac{1}{n!}f^{(n)}(0)(\psi, ..., \psi) + ... \quad (2)$$

which converges uniformly on any ball of $H^{\mathbf{C}}$.

3. Dequantization

3.1. *Classical and quantum statistical models*

We define *"classical statistical models"* in the following way, see Ref. 9 for more detail (and even philosophic considerations):

a). Physical states ω are represented by points of some set Ω (state space).

b). Physical variables are represented by functions $f : \Omega \to \mathbf{R}$ belonging to some functional space $V(\Omega)$.*

c). Statistical states are represented by probability measures on Ω belonging to some class $S(\Omega)$. †

*The choice of a concrete functional space $V(\Omega)$ depends on various physical and mathematical factors.

†It is assumed (by the Kolmogorov axiomatics) that there is given a fixed σ-field of subsets of Ω denoted by F. Probabilities are defined on F It is, of course, assumed that physical variables are represented by random variables – measurable functions. The choice of a concrete space of probability measures $S(\Omega)$ depends on various physical and mathematical factors.

d). The average of a physical variable (which is represented by a function $f \in V(\Omega)$) with respect to a statistical state (which is represented by a probability measure $\rho \in S(\Omega)$) is given by

$$< f >_\rho \equiv \int_\Omega f(\omega) d\rho(\omega). \tag{3}$$

A *classical statistical model* is a pair $M = (S(\Omega), V(\Omega))$. We recall that classical statistical mechanics on the phase space $\Omega_{2n} = \mathbf{R}^n \times \mathbf{R}^n$ gives an example of a classical statistical model. But we shall not be interested in this example in our further considerations. We shall develop a classical statistical model with *an infinite-dimensional phase-space.*

In real Hilbert space H a quantum statistical model is described in the following way:

a). Physical observables are represented by operators $A : H \to H$ belonging to the class of continuous self-adjoint operators $\mathcal{L}_s \equiv \mathcal{L}_s(H)$.

b). Statistical states are represented by density operators, see Ref. 4. The class of such operators is denoted by $\mathcal{D} \equiv \mathcal{D}(H)$.

d). The average of a physical observable (which is represented by the operator $A \in \mathcal{L}_s(H)$) with respect to a statistical state (which is represented by the density operator $D \in \mathcal{D}(H)$) is given by von Neumann's formula:[4]

$$< A >_D \equiv \text{Tr DA} \tag{4}$$

The *quantum statistical model* is the pair $N_{\text{quant}} = (\mathcal{D}(H), \mathcal{L}_s(H))$.

3.2. *Asymptotic Gaussian analysis*

Let us consider a classical statistical model in that the state space $\Omega = H$ (in physical applications $H = L_2(\mathbf{R}^3)$ is the space of classical fields on $\mathbf{R}^3$) and the space of statistical states consists of Gaussian measures with zero mean value and dispersion

$$\sigma^2(\rho) = \int_\Omega \|\psi\|^2 d\rho(\psi) = \alpha, \tag{5}$$

where $\alpha > 0$ is a small real parameter. Denote such a class of Gaussian measures by the symbol $S_G^\alpha(\Omega)$. For $\rho \in S_G^\alpha(\Omega)$, we have

$$\text{Tr cov } \rho = \alpha \tag{6}$$

We remark that any linear transformation (in particular, scaling) preserves the class of Gaussian measures. Let us make the change of variables (scaling):

$$\psi \to \frac{\psi}{\sqrt{\alpha}} \tag{7}$$

(we emphasize that this is a scaling not in the physical space $\mathbf{R}^3$, but in the space of fields on it). To find the covariation operator D of the image ρ_D of the Gaussian measure ρ_B, we compute its Fourier transform:

$$\tilde{\rho}_D(\xi) = \int_\Omega e^{i(\xi,y)} d\rho_D(y) = \int_\Omega e^{i(\xi,\frac{\psi}{\sqrt{\alpha}})} d\rho_B(\psi) = e^{-\frac{1}{2\alpha}(B\xi,\xi)}.$$

Thus

$$D = \frac{B}{\alpha} = \frac{\mathrm{cov}\rho}{\alpha}. \tag{8}$$

We shall use this formula later. We remark that by definition:

$$< f >_{\rho_B} = \int_\Omega f(\psi) d\rho_B(\psi) = \int_\Omega f(\sqrt{\alpha}\psi) d\rho_D(\psi).$$

To make our further considerations mathematically rigorous, we should attract the theory of analytic functions $f : \Omega^{\mathbf{C}} \to \mathbf{C}$. Here $\Omega^{\mathbf{C}} = \Omega \oplus i\Omega$ is the complexification of the real Hilbert space Ω.

Let $b_n : \Omega^{\mathbf{C}} \times ... \times \Omega^{\mathbf{C}} \to \mathbf{C}$ be a continuous n-linear symmetric form. We define its norm by $\|b_n\| = \sup_{\|\psi\| \le 1} |b_n(\psi, ..., \psi)|$. Thus

$$|b_n(\psi, ..., \psi)| \le \|b_n\| \|\psi\|^n \tag{9}$$

Let us consider the space analytic functions of the *exponential growth:*

$$|f(\psi)| \le a e^{b\|\psi\|}, \psi \in \Omega^{\mathbf{C}}. \tag{10}$$

Here constants depend on f : $a = a_f, b = b_f$.

Lemma 3.1. [13] *The space of analytic functions of the exponential growth coincides with the space of analytic functions such that:*

$$\|f^{(n)}(0)\| \le c\, r^n,\ n = 0, 1, 2, ... \tag{11}$$

Here constants $c = c_f$ and $r = r_f$ depend on the function f.

Proof.

A). Let f have the exponential growth. For any $\psi \in \Omega^{\mathbf{C}}$, we consider the function of the complex variable $z \in \mathbf{C} : g_\psi(z) = f(z\psi)$. By the Cauchy integral formula for $g_\psi(z)$ we have: $g_\psi^{(n)}(0) = \frac{n!}{2\pi i} \int_{|z|=R} g_\psi(z) z^{-(n+1)} dz$, where at the moment $R > 0$ is a free parameter. Thus: $|g_\psi^{(n)}(0)| \le n! R^{-n} \sup_{0 \le \theta \le 2\pi} |f(Re^{i\theta}\psi)| \le a_f n! R^{-n} e^{b_f R \|\psi\|}$. By choosing $R = n$ and observing that $g_\psi^{(n)}(0) = f^{(n)}(0)(\psi, ..., \psi)$ we obtain:

$$\|f^{(n)}(0)\| \le a'_f e^{-n} n^{1/2} e^{b_f n}.$$

Thus the derivatives of f satisfy the inequalities (11) with $r_f = e^{b_f}$.

B). Let now derivatives of f satisfy the inequalities (11). Then by the inequalities (9) we have $|f(\psi)| \leq \sum_{n=0}^{\infty} \|f^{(n)}(0)\| \|\psi\|^n/n! \leq c_f \sum_{n=0}^{\infty}(r_f \| \|\psi\|)^n/n! \leq c_f e^{r_f \|\psi\|}$. Thus f has the exponential growth with $b_f = r_f$. □

We denote by the symbol $\mathcal{V}(\Omega)$ the following space of functions $f : \Omega \to \mathbf{R}$. Each $f \in \mathcal{V}(\Omega)$ takes the value zero at the point $\psi = 0$ and it can be extended to the analytic function $f : \Omega^{\mathbf{C}} \to \mathbf{C}$ having the exponential growth.

Example 3.1. In particular, any polynomial on Ω belongs to the space $\mathcal{V}(\Omega)$. For example, let $A_1, ..., A_N$ be continuous linear operators. Then function $f(\psi) = \sum_{n=1}^{N}(A_n\psi, \psi)^n$ belongs to the space $\mathcal{V}(\Omega)$.

Any function $f \in \mathcal{V}(\Omega)$ is integrable with respect to any Gaussian measure on Ω. Let us consider the family of the classical statistical models

$$M^\alpha = (S_G^\alpha(\Omega), \mathcal{V}(\Omega)).$$

Let a variable $f \in \mathcal{V}(\Omega)$ and let a statistical state $\rho_B \in S_G^\alpha(\Omega)$. Our further aim is to find an asymptotic expansion of the (classical) average $< f >_{\rho_B} = \int_\Omega f(\psi) d\rho_B(\psi)$ with respect to the small parameter α.

Lemma 3.2. *Let $f \in \mathcal{V}(\Omega)$ and let $\rho \in S_G^\alpha(\Omega)$. Then the following asymptotic equality holds:*

$$< f >_\rho = \frac{\alpha}{2} \operatorname{Tr} \mathrm{D}\, \mathrm{f}''(0) + \mathrm{o}(\alpha), \ \alpha \to 0, \tag{12}$$

where the operator D is given by (8). Here

$$o(\alpha) = \alpha^2 R(\alpha, f, \rho), \tag{13}$$

where $|R(\alpha, f, \rho)| \leq c_f \int_\Omega e^{r_f \|\psi\|} d\rho_D(\psi)$.

Proof. In the Gaussian integral $\int_\Omega f(\psi) d\rho(\psi)$ we make the scaling (7):

$$< f >_\rho = \int_\Omega f(\sqrt{\alpha}\psi) d\rho_D(\psi) = \frac{\alpha}{2} \int_\Omega (f''(0)\psi, \psi)\, d\rho_D(\psi) + \alpha^2 R(\alpha, f, \rho), \tag{14}$$

where

$$R(\alpha, f, \rho) = \int_\Omega g(\alpha, f; \psi) d\rho_D(\psi), g(\alpha, f; \psi) = \sum_{n=4}^{\infty} \frac{\alpha^{n/2-2}}{n!} f^{(n)}(0)(\psi, ..., \psi).$$

We pay attention that

$$\int_\Omega (f'(0),\psi)d\rho_D(\psi)=0, \quad \int_\Omega f'''(0)(\psi,\psi,\psi)d\rho_D(\psi)=0,$$

because the mean value of ρ (and, hence, of ρ_D) is equal to zero. Since $\rho \in S_G^\alpha(\Omega)$, we have Tr D $= 1$.

The change of variables in (14) can be considered as scaling of the magnitude of statistical (Gaussian) fluctuations. Negligibly small random fluctuations $\sigma(\rho)=\sqrt{\alpha}$ (where α is a small parameter) are considered in the new scale as standard normal fluctuations. If we use the language of probability theory and consider a Gaussian random variables $\xi(\lambda)$, then the transformation (7) is nothing else than the standard normalization of this random variable (which is used, for example, in the central limit theorem): $\eta(\lambda)=\frac{\xi(\lambda)-E\xi}{\sqrt{E(\xi(\lambda)-E\xi)^2}}$ (in our case $E\xi=0$).

We now estimate the rest term $R(\alpha,f,\rho)$. By using the inequality (11) we have for $\alpha\le 1$:

$$|g(\alpha,f;\psi)|=\sum_{n=4}^{\infty}\frac{\|f^{(n)}(0)\|\|\psi\|^n}{n!}\le c_f\sum_{n=4}^{\infty}\frac{r_f^n\|\psi\|^n}{n!}=C_f e^{r_f\|\psi\|}.$$

Thus: $|R(\alpha,f,\rho)|\le c_f\int_\Omega e^{r_f\|\psi\|}d\rho_D(\psi)$. We obtain:

$$<f>_\rho=\frac{\alpha}{2}\int_\Omega (f''(0)\psi,\psi)\;d\rho_D(\psi)+o(\alpha),\;\alpha\to 0. \tag{15}$$

By using the equality (1) we finally come the asymptotic equality (12). □

We see that the classical average (computed in the model $M^\alpha=(S_G^\alpha(\Omega),\mathcal{V}(\Omega))$ by using the measure-theoretic approach) is coupled through (12) to the quantum average (computed in the model $N_{\rm quant}=(\mathcal{D}(\Omega),\mathcal{L}_{\rm s}(\Omega))$ by the von Neumann trace-formula).

The equality (12) can be used as the motivation for defining the following classical $\to$ quantum map T from the classical statistical model $M^\alpha=(S_G^\alpha,\mathcal{V})$ onto the quantum statistical model $N_{\rm quant}=(\mathcal{D},\mathcal{L}_{\rm s})$:

$$T: S_G^\alpha(\Omega)\to\mathcal{D}(\Omega),\;\; D=T(\rho)=\frac{\operatorname{cov}\rho}{\alpha} \tag{16}$$

(the Gaussian measure ρ is represented by the density matrix D which is equal to the covariation operator of this measure normalized by α);

$$T:\mathcal{V}(\Omega)\to\mathcal{L}_{\rm s}(\Omega),\;\; A_{\rm quant}=T(f)=\frac{1}{2}f''(0). \tag{17}$$

Our previous considerations can be presented as

Theorem 3.1. *The map* $T : S_G^{\alpha}(\Omega) \to \mathcal{D}(\Omega)$ *is one-to-one; the map* $T : \mathcal{V}(\Omega) \to \mathcal{L}_s(\Omega)$ *is linear surjection and the classical and quantum averages are coupled by the asymptotic equality (12).*

We emphasize that the correspondence between physical variables $f \in \mathcal{V}(\Omega)$ and physical observables $A \in \mathcal{L}_s(\Omega)$ is not one-to-one.

Let again $\Omega = H$. We consider a classical statistical model such that the class of statistical states consists of Gaussian measures ρ on Ω having zero mean value and unit dispersion

$$\sigma^2(\rho) = \int_\Omega ||x||^2 d\rho(x) = 1.$$

These are Gaussian measures having covariance operators with the unit trace. Denote the class of such probabilities by the symbol $S_G \equiv S_G(\Omega)$. In this model we choose a class of physical variables consisting of quadratic forms $f_A(x) = (Ax, x)$. We denote this class by V_{quad}. We remark that this is a linear space (over $\mathbf{R}$). We consider the following classical statistical model:

$$M_{\text{quad}} = (S_G(\Omega), V_{\text{quad}}(\Omega))$$

As always in a statistical model, we are interested only in averages of physical variables $f \in V_{\text{quad}}$ with respect to statistical states $\rho \in S_G(\Omega)$. We emphasize that by (1):

$$< f_A >_\rho = \text{Tr } BA \tag{18}$$

Let us consider the following map T from the classical statistical model $M_{\text{quad}} = (S_G, V_{\text{quad}})$ to the quantum statistical model $N_{\text{quant}} = (\mathcal{D}, \mathcal{L}_s)$:

$$T : S_G(\Omega) \to \mathcal{D}(\Omega), T(\rho) = \text{cov } \rho \tag{19}$$

(the Gaussian measure ρ_B is represented by the density matrix D which is equal to the covariation operator of this measure), and we define:

$$T : V_{\text{quad}}(\Omega) \to \mathcal{L}_s(\Omega), T(f) = \frac{1}{2} f''(0) \tag{20}$$

(thus a variable $f \in V_{\text{quad}}(\Omega)$ is represented by its second derivative).

Theorem 3.2. *The map* T *provides one-to-one correspondence between the classical statistical model* M_{quad} *and the quantum model* N_{quant}.

4. Gaussian underground for pure states

In quantum mechanics a pure quantum state is given by a normalized vector $\Psi \in H : \|\Psi\| = 1$. In our model such a state is not pure at all (in the sense that such a vector Ψ does not provide a description of an individual system). Such a normalized vector Ψ is the label of a Gaussian statistical mixture. The corresponding quantum statistical state is represented by the density operator: $D_\Psi = \Psi \otimes \Psi$. In particular, the von Neumann's trace-formula for expectation has the form: $\mathrm{Tr}\, \mathrm{D}_\Psi \mathrm{A} = (\mathrm{A}\Psi, \Psi)$. Let us consider the correspondence map T for statistical states for the classical statistical model $M^\alpha = (S_G^\alpha, \mathcal{V})$, see (16). A pure quantum state Ψ (i.e., the state with the density operator D_Ψ) is the image of the Gaussian statistical mixture ρ_Ψ of states $\psi \in H$. We use the capital Ψ to denote a quantum pure state. This is just the special system of labeling of the Gaussian measure ρ_Ψ by the normalized vector Ψ of Hilbert space. Points of the sample space on that this measure is defined we denote by the low ψ. The measure ρ_Ψ has the covariation operator $B_\Psi = \alpha D_\Psi$. This means that the measure ρ_Ψ is concentrated on the one-dimensional subspace $H_\Psi = \{x \in H : x = s\Psi, s \in \mathbf{R}\}$. This is one-dimensional Gaussian distribution.

5. Pure states as one-dimensional projections of spatial white-noise

In section 4 we showed that so called pure states of quantum mechanics have the natural classical statistical interpretation as Gaussian measures concentrated on one-dimensional subspaces of the Hilbert space H. On the other hand, it is well known that *any Gaussian measure on H is determined by its one-dimensional projections.* To determine a Gaussian random variable $\xi(\omega) \in H$, it is sufficient to determine all its one-dimensional projections: $\xi_\Psi(\omega) = (\Psi, \xi(\omega)), \Psi \in H$. The covariation operator B of ξ (having the zero mean value) is defined by $(B\Psi, \Psi) = E\xi_\Psi^2$. We are interested in the following problem:

Is it possible to construct a Gaussian distribution on H such that its one-dimensional projections will give us all pure quantum states, $\Psi \in H, \|\Psi\| = 1$?

We recall that in our approach a pure quantum state Ψ is just the label for a Gaussian random variable ξ_Ψ such that $E\xi_\Psi^2 = \alpha||\Psi||^2 = \alpha$. Thus the answer to our question is positive and pure quantum states can be considered as one-dimensional projections of the $\sqrt{\alpha}$-scaling of the standard Gaussian distribution on H. The standard Gaussian distribution μ on H (so the average of μ is equal to zero and cov $\mu = I$, where I is the unit operator) is nothing else than the *white noise* on $\mathbf{R}^3$ (if one chooses $H = L_2(\mathbf{R}^3)$), see [12] for details. Thus pure quantum states are simply one-dimensional projections of the *spatial white noise*. It is well known, see, e.g., [12], that the μ *is not σ-additive on the σ-field of Borel subsets of H.*

To escape mathematical difficulties and concentrate on the dequantization of quantum mechanics, we start with consideration of the finite-dimensional case.

5.1. *The finite-dimensional case*

We consider the family of Gaussian random variables $\xi_\Psi, \Psi \in \mathbf{R}^n$, $E\xi_\Psi = 0, E\xi_\Psi^2 = \alpha||\Psi||^2$. This family can be realized as $\xi_\Psi(\omega) = (\Psi, \xi(\omega))$ where $\xi(\omega) = \sqrt{\alpha}\eta(\omega)$ and $\eta(\omega) \in \mathbf{R}^n$ is standard Gaussian random variable (so $E\eta = 0, \text{cov } \eta = I$). For any $\Psi \in \mathbf{R}^n$, we define the projection P_Ψ to this vector: $P_\Psi(k) = (\Psi, k)\Psi$.

Denote by the symbol $\mathcal{V}(\mathbf{R}^n)$ the class of functions $f : \mathbf{R}^n \to \mathbf{R}$ such that $f(0) = 0$ and f can be continued analytically onto $\mathbf{C}^n$ and this continuation $f(z)$ has the exponential growth.

Proposition 5.1. *Let $f \in \mathcal{V}(\mathbf{R}^n)$. Then we have:*

$$Ef(P_\Psi\xi(\omega)) = \frac{\alpha||\Psi||^2}{2}(f''(0)\Psi, \Psi) + o(\alpha), \alpha \to 0. \tag{21}$$

Proof. By using the Taylor expansion of f we obtain:

$$Ef(P_\Psi\xi(\omega)) = \frac{1}{2}E(\Psi, \xi(\omega))^2(f''(0)\Psi, \Psi) + o(\alpha), \alpha \to 0.$$

By setting into this asymptotic equality the dispersion of the random variable $P_\Psi\xi(\omega)$ we obtain (21). □

If $||\Psi|| = 1$ (a pure quantum state), then we get:

$$Ef(P_\Psi\xi(\omega)) = \frac{\alpha}{2}(f''(0)\Psi, \Psi) + o(\alpha), \alpha \to 0. \tag{22}$$

Here $A = f''(0)$ is a symmetric linear operator. We "quantized" the classical variable $f(x), x \in \mathbf{R}^n$, by mapping it to the operator $A = \frac{1}{2}f''(0)$, see Theorem 3.1. The Gaussian random variable $\xi_\Psi, ||\Psi|| = 1$ is "quantize" by mapping it into the pure quantum state Ψ.

Theorem 5.1. *There exists a Kolmogorov probability space such that all pure quantum states can be represented by Gaussian random variables on* *this space. The correspondence* $\Psi \to \xi_\Psi(\omega)$ *is linear:*

$$\lambda_1\Psi_1 + \lambda_2\Psi_2 \to \lambda_1\xi_{\Psi_1}(\omega) + \lambda_2\xi_{\Psi_2}(\omega), \tag{23}$$

where $\lambda_1, \lambda_2 \in \mathbf{R}$.

Proof. We choose $\Omega = \mathbf{R}^n$ as the space of elementary events, the σ-field of Borel subsets is the space of events and the standard Gaussian measure μ as the probability measure. Then for $\Psi \to \xi_\Psi(\omega) = \sqrt{\alpha}(\Psi, \omega), \omega \in \mathbf{R}^n$, we have: $\lambda_1\xi_{\Psi_1}(\omega) + \lambda_2\xi_{\Psi_2}(\omega) = \lambda_1(\Psi_1, \omega) + \lambda_2(\Psi_2, \omega) = (\lambda_1\Psi_1 + \lambda_2\Psi_2, \omega) = \xi_{\lambda_1\Psi_1+\lambda_2\Psi_2}(\omega)$. □

This theorem is rather surprising from the common viewpoint (by that essentially nonclassical probabilistic features of quantum states are consequences of the *non-Kolmogorovian structure* of the quantum probabilistic model).

We pay attention that physical variables $\xi_\Psi(\omega) = P_\Psi\xi(\omega), \Psi \in \mathbf{R}^n$, (one-dimensional projections of the scaling $\xi(\omega)$ of the standard Gaussian random variable $\eta(\omega) \in \mathbf{R}^n$) *cannot be mapped onto nontrivial quantum observables.* Prequantum classical physical variables $\xi_\Psi(\omega) = (\Psi, \omega)$ are linear functionals of ω. Therefore $T(\xi_\Psi) = \xi_\Psi''(0) = 0$. Nevertheless, quantum mechanics contains images of ξ_Ψ given by quantum states Ψ, but only for Ψ with $||\Psi|| = 1$!

We call $\xi(\omega)$ a *background random field.* All pure states could be extracted from the the background random field by projecting it to one dimensional subspaces. PCSFT explains the origin of the scalar product on the set of pure quantum states. We consider the $1/\alpha$-amplification of the covariation of two Gaussian (prequantum) random variables $\xi_{\Psi_1}(\omega)$ and $\xi_{\Psi_2}(\omega)$. We have:

$$\frac{1}{\alpha} E\xi_{\Psi_1}(\omega)\xi_{\Psi_2}(\omega) = (\Psi_1, \Psi_2). \tag{24}$$

Conclusion. *The Hilbert space structure of quantum mechanics is induced by the (prequantum) Gaussian random field (the background field* $\xi(\omega)$*) through the* $\alpha \to 0$ *asymptotic.*

At the moment we proved this only in the finite-dimensional case. In subsection 5.2 we shall do this in the infinite-dimensional case. Finally, we pay attention to the fact that, for quadratic physical variables $f(x) = \frac{1}{2}(Ax, x)$, where $A : H \to H$ is a symmetric operator, the asymptotic equality (22) is reduced to the precise equality of averages. By considering directly the standard Gaussian random variable $\eta(\omega)$ (instead of the background random field $\xi(\omega) = \sqrt{\alpha}\eta(\omega)$) we come to the following classical probabilistic representation of the quantum average: $Ef(P_\Psi \eta(\omega)) = (A\Psi, \Psi) \equiv < A >_\Psi$.

5.2. *Prequantum white noise field*

To repeat consideration of susection 5.1 for the infinite-dimensional case, we consider measures on the so called rigged Hilbert spaces. We apply some rather abstract mathematical constructions. However, finally we shall consider a simple concrete example which will be then used as the basis of our prequantum classical statistical model.

Let Ω be a nuclear Frechet[‡] topological linear space and Ω' its dual space. Suppose that Ω is densely and continuously embedded into a Hilbert space H, so $\Omega \subset H$. Thus the dual space H' is densely embedded into Ω'. By identifying H and H' we obtain the rigged Hilbert space:

$$\Omega \subset H \subset \Omega' \tag{25}$$

In our final application we shall set $\Omega = \mathcal{S}(\mathbf{R}^3)$. This is the space of Schwartz test functions on $\mathbf{R}^3$. Here $\Omega' = \mathcal{S}'(\mathbf{R}^3)$ is the space of Schwartz distributions. In this case we choose $H = L_2(\mathbf{R}^3)$ and we shall consider the rigged Hilbert space:

$$\mathcal{S}(\mathbf{R}^3) \subset L_2(\mathbf{R}^3) \subset \mathcal{S}'(\mathbf{R}^3) \tag{26}$$

Readers who are not so much interested in general theory of topological linear spaces can consider this rigged Hilbert space throughout this section.

A Gaussian measure ρ on Ω' is determined by its characteristic functional (Fourier transform) $\tilde{\rho}$ which is defined on $\Omega : \tilde{\rho}(\Psi) = e^{-\frac{1}{2}b(\Psi,\Psi)}$, where $b : \Omega \times \Omega \to \mathbf{R}$ is a continuous positively defined quadratic form. By the well known theorem of Minlos-Sazonov, see e.g., [], ρ is σ-additive on Ω' and its covariation functional is equal to b. Here $b(\Psi_1, \Psi_2) = \int_{\Omega'} (\phi, \Psi_1)(\phi, \Psi_2) d\rho(\phi)$, where $\Psi_1, \Psi_2 \in \Omega$. This functional defines the covariation operator $B : \Omega \to \Omega'$ by $(B\Psi_1, \Psi_2) = b(\Psi_1, \Psi_2)$. This operator is self-adjoint in the following sense. The dual operator $B' : \Omega'' \to \Omega'$. But, since the topological linear space Ω is a nuclear Frechet space, it is *reflexive*. Hence, $\Omega'' = \Omega$. Thus the operator $B' : \Omega \to \Omega'$. Thus it is meaningful to speak about self-adjoint operators in this framework (by extending the ordinary theory of self-adjoint operators in Hilbert space). We also pay attention to the fact that the covariation operator B is positively defined.

[‡]so complete metrizable and locally convex

Let us consider the standard Gaussian distribution μ on H that is defined by its covariation functional:

$$b(\Psi_1, \Psi_2) = (\Psi_1, \Psi_2).$$

The corresponding covariation operator $B = I : \Omega \to \Omega'$ is the canonical embedding operator. Since the embedding $\Omega \subset H$ is continuous, $b : \Omega \times \Omega \to \mathbf{R}$ is continuous and, hence, the measure μ is σ-additive on Ω'. Therefore there is well defined the corresponding Gaussian random variable $\eta(\phi) \in \Omega'$.

In the case of the rigged Hilbert space (26) the Gaussian random field $\eta(\phi) \in \mathcal{S}'(\mathbf{R}^3)$ is nothing else than the *spatial white noise.* We extend this terminology and we shall call $\eta(\phi)$ *white noise* even in the abstract framework. Let us consider $\sqrt{\alpha}$-scaling of white noise

$$\xi(\phi) = \sqrt{\alpha}\eta(\phi)$$

and its one-dimensional projections: $\xi_\Psi(\phi) = (\xi(\phi), \Psi), \Psi \in \Omega$. We have $E\xi_\Psi = 0, E\xi_\Psi^2 = \alpha||\Psi||^2$. The $\xi(\phi)$ is the *background field* in our prequantum model (PCSFT).

For any $\Psi \in \Omega$, we consider the one-dimensional projector $P_\Psi(\phi) = (\phi, \Psi)\Psi, \phi \in \Omega'$, and the Ω-valued random variable $P_\Psi\xi(\phi) = \xi_\Psi(\phi)\Psi$. If $||\Psi|| = 1$, then the T-image of the corresponding Gaussian distribution ρ_Ψ is nothing else than the pure state Ψ.

This correspondence can be extended form the space Ω to the Hilbert space H. If $\Psi \in H$, then $\xi_\Psi(\phi) = (\phi, \Psi)$ is also well defined, but is is not a continuous linear functional on the space Ω'. The $\xi_\Psi(\phi)$ is defined as an element of the space of square integrable functionals of the white noise: $\xi_\Psi \in L_2(\Omega', d\mu)$. To define ξ_Ψ, we approximate $\Psi \in H$ by elements Ψ_n of $\Omega, \Psi_n \to \Psi$ in H (we recall that Ω is dense in H). Then $\xi_\Psi = \lim_{n\to\infty} \xi_{\Psi_n}$ in $L_2(\Omega', d\mu)$.

Lemma 5.1. *Let $f : H \to \mathbf{R}$ be a polynomial and let $f(0) = 0$. Then, for any $\Psi \in H$, the asymptotic equality (21) holds.*

Proof. Here the main difference from consideration in section 3 is that the measure μ is not concentrated on Hilbert space H on which the function f is defined (and continuous). Therefore even the exponential growth of f on H would not help so much, because $\int_{\Omega'} e^{a\|\phi\|} d\mu(\phi) = \infty$ (since even $\int_{\Omega'} \|\phi\| d\mu(\phi) = \infty$). We have for a polynomial f:

$$Ef(P_\Psi\xi(\phi)) = \sum_{k=1}^{N} \frac{f^{(2k)}(0)(\Psi, ..., \Psi)}{2k!} E\xi_\Psi^{2k}(\phi)$$

$$= \sum_{k=1}^{N} \frac{f^{(2k)}(0)(\Psi, ..., \Psi)}{(2k)!} \frac{\alpha^{2k} \|\Psi\|^{2k} (2k)!}{2^k k!}.$$

Since the sum is finite and derivatives of f are continuous forms on H, we obtain (21). □

We "quantize" $f(u)$ by mapping it into $\frac{f''(0)}{2}$. For quadratic functionals $f(u) = \frac{1}{2}(Au, u), A \in \mathcal{L}_s(H)$, we have the precise equality and we can directly use the average with respect to the canonical Gaussian random variable $\eta(\omega)$. Here

$$Ef(P_\Psi \eta(\omega)) = \frac{1}{2}(f''(0)\Psi, \Psi).$$

6. Unbounded operators

Let $f : \Omega \to \mathbf{R}$ be a smooth function. Then, at any point $\psi_0 \in \Omega$, $f''(\psi_0) : \Omega \to \Omega'$. Therefore $f''(0)$ is in general unbounded operator in H.

Moreover, in this way (i.e., starting with PCSFT) we obtain the class of linear operators (quantum observables) that is even essentially larger than in the conventional quantum formalism. In general, $A = f''(0)$ maps Ω not into H, but into Ω'.

Example 6.1. Let us consider the rigged Hilbert space (26). We consider the map $f : \mathcal{S}(\mathbf{R}^3) \to \mathbf{R}$ determined by a fixed point $x_0 \in \mathbf{R}^3$:

$$f(\psi) = \frac{1}{2}\psi^2(x_0).$$

(For example, the classical field $\psi(x) = e^{-x^2}$ is mapped into the real number $e^{-x_0^2}$). Then $(f''(0)\psi_1, \psi_2) = \psi_1(x_0)\psi_2(x_0)$. Thus

$$A\psi(x) = \frac{1}{2} f''(0)\psi(x) = \frac{1}{2}\delta(x - x_0)\psi(x)$$

is the operator of multiplication by the δ-function $\delta(x - x_0)$. Hence

$$f''(0)(\mathcal{S}(\mathbf{R}^3)) \not\subset L_2(\mathbf{R}^3).$$

For any $\Psi \in \mathcal{S}(\mathbf{R}^3)$, we have

$$Ef(P_\Psi \eta(\omega)) = \frac{1}{2}\Psi^2(x_0) = (A\Psi, \Psi) \equiv < A >_\Psi .$$

However, in general for $\Psi \in L_2(\mathbf{R}^3)$ the average $< A >_\Psi$ is not well defined.

We can consider not only pure states, but general density operators. Let us now consider a Gaussian measure $\rho \in S_G^\alpha(H)$ which has the support on the space Ω. Thus ρ can be considered as a measure on Ω. For such a measure ρ its covariation operator $B : \Omega' \to \Omega$ and its Fourier transform $\tilde{\rho}$ is defined on Ω'. We denote this class of statistical states by the symbol $S_G^\alpha(\Omega)$. We remark that $S_G^\alpha(\Omega) \subset S_G^\alpha(H)$.

Let E be a complex locally convex topological linear space. We recall that the topology of E can be determined by a system of semi-norms (the notion of a semi-norm p generalizes the notion of a norm $\|\cdot\|$; the only difference is that $p(\psi)$ can be equal to zero even for a nonzero vector ψ). Let $b_n : E \times ... \times E \to \mathbf{C}$ be a continuous n-linear symmetric form. There exits a continuous semi-norm p on E such that

$$\|b_n\|_p = \sup_{p(\psi)\leq 1} |b_n(\psi, ..., \psi)| < \infty$$

(here $p \equiv p_{b_n}$). Thus

$$|b_n(\psi, ..., \psi)| \leq \|b_n\| \, p^n(\psi) \tag{27}$$

An analytic function, see, e.g., [11] for details, $f : E \to \mathbf{C}$ has the exponential growth if there exits a continuous semi-norm p on E on such that:

$$|f(\psi)| \leq a e^{bp(\psi)}, \psi \in E. \tag{28}$$

Here the constants and the semi-norm depend on $f : a \equiv a_f, b \equiv b_f, p \equiv p_f$.

Lemma 6.1. [13] *The space of analytic functions of the exponential growth coincides with the space of analytic functions such that there exists a continuous semi-norm $p = p_f$:*

$$\|f^{(n)}(0)\|_p \leq c\, r^n, \; n = 0, 1, 2, ... \tag{29}$$

Here constants $c = c_f$ and $r = r_f$ depend on the function f.

Proof.

A). Let f have the exponential growth. For any $\psi \in E$, we consider the function of the complex variable $z \in \mathbf{C} : g_\psi(z) = f(z\psi)$. As in Lemma 3.1, we have: $|g_\psi^{(n)}(0)| \leq n! R^{-n} \sup_{0\leq\theta\leq 2\pi} |f(Re^{i\theta}\psi)| \leq a_f n! R^{-n} e^{b_f R p(\psi)}$. We obtain:

$$\|f^{(n)}(0)\|_p \leq a'_f e^{-n} n^{1/2} e^{b_f n}.$$

Thus the derivatives of f satisfy the inequalities (29) with $r_f = e^{b_f}$.

B). Let now derivatives of f satisfy the inequalities (29) for some continuous semi-norm p. Then by the inequalities (27) we have

$$|f(\psi)| \leq \sum_{n=0}^{\infty} \|f^{(n)}(0)\|_p p^n(\psi)/n! \leq c_f e^{r_f p(\psi)}.$$

Thus f has the exponential growth with $b_f = r_f$ and the same continuous semi-norm p as in (29). □

We denote by $\Omega^{\mathbf{C}}$ the complexification of $\Omega : \Omega^{\mathbf{C}} = \Omega \oplus i\Omega$. We denote by $\mathcal{V}(\Omega)$ the class of functions $f : \Omega \to \mathbf{R}$, $f(0) = 0$, which can be analytically continued onto $\Omega^{\mathbf{C}}$ and they have the exponential growth.

Lemma 6.2. *Let $\rho \in S_G^{\alpha}(\Omega)$. Then, for any function $f \in \mathcal{V}(\Omega)$, the following asymptotic equality holds:*

$$< f >_\rho \equiv \int_\Omega f(\psi) d\rho(\psi) = \frac{\alpha}{2} \int_\Omega (f''(0)\psi, \psi) d\rho_D(u) + o(\alpha), \; \alpha \to 0, \quad (30)$$

where $D = \frac{\mathrm{cov}\rho}{\alpha}$. Here

$$o(\alpha) = \alpha^2 R(\alpha, f, \rho), \quad (31)$$

where

$$|R(\alpha, f, \rho)| \leq c_f \int_\Omega e^{r_f p(\psi)} d\rho_D(\psi). \quad (32)$$

The semi-norm p is determined by the inequality (28).

The proof of this Theorem repeats the proof of Lemma 3.2. Instead of Lemma 3.1, we apply its generalization to the case of an arbitrary locally convex topological linear space, see Lemma 6.1.

We pay attention that $D : \Omega' \to \Omega$, and $A = \frac{f''(0)}{2} : \Omega \to \Omega'$, so $C = DA : \Omega \to \Omega$. In general, this operator can not be extended to a continuous operator in H. We would like to obtain an analogue of the formula (1) for linear continuous operators $A : \Omega \to \Omega'$:

$$\int_\Omega (A\psi, \psi) d\rho_D(u) = \mathrm{Tr}\, DA \quad (33)$$

The main mathematical problem is that in general the operator $C = DA$ is not even continuous in H, so it is not a trace class operator in the Hilbert space H. Nevertheless, we can introduce the notion of trace even in such a framework.

We recall that systems of vectors $\{e_j\}_{j=1}^{\infty}, e_j \in \Omega$, and $\{e'_j\}_{j=1}^{\infty}, e'_j \in \Omega'$, are called *biorthogonal topological bases* in Ω and Ω' if

$$(e'_j, e_i) = \delta_{ij}, \text{ and } \psi = \sum_{j=1}^{\infty} (e'_j, \psi) e_j, \psi \in \Omega, \phi = \sum_{j=1}^{\infty} (\phi, e_j) e'_j, \phi \in \Omega',$$

where the series converge in Ω and Ω', respectively.

Definition 6.1. A linear continuous operator $C : \Omega \to \Omega$ is called trace-class operator if, for any pair of biorthogonal topological bases, the series

$$\mathrm{Tr}\, C = \sum_{j=1}^{\infty} (e'_j, Ce_j)$$

converges and its sum does not depend on bases.

Lemma 6.3. *Let ρ be a Gaussian measure on Ω and let $A : \Omega \to \Omega'$ be a continuous operator. Then the operator $C = DA$, where $D = \mathrm{cov}\rho$, belongs to the trace class and the equality (33) holds.*

As a consequence of Lemmas 6.2 and 6.3, we obtain:

Theorem 6.1. *Let $\rho \in S_G^{\alpha}(\Omega)$. Then, for any function $f \in \mathcal{V}(\Omega)$, the following asymptotic equality holds:*

$$< f >_{\rho} \equiv \int_{\Omega} f(\psi) d\rho(\psi) = \mathrm{Tr\, D} f''(0)/2 + o(\alpha), \ \alpha \to 0, \tag{34}$$

where $D = \frac{\mathrm{cov}\rho}{\alpha}$.

Thus our prequantum model, PCSFT, provides the motivation to extend the set of quantum observables and consider all continuous operators $A : \Omega \to \Omega'$. Operators should be self-adjoint in the ordinary sense: $A' = A$. We recall that here $A' : \Omega'' \to \Omega'$, but $\Omega'' \equiv \Omega$, since Ω is a nuclear Frechet space and hence it is reflexive. Denote the set of such operators by the symbol $\mathcal{L}_s(\Omega, \Omega')$. Denote the set of covariation operators of Gaussian measures belonging the space $S_G^1(\Omega)$ by the symbol $\mathcal{D}(\Omega', \Omega)$.

Definition 6.2. A statistical quantum model corresponding to a rigged Hilbert space $\mathcal{T}$ given by (25) is the pair $N_{\text{quant}}(\mathcal{T}) = (\mathcal{D}(\Omega', \Omega), \mathcal{L}_s(\Omega, \Omega'))$. A generalized density operators $D \in \mathcal{D}(\Omega', \Omega)$ represents a statistical state; a linear operator $A \in \mathcal{L}_s(\Omega, \Omega')$ represents a quantum observable. The average of such an observable with respect to such a statistical state is given by the following generalization of the von Neumann trace-formula:

$$< A >_D = \mathrm{Tr\, DA} \tag{35}$$

We choose the state space Ω – a nuclear Frechet space. For a rigged Hilbert space $\mathcal{T}$ given by (25), we consider the classical statistical model $M^\alpha(\mathcal{T}) = (S^\alpha_G(\Omega), \mathcal{V}(\Omega))$. Here as always $< f >_\rho = \int_\Omega f(\psi)d\rho(\psi)$.

The equality (34) can be used as the motivation for defining the following classical $\rightarrow$ quantum map T from the classical statistical model $M^\alpha(\mathcal{T}) = (S^\alpha_G(\Omega), \mathcal{V}(\Omega))$ onto the quantum statistical model $N_{\text{quant}}(\mathcal{T}) = (\mathcal{D}(\Omega', \Omega), \mathcal{L}_s(\Omega, \Omega'))$ by (16), (17). Our previous considerations can be presented as

Theorem 6.2. *The map $T : S^\alpha_G(\Omega) \rightarrow \mathcal{D}(\Omega', \Omega)$ is one-to-one; the map $T : \mathcal{V}(\Omega) \rightarrow \mathcal{L}_s(\Omega, \Omega')$ is linear surjection and the classical and quantum averages are coupled by the asymptotic equality (34).*

Example 6.2. The position operators $\hat{x}_j, j = 1, 2, 3$ can be obtained as $\hat{x}_j = \frac{1}{2} f''_{x_j}(0)$, where

$$f_{x_j}(\psi) = \int_{\mathbf{R}^3} x_j \psi^2(x) dx.$$

Here the operator of multiplication $\hat{x}_j : \mathcal{S}(\mathbf{R}^3) \rightarrow \mathcal{S}(\mathbf{R}^3), \psi \rightarrow x_j\psi$, is continuous. Hence $\hat{x}_j : \mathcal{S}(\mathbf{R}^3) \rightarrow \mathcal{S}'(\mathbf{R}^3)$ is also continuous. Thus, for any measure $\rho \in S^\alpha_G(\mathcal{S}(\mathbf{R}^3))$, we have

$$< f_{x_j} >_\rho \equiv \int_{\mathcal{S}(\mathbf{R}^3)} \int_{\mathbf{R}^3} x\psi^2(x) dx d\rho(\psi) = \alpha \text{Tr } D\hat{x}_j,$$

$D = \text{cov } \rho/\alpha$ (here the trace of the composition $D\hat{x}_j$ is well defined).

Example 6.3. Let x_0 be a fixed point in $\mathbf{R}^3$. Let now $A\psi(x) = \delta(x - x_0)\psi(x), \psi \in \mathcal{S}(\mathbf{R}^3)$. This operator does not belong to the domain of the conventional quantum formalism. It could not be represented as an unbounded operator in $H = L_2(\mathbf{R}^3)$ with a dense domain of definition. Nevertheless,

$$\int_{\mathcal{S}(\mathbf{R}^3)} (A\psi, \psi) d\rho(\psi) = \int_{\mathcal{S}(\mathbf{R}^3)} \psi^2(x_0) d\rho(\psi) = \alpha \text{Tr } DA$$

and the trace of the composition DA is well defined.

Example 6.4. The momentum operators $\hat{p}_j, j = 1, 2, 3$, can be obtained as $\hat{p}_j = \frac{1}{2} f''_{p_j}(0)$, where

$$f_{p_j}(\psi) = -i \int_{\mathbf{R}^3} \frac{\partial \psi}{\partial x_j}(x) \overline{\psi(x)} dx.$$

Here the operator $\hat{p}_j : \mathcal{S}(\mathbf{R}^3) \to \mathcal{S}(\mathbf{R}^3)$ is continuous. Hence, $\hat{p}_j : \mathcal{S}(\mathbf{R}^3) \to \mathcal{S}'(\mathbf{R}^3)$ is also continuous. Thus for any measure $\rho \in S^\alpha(\mathcal{S}(\mathbf{R}^3) \times \mathcal{S}(\mathbf{R}^3))$, we have (for $\psi(x) = q(x) + ip(x)$):

$$< f_{p_j} >_\rho \equiv -i \int_{\mathcal{S}(\mathbf{R}^3)\times\mathcal{S}(\mathbf{R}^3)} \int \frac{\partial\psi}{\partial x_j}(x)\overline{\psi(x)}dx d\rho(\psi) = \alpha \mathrm{Tr}\, D\, \hat{p}_j,$$

where $D = \mathrm{cov}\ \rho/\alpha$. Here $D\hat{p}_j : \mathcal{S}(\mathbf{R}^3) \to \mathcal{S}(\mathbf{R}^3)$ is the trace class operator. Similar considerations can be done for angular momentum operators.

This paper was partially supported by EU-network "Quantum Probability and Applications."

References

1. D. Hilbert, J. von Neumann, L. Nordheim, *Math. Ann.*, **98**, 1 (1927).
2. P. A. M. Dirac, *The Principles of Quantum Mechanics,* (Oxford Univ. Press, 1930).
3. W. Heisenberg, *Physical principles of quantum theory,* (Chicago Univ. Press, 1930).
4. J. von Neumann, *Mathematical foundations of quantum mechanics,* (Princeton Univ. Press, Princeton, N. J., 1955).
5. A. Yu. Khrennikov (editor), *Foundations of Probability and Physics,* Q. Prob. White Noise Anal., Vol. 13, (WSP, Singapore, 2001).
6. A. Yu. Khrennikov (editor), *Quantum Theory: Reconsideration of Foundations,* Ser. Math. Modeling, 2, (Växjö Univ. Press, 2002).
7. A. Yu. Khrennikov (editor), *Foundations of Probability and Physics*-2, Ser. Math. Modeling, 5, (Växjö Univ. Press, 2003).
8. A. Yu. Khrennikov (editor), *Quantum Theory: Reconsideration of Foundations*-2, Ser. Math. Modeling, 10, (Växjö Univ. Press, 2004).
9. A. Yu. Khrennikov, A pre-quantum classical statistical model with infinite-dimensional phase space. *J. Phys. A: Math. Gen.*, **38**, 9051 (2005).
10. A. Yu. Khrennikov, Quantum mechanics as an asymptotic projection of statistical mechanics of classical fields: derivation of Schrödinger's, Heisenberg's and von Neumann's equations. http://www.arxiv.org/abs/quant-ph/0511074
11. A. Yu. Khrennikov, Infinite-dimensional pseudo-differential operators. *Izvestia Akademii Nauk USSR, ser.Math.*, **51**, 46 (1987).
12. T. Hida, M. Hitsuda *Gaussian Processes,* Translations of Mathematical Monographs, **120**, (American Mathematical Society, 1993).
13. A. Yu. Khrennikov, Equations with infinite-dimensional pseudo-differential operators. Dissertation for the degree of candidate of phys-math. sc., (Dept. Mechanics-Mathematics, Moscow State University, Moscow, 1983).

Advances in Deterministic and Stochastic Analysis
Eds. N. M. Chuong *et al.* (pp. 341–360)

§18. NONCOMMUTATIVE TRIGONOMETRY AND QUANTUM MECHANICS

Karl Gustafson*

Department of Mathematics,
University of Colorado, Boulder, Colorado
E-mail: gustafs@euclid.colorado.edu

A noncommutative operator trigonometry originated by this author in 1966 will be summarized. Some important applications that have taken place during the intervening years will be recalled. Then we will turn to the recent application of this noncommutative operator trigonometry to Bell's inequality in quantum mechanics. Our geometrization of this celebrated inequality clarifies the proper meaning of several physical issues in the Einstein-Podolsky-Rosen Paradox. Those clarifications lead us to look more closely at Von Neumann's formulation of the collapse of the quantum mechanical wave packet. These investigations give us a new clarification of the quantum Zeno Paradox. We also announce a fundamental new theorem on quantum mechanical reversibility and we take note of some of its physical implications.

Keywords: noncommutative operators; antieigenvectors; Bell Inequality; Einstein-Podolsky-Rosen Paradox; Zeno Paradox; reversibility

1. Introduction, Background, and Summary

Starting from a purely operator-theoretic question which this author had formulated in 1966 within the Hille-Yosida-Phillips-Lumer theory of abstract operator semigroups, this author developed in the period 1966-1972 the essentials of a noncommutative operator trigonometry. In this beginning period, I called this a theory of *antieigenvalues* and *antieigenvectors*, because intuitively it compares and contrasts in a natural manner to the Rayleigh-Ritz variational theory of eigenvalues and eigenvectors. From 1973

*This paper is a summary of the two lectures, 'Noncommutative Trigonometry' presented at The Second International Conference on Abstract and Applied Analysis 2005–Quy Nhon, Vietnam, June 4–9, 2005, and 'Noncommutative Operator Algebra and Bell Inequality' presented at the Institute of Mathematics, Vietnamese Academy of Science and Technology, Hanoi, Vietnam, June 15, 2005. In addition, some new results are included.

to 1993, this author was chiefly involved with other research, including scattering theory, mathematical physics, computational fluid dynamics, optical computing systems, neural networks. However, a few further results for the noncommutative trigonometry were obtained. Then from 1994 to the present I returned to the noncommutative trigonometry and successfully applied it to numerical linear algebra, wavelets and control theory, statistical estimation and efficiency, and quantum mechanics. In 1997 I published two books[1,2] which contain chapters on this noncommutative trigonometry. At that time I decided to call this theory *operator trigonometry.* This, because I was beginning to realize its importance beyond the original context in which I had viewed it variationally, juxtaposed against the Rayleigh-Ritz variational theory. More recently, I have settled upon the name *noncommutative trigonometry.* This, because the noncommutative trigonometry now seems to me to convey a spirit much like the theory of A. Connes, which now carries the name *noncommutative geometry.* Suffice it to say that this author has always viewed operator theory as primarily noncommutative. It is time to emphasize that fact with the name noncommutative trigonometry for the noncommutative operator trigonometry I will describe herein.

A bibliography of 62 papers written by this author that deal wholely or in part with this noncommutative trigonometry may be found in the recent survey.[3] Thus, the reader may find more bibliography, facts, and history in Refs. 1–3. Although this noncommutative operator trigonometry has had contact with related contributions by a few other notable mathematicians, including M. G. Krein, H. Wielandt, V. Ptak, C. Davis, B. Mirman, L. Kantorovich, C. R. Rao, for the most part it is my creation.

In Section 2 we quickly recall the key essentials of the noncommutative trigonometry. In Section 3 we quickly recount some of its main application to date. More details about these may be found in Ref. 1–3 and especially in the 62 citations of Ref. 3. However, the recent survey[3] could not present the recent results of this author in which the noncommutative trigonometry fundamentally clarifies key issues in the celebrated Bell and Zeno problems of quantum mechanics. So I go beyond[3] here and in Sections 4 and 5, respectively, I treat those two problems from the viewpoint of the noncommutative trigonometry.

2. Essentials of the Noncommutative Trigonometry

To fix ideas, let us restrict attention to A a symmetric $n \times n$ invertible positive definite (SPD) matrix. The theory extends to arbitrary positive selfadjoint operators in a Hilbert space and to arbitrary invertible matrices and even to accretive operators in a Banach space and we refer the reader to Refs. 1–3 for those results. Let A have eigenvalues $0 < \lambda_1 < \lambda_2 < \cdots < \lambda_n$ where again for simplicity we assume them to be all distinct, although the theory does not need that. Then the essential facts of the noncommutative trigonometry depend upon the following fundamental four entities:

$$\mu_1(A) \equiv \inf_{x \neq 0} \frac{\langle Ax, x \rangle}{||Ax|| ||x||} \equiv \cos\phi(A) = \frac{2\sqrt{\lambda_1 \lambda_n}}{\lambda_1 + \lambda_n} \tag{1}$$

$$\nu_1(A) \equiv \inf_{\epsilon > 0} ||\epsilon A - I|| \equiv \sin\phi(A) = \frac{\lambda_n - \lambda_1}{\lambda_n + \lambda_1} \tag{2}$$

$$\phi(A) \equiv \text{the maximum turning angle of } A \tag{3}$$

$$x_\pm = \pm \left(\frac{\lambda_n}{\lambda_1 + \lambda_n} \right)^{1/2} x_1 + \left(\frac{\lambda_1}{\lambda_1 + \lambda_n} \right)^{1/2} x_n \tag{4}$$

I named μ_1 the first antieigenvalue of A. The entity ν_1 is equally important in applications. I called $\phi(A)$ as defined from either μ_1 or ν_1, the angle of A. I called the two vectors $x_\pm$ the first antieigenvectors of A. They are the vectors (normalized to norm 1 in (4)) most turned by A. The x_1 and x_n are normalized eigenvectors for λ_1 and λ_n respectively.

A good example is $A = \begin{bmatrix} 9 & 0 \\ 0 & 16 \end{bmatrix}$. Then $\phi(A) = 16.2602°$ and the two antieigenvectors are $x_+ = (4, 3)/5$ and $x_- = (-4, 3)/5$.

Early on I realized that these entities form the basis of an (angular) spectral theory as rich as the usual (dilation) spectral theory based upon eigenvalues. Moreover I showed that the Euler equation

$$2||Ax||^2||x||^2(Re\, A)x - ||x||^2 Re\langle Ax, x\rangle A^*Ax - ||Ax||^2 Re\langle Ax, x\rangle x = 0 \tag{5}$$

generalizes the usual Rayleigh-Ritz variational theory, and moreover, contains it: for A symmetric or normal, A's antieigenvectors, and as well all eigenvectors, satisfy (5). I defined the sequence of higher antieigenvalues and corresponding higher antieigenvectors on successive reducing subspaces of A, more or less in analogy with the Rayleigh-Ritz theory. All of that was done in the period 1966–1972. See Refs. 1–3 for more details.

A couple of remarks here may be helpful. The two antieigenvectors may not be linearly combined to form an antieigenspace. However, their span is of course the reducing subspace $sp\{x_1, x_n\}$ formed by the 'smallest' and 'largest' eigenvectors. Thus the antiegenvectors $x_\pm$ are rather special vectors contained within that spectral subspace. Specifically[4] they have relative angle $\langle x_+, x_- \rangle = -\sin\phi(A)$, and thus the angle between them is always $\phi(A) + \pi/2$.

Second, in applications I often must use the operator trigonometry and the usual trigonometry in conjunction. However, the operator trigonometry generally does not satisfy the identities of the usual trigonometry. Moreover, it has some of its own interesting identities. A useful example is:[5] $\sin\phi(A^{1/2}) = \sin\phi(A)/[1+\cos\phi(A)]$. Here $A^{1/2}$ denotes the positive square root of the operator A. Generally we have the operator angle inequality: $\phi(AB) \leqq \phi(A) + \phi(B)$.

Third, although I developed some rudiments of the noncommutative operator trigonometry for operators A on a Banach space, it is better to stay in Hilbert space, because the essential identity $\sin^2\phi(A) + \cos^2\phi(A) = 1$ does not hold generally in Banach space.

3. Quick Summary of Applications to Date

Applications of this noncommutative trigonometry to date include, roughly and chronologically: perturbation of operator semigroups, positivity of operator products, Markov processes, Rayleigh-Ritz theory, numerical range, normal operators, convexity theory, minimum residual and conjugate gradient solvers, Richardson and relaxation schemes, wavelets, domain decomposition and multilevel methods, control theory, scattering theory, preconditioning and condition number theory, statistical estimation and efficiency, canonical correlations, Bell's inequalities, quantum spin systems, and quantum computing. Here is a quick summary.

In the initial development period 1966–1989 one could say that the research emphasis was functional-analytic and theoretical. However, I had turned to computational mathematics in the 1980's, working with several engineering groups. An example is in Ref. 6. In particular, in such research I had used some steepest descent and conjugate gradient scheme computations and had learned of the basic convergence rate bound

$$E(x_{k+1}) \leqq \left(1 - \frac{4\lambda_1\lambda_n}{(\lambda_1 + \lambda_n)^2}\right) E(x_k) \tag{6}$$

Here $E(x)$ is the error $\langle (x - x^*), A(x - x^*)\rangle/2$, x^* the true solution. This bound also occurs in a key way in some parts of optimization theory. Immediately from my operator trigonometry I knew that what (6) really meant, geometrically, was trigonometric:

$$E(x_{k+1}) \leqq \sin^2\phi(A) E(x_k) \tag{7}$$

Although (6) was known as a form of the Kantorovich inequality, nowhere in the numerical nor in the optimization literature did I find this important geometrical result: that gradient and conjugate gradient convergence rates are fundamentally trigonometric. That is, they are a direct reflection of the elemental fact of A's maximum turning angle $\phi(A)$. The known conjugate gradient convergence rate, by the way, becomes, trigonometrically written now,

$$E(x_k) \leqq 4(\sin\phi(A^{1/2}))^{2k} E(x_0). \tag{8}$$

See Ref. 7 where I published this result (8) for conjugate gradient convergence. And I first announced the steepest descent result (7) at the 1990 Dubrovnik conference.[8]

Many other important iterative computational linear solvers for large problems $Ax = b$ were shown by the noncommutative trigonometry to have trigonometric convergence rates. I obtained these results in the period 1994–2004. All were new. It is really quite astounding that these fundamentally trigonometric convergence geometries had lain undiscovered all those years, going back to Richardson (1910), Kantorovich (1948), other notable computational mathematicians. However, without the $\cos\phi(A)$ and $\sin\phi(A)$ of the noncommutative trigonometry at your fingertips, one cannot see the geometry. See Ref. 9 to come somewhat up-to-date on these results.

Another important application of the noncommutative trigonometry has been to statistics. For example, there is the famous Durbin-Watson-Bloomfield (1955, 1975) lower bound for least squares relative efficiency

$$RE(\hat{\beta}) = \frac{|Cov\beta^*|}{|Cov\hat{\beta}|} = \frac{1}{|X'VX||X'V^{-1}X|} \tag{9}$$

of estimators, namely,

$$RE(\hat{\beta}) \geqq \prod_{i=1}^{p} \frac{4\lambda_i\lambda_{n-i+1}}{(\lambda_i + \lambda_{n-i+1})^2} \tag{10}$$

where the λ's are the eigenvalues of the covariance matrix V. In (9) the β^* is the best possible estimator and X is an $n \times p$ regression design matrix. Coincidentally, I had recently[4] extended my operator trigonometry to arbitrary matrices and there (also in earlier papers) I had stressed the alternate definition of higher antieigenvalues and their corresponding higher antieigenvectors as defined combinatorially in accordance with (1)-(4) but with successively smaller operator turning angles now defined by corresponding pairs of eigenvectors. It follows readily that the geometric meaning of statistical efficiency then becomes clear:[10]

$$RE(\hat{\beta}) \geqq \prod_{i=1}^{p} \cos^2 \phi_i(V) \tag{11}$$

It is that of operator turning angles.

Other intimate connections between statistical efficiency, statistical estimation, and canonical correlations may be found in Ref. 10. For example, LaGrange multiplier techniques used in statistics lead to equations for the columns x_i of the regression design matrices X to satisfy

$$\frac{V^2 x_i}{\langle V x_i, x_i \rangle} + \frac{x_i}{\langle V^{-1} x_i, x_i \rangle} = 2V x_i, \quad i = 1, \ldots, p \tag{12}$$

I call this equation the Inefficiency Equation. In the antieigenvalue theory, the Euler equation (5) becomes in the case when A is symmetric positive definite,

$$\frac{A^2 x}{\langle A^2 x, x \rangle} - \frac{2Ax}{\langle Ax, x \rangle} + x = 0 \tag{13}$$

The combined result is the following. All eigenvectors x_j satisfy both the Inefficiency Equation and the Euler Equation. The only other (normalized) vectors satisfying the Inefficiency Equation are the "inefficiency vectors" (my terminology)

$$x_{\pm}^{j+k} = \pm \frac{1}{\sqrt{2}} x_j + \frac{1}{\sqrt{2}} x_k. \tag{14}$$

The only other vectors satisfying the Euler equation are the antieigenvectors

$$x_{\pm}^{jk} = \pm \left(\frac{\lambda_k}{\lambda_j + \lambda_k} \right)^{1/2} x_j + \left(\frac{\lambda_j}{\lambda_j + \lambda_k} \right)^{1/2} x_k. \tag{15}$$

Further recent results from the application of the noncommutative trigonometry to matrix statistics may be found in Refs. 11,12.

For other applications of the noncommutative trigonometry, see the citations of Ref. 3. These include applications to quantum computing and to elementary particle theory, which I will not discuss in this paper. In the next two sections I will concentrate on the Bell and Zeno problems from quantum theory.

4. The Bell Inequalities of Quantum Theory

Addressing fundamental issues raised by Einstein, Podolsky, Rosen,[13] concerning the foundations of quantum mechanics of Von Neumann,[14] Bell[15] presented his famous inequality

$$|P(\mathbf{a},\mathbf{b}) - P(\mathbf{a},\mathbf{c})| \leqq 1 + P(\mathbf{b},\mathbf{c}) \tag{16}$$

and exhibited quantum spin measurement configurations whose quantum expectation values could not satisfy his inequality. Bell's analysis assumed that two measuring apparatuses could be regarded as physically totally separated, and free from any effects from the other. Thus his inequality could provide a test which could be failed by measurements performed on correlated quantum systems. It was therefore argued that local realistic hidden variable theories could not hold in quantum mechanics if future physical experiments would violate Bell's inequality. Later Aspect *et al.*[16] indeed demonstrated violation of Bell's inequality in their laboratory experiments. But the controversy about the Bell inequality and related inequalities to be mentioned below and their physical implications, continues to this day.

Bell's arguments in arriving at his inequality were classical probabilistic correlation arguments. However, it is known and easy to prove that this inequality holds for any real numbers a, b, c in the interval $[-1, 1]$: then $ab - bc + ac \leqq 1$. Here is a proof. From $b^2 \leqq 1$ and $c^2 \leqq 1$ we have $b^2(1 - c^2) \leqq 1 - c^2$ and hence $b^2 + c^2 \leqq 1 + b^2c^2$. Adding $2bc$ to both sides and multiplying by $a^2 \leqq 1$ we therefore have $a^2(b^2 + c^2 + 2bc) \leqq b^2 + c^2 + 2bc \leqq 1 + b^2c^2 + 2bc$, that is, $a^2(b + c)^2 \leqq (1 + bc)^2$. Taking the positive square root yields $a(b + c) \leqq |a||b + c| \leqq 1 + bc$.

Wigner[17] presented his own version, making more clear the issues of locality and socalled realism. Furthermore, Wigner was sure to use quantum mechanical probabilistic correlations. His version of Bell's theory then becomes the inequality

$$\frac{1}{2}\sin^2\frac{1}{2}\theta_{23} + \frac{1}{2}\sin^2\frac{1}{2}\theta_{12} \geqq \frac{1}{2}\sin^2\frac{1}{2}\theta_{31} \tag{17}$$

where the θ_{ij} are angles between spin directions ω_i and ω_k.

Another important Bell-type inequality is that of Clauser, Horne, Shimony, Holt.[18] Let $\mathbf{a}, \mathbf{b}, \mathbf{c}, \mathbf{d}$ be four arbitrary chosen unit vector directions in plane orthogonal to the two beams produced by the source. Let $v_i(\mathbf{a})$ and $v_i(\mathbf{d})$ be the "hidden" predetermined values ± 1 of the spin components along $\mathbf{a}$ and $\mathbf{d}$, respectively, of particle 1 of the ith pair, similarly $w_i(\mathbf{b})$ and $w_i(\mathbf{c})$ for particle 2 values along directions $\mathbf{b}$ and $\mathbf{c}$. Then the average correlation value for particle 1 spins measured along $\mathbf{a}$ and particle 2 spins measaured along $\mathbf{b}$ is $E(\mathbf{a},\mathbf{b}) = \sum_{i=1}^{N} v_i(\mathbf{a})w_i(\mathbf{b})/N$. Taking into account in the same way the average correlation values $E(\mathbf{a},\mathbf{c})$, $E(\mathbf{d},\mathbf{b})$, $E(\mathbf{d},\mathbf{c})$

and adding up all pairs, one arrives at the CHSH inequality

$$|E(\mathbf{a},\mathbf{b}) + E(\mathbf{a},\mathbf{c}) + E(\mathbf{d},\mathbf{b}) - E(\mathbf{d},\mathbf{c})| \leqq 2. \tag{18}$$

In a series of papers starting with Ref. 19, see Ref. 3 for citations to later papers, I placed all of the inequalities of the Bell theory, into my noncommutative trigonometry. From my results, it may be argued that many important issues in the Bell theory, about which there are at times furious arguments about physical and probabilistic meaning among physicists, are really better seen as new mathematical quantum geometry from the noncommutative trigonometry.

As a first example, consider Wigner's version (17) of the Bell inequality discussed above. For the three coplanar directions, our corresponding inequality equality[20] becomes in Wigner's quantum spin setting terminology

$$\begin{aligned}&\sin^2\left(\frac{1}{2}\theta_{12}\right) + \sin^2\left(\frac{1}{2}\theta_{23}\right) - \sin^2\left(\frac{1}{2}\theta_{13}\right)\\ &= 2\cos\left(\frac{1}{2}\theta_{13}\right)\left[\cos\left(\frac{1}{2}\theta_{13}\right) - \cos\left(\frac{1}{2}\theta_{12}\right)\cos\left(\frac{1}{2}\theta_{23}\right)\right]\end{aligned} \tag{19}$$

Violation of the conventionally assumed quantum probability rule $|\langle u,v\rangle|^2 \equiv \cos^2\theta_{u,v}$ for unit vectors u and v representing prepared state u to be measured as state v, is equivalent according to Wigner to the right side of this identity being negative. This is his Bell "violation" test. However, from our point of view, there is no violation, there is just a quantum trigonometric identity, valid for certain formulations of measurement of certain spin systems.

As a second example, let us consider the important CHSH inequality (18) given above. Wishing now to preserve equality, we write

$$\begin{aligned}|\mathbf{a}\cdot\mathbf{b} + \mathbf{a}\cdot\mathbf{c} + \mathbf{d}\cdot\mathbf{b} - \mathbf{d}\cdot\mathbf{c}| &= |\mathbf{a}\cdot(\mathbf{b}+\mathbf{c}) + \mathbf{d}\cdot(\mathbf{b}-\mathbf{c})|\\ &= |\|b+c\|\cos\theta_{a,b+c} + \|b-c\|\cos\theta_{d,b-c}|\\ &= |(2+2\cos\theta_{bc})^{1/2}\cos\theta_{a,b+c}\\ &\quad + (2-2\cos\theta_{bc})^{1/2}\cos\theta_{d,b-c}|\end{aligned}$$

Squaring this expression and writing everything quantum trigonometrically,

$$\begin{aligned}|\mathbf{a}\cdot\mathbf{b}+\mathbf{a}\cdot\mathbf{c}+\mathbf{d}\cdot\mathbf{b}-\mathbf{d}\cdot\mathbf{c}|^2 &= (2+2\cos\theta_{bc})\cos^2\theta_{a,b+c}\\ &\quad+(2-2\cos\theta_{bc})\cos^2\theta_{d,b-c)}\\ &\quad+2(4-4\cos^2\theta_{bc})^{1/2}\cos\theta_{a,b+c}\cos\theta_{d,b-c}\\ &= 4\cos^2(\theta_{bc}/2)\cos^2\theta_{a,b+c}\\ &\quad+4\sin^2(\theta_{bc}/2)\cos^2\theta_{d,b-c}\\ &\quad+4\sin\theta_{bc}\cos\theta_{a,b+c}\cos\theta_{d,b-c}.\end{aligned}$$

In the above we used two standard trigonometric half angle formulas. Now substituting the double angle formula $\sin\theta_{bc} = 2\sin(\theta_{bc}/2)\cos(\theta_{bc}/2)$ into the above we arrive at

$$|\mathbf{a}\cdot\mathbf{b}+\mathbf{a}\cdot\mathbf{c}+\mathbf{d}\cdot\mathbf{b}-\mathbf{d}\cdot\mathbf{c}|^2 = 4[\cos\theta_{bc}/2)\cos\theta_{a,b+c}+\sin(\theta_{bc}/2)\cos\theta_{d,b-c}]^2$$

and hence we have a new quantum *CHSH equality*

$$|\mathbf{a}\cdot\mathbf{b}+\mathbf{a}\cdot\mathbf{c}+\mathbf{d}\cdot\mathbf{b}-\mathbf{d}\cdot\mathbf{c}| = 2|\cos(\theta_{bc}/2)\cos\theta_{a,b+c}+\sin(\theta_{bc}/2)\cos\theta_{d,b-c}|$$

We may also write the righthand side as twice the absolute value of the two-vector inner product

$$\mathbf{u}_1\cdot\mathbf{u}_2 \equiv (\cos(\theta_{bc}/2),\sin(\theta_{bc}/2))\cdot(\cos\theta_{a,b+c},\cos\theta_{d,b-c})$$

to arrive at the quantum trigonometric identity

$$|\mathbf{a}\cdot\mathbf{b}+\mathbf{a}\cdot\mathbf{c}+\mathbf{d}\cdot\mathbf{b}-\mathbf{d}\cdot\mathbf{c}| = 2(\cos^2\theta_{a,b+c}+\cos^2\theta_{d,b-c})^{1/2}|\cos\theta_{u_1,u_2}| \tag{20}$$

The right sides of these equalities isolate the "classical limiting probability factor" 2 from the second factor, which may achieve its maximum $\sqrt{2}$. Fix any directions **b** and **c**. Then choose **a** relative to **b**+**c** and choose **d** relative to **b** − **c** so that $\cos^2\theta_{a,b+c} = 1$ and $\cos^2\theta_{d,b-c} = 1$, respectively. Now we may choose the free directions **b** and **c** to maximize the third factor to $\cos\theta_{u_1,u_2} = \pm 1$. But that means the two-vectors $\mathbf{u}_1$ and $\mathbf{u}_2$ are colinear and hence

$$\begin{aligned}\mathbf{u}_1 &= (\cos(\theta_{bc}/2),\sin(\theta_{bc}/2)) = 2^{-1/2}(\cos\theta_{a,b+c},\cos\theta_{d,b-c})\\ &= 2^{-1/2}(\pm 1,\pm 1)\end{aligned}$$

and thus the important angle θ_{bc} is seen to be $\pm\pi/2$. More to the point, our identity allows one to exactly trace out the "violation regions" analytically in terms of the trigonometric inner product condition $1 \leqq |\mathbf{u}_1\cdot\mathbf{u}_2| \leqq \sqrt{2}$.

Thus, and we have only presented two examples here, the noncommutative trigonometry applied to the celebrated inequalities of the Bell theory has shown that those inequalities may be made special cases of identities for the noncommutative trigonometry. This enables[20] much clearer delineations of physical, probabilistic, and geometrical implications of those inequalities. Further analysis[20] leads to the opinion that one cannot claim quantum locality or nonlocality on the basis of satisfaction or violation of Bell's inequalities. Rather, one is led[20] to the assertion that the real issue as concerns nonlocality is the Von Neumann projection rule.[14] That rule, under which measurements of quantum probabilities correspond to projections on Hilbert space subspaces, is considered next, in the last section of this paper.

5. The Zeno Problem of Quantum Theory

The term Zeno Paradox, i.e., "a watched pot never boils", was introduced in Misra, Sudarshan[21] to highlight certain fundamental issues in quantum mechanical measurement theory. There is a long history to such quantum measurement problems. For example, Von Neumann,[14] when he created his Hilbert space model of quantum mechanics, proved that any given state ϕ of a quantum mechanical system can be steered into any other state ψ in the Hilbert space by an appropriate sequence of very frequent measurements. Thus, in particular, you can freeze a quantum evolution in time by "continually watching it". The allusion to the greek Zeno refers to a debate[22] which he and Parmenides had with Socrates in Athens approximately 445 B.C.. Parmenides, an elderly philosopher, presented to Socrates an interesting proposition to start the debate: reality never changes, therefore motion is not possible. Although Parmenides relied on the younger Zeno for argumentive support, the originating thesis was that of Parmenides. Thus is might be more accurate to call the quantum versions Parmenides Paradox and Parmenides theory.

What is at issue is not a matter of names, it is the fundamental physical question of what happens when a quantum mechanical wave function $\psi(t)$ is measured by some classical measuring instrument. The act of measuring is seen as a 'collapse of the wave function,' which according to the Von Neumann postulates is implemented by an orthogonal projection. This gives rise to numerous mathematical issues. I was involved, going back to 1974, in some of these questions, and left them unresolved, and hence as concerns my partial results, unpublished. Misra and Sudarshan[21] went from the Schrodinger picture to the Heisenberg picture to partially move beyond these issues, but a close reading of Ref. 21 shows that the issues are not removed, they are only assumed away. Recently I returned to some of these questions and some of the following results are taken from Ref. 23. Because there is now a large literature on the Zeno Problem, I must refer the reader to Ref. 23 for further bibliography. Also one should look at several articles in the 2003 volume in which Ref. 20 appeared, a volume dedicated in significant part to the Zeno problem, and its relation to quantum computing.

At issue both physically and mathematically is the nature of the interaction of the measuring instrument or subspace with the evolving wave packet $\psi_t = U_t\psi_0$ where ψ_0 was the initial state vector and where $U_t = e^{itH}$ is the unitary evolution group generated by the governing self-adjoint Hamiltonian. One also needs to look at projected evolutions $Z_t = PU_tP$. Here P is an orthogonal projection. It turns out that commutativity or noncommutativity play critical roles in both the mathematical considerations and in the physical outcomes.

The following was known early. Let $U(t)$ be an arbitrary unitary evolution with selfadjoint infinitesimal generator H and let P be an arbitrary selfadjoint bounded projection onto closed subspace M in Hilbert space $\mathcal{H}$. Let $D(T)$ denote the domain of an operator T in $\mathcal{H}$, and $R(T)$ its range. We recall that $PU_t = U_tP$ iff M is a reducing subspace for U_t, and more generally $PU_tP = U_tP$ iff M is an invariant subspace for U_t. More general yet, we have (see Refs. 24–27)

Lemma 5.1. *The projected evolution $Z_t = PU_tP$ is a semigroup for all $t \geqq 0$ iff M is a proper subspace without regeneration for U_t, i.e., $PU_sP^{\perp}U_tP = 0$ for all $t, s \geqq 0$.*

We remark that when M is a reducing subspace then Z_t is a unitary group, when M is an invariant subspace then Z_t is a semigroup of partial isometries, and if M fails to meet the requirement of Lemma 5.1, the Z_t loses the semigroup property entirely.

The possibility of retaining a unitary, hence reversible, evolution on socalled Zeno subspaces has been investigated in a number of recent books/papers.[28–31] For a better understanding of what is involved in such formulations, we wish to take note of certain unbounded operator-theoretic considerations. The following was clear early,[25,26] but not published due to the general incompleteness of my study of the Zeno issues.

Lemma 5.2. *Assume $D(HP)$ is dense. Then PHP is symmetric in $\mathcal{H}$, and PHP is selfadjoint if PH is closed in $\mathcal{H}$.*

Proof. Since $D(HP) \equiv D(PHP)$ is dense, $(HP)^*$ and $(PHP)^*$ exist and $(PHP)^* = (HP)^*P \supset (PH)P = PHP$ so PHP is symmetric. If PH

is closed, then $PH = (PH)^{**} = (HP)^*$, so one obtains equality in the previous sentence. □

Before looking more closely at the fundamental underlying issue of $D(HP)$ dense, we wish at this point to recall a few operator-theoretic facts which we will use in the remainder of this paper. For more details see the books [32,33] among others. In particular, the notion of (closed) invariant subspace M for an unbounded operator T is better thought of in the more general context as a decomposition of T by a direct sum $\mathcal{H} = M \oplus N$ of a pair of subspaces with the requirements that the projection P on M map $D(T)$ into $D(T)$, T maps M into M, T maps N into N. Here P is the (generally oblique) projection of M along N. Such decomposition of T is equivalent to T commuting with $P : PT \subset TP$. Then $TP = PTP = PT$ on $D(T)$. When T is a selfadjoint operator H and P an orthogonal projection and $U_t = e^{iHt}$, then when one says that M reduces H one is saying all of the following: $PH \subset HP$, $P : D(H)$ into $D(H)$, $(H - zI)^{-1}P = P(H - zI)^{-1}$ for all z with $Imz \neq 0$, $PE(s) = E(s)P$ for all real s and all spectral family projectors of H, and $PU_t = U_tP$ for all real t. Our point-of-view is that such a simplified situation is not that of the Zeno issues. A measuring projection already within H's spectral calculus is a far oversimplified situation, although certain proposed spin measurers may have the property.

A second set of facts to remember is that T^* exists iff $D(T)$ is dense, that $D(T^*)$ need not be dense, but $D(T^*)$ is dense iff T is closeable, and then its closure satisfies $\overline{T} = T^{**}$. For any two densely defined operators A and B and if AB is densely defined, then $(AB)^* \supset B^*A^*$, with equality if $A \in \mathcal{B}(\mathcal{H})$. Other conditions for equality when A and B are both unbounded may be found in [34]. We also will use other unbounded operator theory, such as the associativity $T_1(T_2T_3) = (T_1T_2)T_3$, without comment.

We may now sharpen Lemma 5.2 to further clarify the situation.

Theorem 5.1. *$D(HP)$ is dense iff PH is closeable. Then $(HP)^*$ is defined and $(HP)^* = \overline{PH} \supset PH$ has domain at least as large as $D(H)$. Thus generally $(PHP)^* = \overline{PH}P$ whenever PH is closeable. Furthermore the polar factors satisfy $|\overline{PH}| = (HP^2H)^{1/2}$ and $|HP| \supset (PH^2P)^{1/2}$.*

Proof. Because PH is densely defined, its adjoint exists and is $(PH)^* = HP$. This operator is densely defined iff PH is a closeable operator. Then HP is a closed densely defined operator and $(HP)^* = (PH)^{**} = \overline{PH} \supset PH$. To obtain the polar factors we form $(\overline{PH})^*PH = HP^2H$ and $(HP)^*HP = \overline{PH}HP \supset PH^2P$. □

The key assumption that HP be densely defined is made throughout the recent treatment [35] of quantum Zeno dynamics. Their approach of quadratic forms follows that of [36] and one is concerned with the form $||H^{1/2}Pu||^2$ with form domain $D(H^{1/2}P)$. The operator $H_P :=$ $(H^{1/2}P)^*(H^{1/2}P)$ is associated with this form. It is noted in [35] that H_P may not be densely defined but they then assert that it will be a selfadjoint operator in some closed subspace of $\mathcal{H}$. From our analysis here we would like to defer slightly, or at least point out some ambiguity in such a conclusion. When $H^{1/2}P$ is not densely defined in $\mathcal{H}$, one cannot even speak of an operator $(H^{1/2}P)^*(H^{1/2}P)$. Of course one can then reduce one's considerations to the smaller Hilbert space M. But if $H^{1/2}P$ was not densely defined in $\mathcal{H}$, then $H^{1/2}$ will not be densely defined in M either. Moreover, whatever its domain there, the range $R(H^{1/2}\mid_M)$ will generally fall at least partially outside of M.

The same reservation applies to the analysis of Ref. 37. The interesting formulation there combines continuous measurement with a coupling limit to force the system to evolve in a set of orthogonal subspaces of the parent Hilbert space. These socalled quantum Zeno subspaces are the eigenspaces of a Hamiltonian which is supposed to represent the interaction between the evolving quantum dynamical system and the measurement apparatus. The use of a superselection rule and an adiabatic theorem are assumed to determine "the subspaces that the apparatus is able to distinguish". Thus the physical description is now that of a dynamical evolution allowing changing Zeno subspaces. However, from our point of view, since the modelling and analysis is carried out in the density matrix formulation, its underlying rigorous validity, e.g., the denseness of the domains of the effective Hamiltonians in the individual Zeno subspace evolutions, has not been considered. A similar comment applies to the Von Neumann Subalgebras Zeno formulation [38], especially if one does not want the measuring projection E to have to be within the Hamiltonian's functional calculus.

Finally, I would like to announce here a fundamental new theorem with I believe fundamental implications to quantum mechanics in general. The reversibility of a quantum mechanical Schrödinger evolution corresponds just to the unitarity of its semigroup e^{itH}. The following result [23] grew out of such thoughts combined with the considerations about domains elsewhere in this paper.

Theorem 5.2. *A unitary group U_t necessarily exactly preserves its infinitesimal generator's domain $D(H)$. That is, U_t maps $D(H)$ one to one onto itself, for all $-\infty < t < \infty$.*

Proof. For every $-\infty < t < \infty$ the linear isometry $||U_t x|| = ||x||$ property and the commutativity property $U_t H \subset H U_t$ guarantee that U_t maps $D(H)$ one to one into $D(H)$. Is it onto $D(H)$? Suppose not. Then for some t there exists an x in $D(H)$ which is not in the image $U_t(D(H))$. Apply U_t^{-1} to x. By the commutativity property again, we know $z = U_{-t}x$ is in $D(H)$. But then $U_t z = x$ must have been in $D(H)$. □

The result is quite evident and natural once one sees it. Therefore it may exist elsewhere in the literature. But in a limited search, I did not find it. It importantly distinguishes the special action of U_t on $D(H)$ from its one to one onto action on the whole Hilbert space. Remember that $D(H)$, although dense, is a Baire Category 1 subspace. In particular, a quantum mechanical Schrodinger evolution $U_t = e^{itH}$ where H is the atomic Schrodinger partial differential equation must take this Baire Category 1 subspace $D(H)$ of all prepared states exactly onto itself. The rest of the Hilbert space, a Baire Category 2 set which is not a subspace, is separately taken onto itself. Thus Theorem 5.2 expresses a 'global' regularity of exact domain preservation.

Contrast that with the heat equation evolution $Z_t = e^{-t\Delta}$, which instantly takes the whole Hilbert space (e.g., $\mathcal{L}^2(-\infty, \infty)$) into $C^\infty(-\infty, \infty)$. See for example the rigorous proof in Ref. 39, pp. 128–131. The Heat evolution operator Z_t is an infinitely smoothing operator. Its smoothing action is a 'local' regularizing of each initial given heat distribution $f \in \mathcal{L}^2(-\infty, \infty)$, whether f be in the domain $W^{2,2}(-\infty, \infty)$ of the Laplacian $-\Delta$, or not. Instantly, information in f is spread out and lost. In particular, the domain $W^{2,2}$ is immediately lost. Not so in the Schrodinger evolution: its unitarity requires that $W^{2,2}$ be mapped exactly onto $W^{2,2}$ by U_t at every instant t, according to Theorem 5.2.

To express our interpretation of Theorem 5.2 in dynamical system terms, the Schrodinger partial differential equation initial value problem

$$\begin{cases} \dot{\psi}(t) = iH\psi(t), & t > 0 \\ \psi(0) = \psi_0 & \text{given} \end{cases} \tag{21}$$

has two classes of states: ψ_0 in the domain $D(H)$, and the rest. Only those ψ_0 in $D(H)$ are 'legally' allowed wave packets from the differential operator theory, in which it is essential that the bounded evolution operator U_t commute with the infinitesimal generator H. Stated in elementary terms: otherwise you cannot substitute the solution back into the differential equation.

One can prove Theorem 5.2 in another way which is more useful when one wants to consider also contraction semigroups Z_t with infinitesimal generator A. To that end we first establish two lemmas.

Lemma 5.3. *Let T be a closed densely defined operator in a Hilbert space $\mathcal{H}$ and let $U_t = e^{itH}$ be a unitary evolution there which commutes with T. Then U_t remains unitary on the graph-norm Hilbert space $\mathcal{H}_T$.*

Proof. $\mathcal{H}_T$ is the Hilbert space $D(T)$ equipped with the inner product $\langle x, y\rangle_T = \langle x, y\rangle + \langle Tx, Ty\rangle$. By the commutativity $U_t T \subseteq TU_t$ we have the isometry property retained,

$$||U_t x||_T^2 = ||U_t x||^2 + ||TU_t x||^2 = ||x||^2 + ||U_t Tx||^2 = ||x||_T^2 \qquad (22)$$

Also the adjoint U_t^* considered in $\mathcal{H}_T$ is the same as the original adjoint, using again the commutativity,

$$\begin{aligned}\langle U_t x, y\rangle_T &= \langle U_t x, y\rangle + \langle U_t Tx, Ty\rangle \\ &= \langle x, U_t^* y\rangle + \langle Tx, TU_t^* y\rangle = \langle x, U_t^* y\rangle_T \end{aligned} \qquad (23)$$

Thus $U_T^* = U_t^{-1}$ in the original space carries over to $\mathcal{H}_T$. □

In particular, Lemma 5.3 provides an alternate proof of Theorem 5.2. Let T be H. Then U_t and U_t^* are both onto $\mathcal{H}_T = D(H)$ and the rest follows easily from bounded operator theory.

Lemma 5.4. *Under the conditions of Lemma 5.3, a contraction semigroup evolution $Z_t = e^{tA}$ remains a contraction semigroup on $\mathcal{H}_T$.*

Proof. As above. Note that the assumed semigroup commutativity property $Z_t T \subseteq TZ_t$ is essential, as is its subproperty that Z_t map $D(T)$ into $D(T)$ In particular, one may take $T = A$, the infinitesimal generator of Z_t □

Now we can provide a partial converse to Theorem 5.2 in the sense of asking, suppose a contraction semigroup Z_t exhibits the regularity preservation property that it map the domain $D(A)$ of its infinitesimal generator one to one onto itself. What 'unitarity' properties does Z_t exhibit for all $t \geqq 0$?

Proposition 5.1. *Let $Z_t = e^{tA}$ be a contraction semigroup on a Hilbert space $\mathcal{H}$ such that Z_t maps $D(A)$ one to one onto $D(A)$. Then Z_t on $\mathcal{H}_A$ can be extended to a group $Z_{-t} = Z_t^{-1}$.*

Proof. By Lemma 5.4 above with $A = T$, we know Z_t remains a contraction semigroup on $\mathcal{H}_A$. Because Z_t maps $\mathcal{H}_A$ 1–1 onto itself, we know Z_t^{-1} is bounded and also maps $\mathcal{H}_A$ 1–1 onto itself. By known results in semigroup theory (e.g., see Ref. 33, p. 393), we may extend Z_t to a group by defining $Z_{-t} = Z_t^{-1}$. We mention that Z_t^* remains the same in $\mathcal{H}_A$ as it was in $\mathcal{H}$, so that one can relate the adjoint semigroups by $(Z_t^*)^{-1} = (Z_t^{-1})^*$, pursue further converse statements for the original Hilbert space $\mathcal{H}$, etc., which we will not do here. One reason we present these results here is in hopes that they be used elsewhere by those working in Zeno theory. □

What Theorem 5.2 states within the context of unitary quantum mechanical evolutions is that one cannot lose any of the totality of detail embodied in the totality of wave functions ψ in $D(H)$. It is quite interesting to think about how much 'mixing around' of $D(H)$ the evolution U_t can do. Perhaps one will investigate such questions elsewhere, e.g., from the viewpoint of ergodic theory. However, just by itself, Theorem 5.2 says that to have reversibility in quantum mechanics, you cannot lose a single wave function from $D(H)$ as you proceed forward in time. In other words, you must be able to account for all evolving domain probabilities, future and past, all of the time.

This finding also gives renewed importance to the role of preparation of states in quantum mechanics. The ensemble $\{\psi_0\}$ of experimental initial states which are prepared to simulate *a posteriori* in an experiment what you could expect *a priori* from all possibilities from $D(H)$, must be sufficiently extensive within $D(H)$ in such a way that their trajectories continue to predict properly during evolution. Ideally, you need a dense subset from $D(H)$. Experimentally of course, one must settle for less.

To emphasize the above discussion, which to us seems important, we may state, at the risk of sounding repetitive, the following.

Corollary 5.1. *A quantum mechanical evolution U_t must and does continuously and simultaneously account for all probabilities $|\psi(t)|^2$ and all wave functions $\psi(t)$ in the domain of its Hamiltonian. No probabilities may be lost.*

It is important to note that Theorem 5.2 is of wider interest, i.e., it will apply to any unitary evolution $U_t = e^{itH}$ in any context. For example, even if one has "resorted" to the Heisenberg interaction picture and to unitary evolutions $\rho(t) = U_t \rho_0 U_t^*$, one knows from Theorem 5.2 that the underlying Hamiltonian must have the domain regularity preservation property. Perhaps more interestingly, effective "interaction Hamiltonians" such as those assumed in quantum dynamical decoupling[31,40] must also enjoy and respect the domain regularity preservation property of Theorem 5.2. Our point of view is that in such situations, this is a physical property that one should try to understand mathematically, in order to obtain a deeper understanding of the physics.

There is another way to place Theorem 5.2 in context. There is a school of thought (see Refs. 41,42) in physics which views irreversibility as "going out of the Hilbert space." This point of view certainly has some merit and goes back in physics to the notion of Rigged Hilbert space, among others. However, as I pointed out in Ref. 43, one is already out of the Hilbert space whenever one needs to consider weak solutions in partial differential equations, a concept going back to the 1930's. Also, and in quantum mechanics, the radiation eigenfunctions over the positive (scattering) spectrum of the Schrodinger Hydrogen operator are already non-L^2 functions. So we construct L^2 wave packets, or work in some Banach function space, or a Rigged Hilbert space.

What Theorem 5.2 says, within this context, is that you do not even need to go out of the Hilbert space to get irreversibility. You need only lose *one* wave function ψ from the domain $D(H)$ of the Hamiltonian. Then the unitarity of the dynamical system has been lost. The wave function ψ can remain in the Hilbert space. Should U_t *immediately* reconstitute its unitarity, the lost wave function ψ would continue its trajectory under the dynamical system U_t but it could not re-enter the 'eligible' $D(H)$ prepared states until the evolution operator U_t again lost its unitarity for an instant and somehow switched ψ back into $D(H)$ as another wave function gets thrown out. I have not seen this conceptual picture anywhere in the quantum theory literature. I would assert that "going out of the operator domain" is a much sharper characterization of irreversibility.

Elsewhere recently[44] I showed that a number of Nobel Prizes depended in a fine analysis on an assumption of *detailed balance* at a microscopic physical level. This then leads to a selfadjoint operator and hence to an implicit unitary evolution. The detailed balance was usually at the level of some, e.g., microscopic reaction operators. I may here interpret Theorem 5.2 as a more precise detailed balance statement. It is at the wave function level and says that U_t must preserve the balance between any two wave functions, and not lose any of them.

From the above considerations, which are of course important in themselves, there now enters the question of how to bring the operator trigonometry to further enrich our understanding of the microscopic dynamics taking place when quantum measurements are performed. The very act of a quantum measurement, except in the case when measuring instrument and the dynamics 'commute', involves an interaction which is, intuitively, irreversible. This must be, in my opinion, the key ansatz in the Zeno problem, and we asserted it in our discussion at the beginning of this Section. On the other hand, Theorem 5.2 (and the results following) involve only a commuting situation. That gave the important new results about micro-irreversibility as the loss of a single wave function. Can the noncommutative operator trigonometry now address the measurement (noncommutative) dynamics? It should be remembered that to date, in my opinion, no-one has successfully treated the quantum measurement interaction dynamics in a mathematically rigorous way.

In Ref. 45 a beginning may have been made. There I showed that under a multiplicative perturbation $H \to BH$, although the direction of time is not changed, the speed of evolving dynamics has been altered. Moreover, there always exists a small disturbance such that a unitary evolution U_t becomes a completely nonunitary contraction semigroup. In [46] a preliminary theory of interaction antieigenvalues for interacting operators A and B has been worked out. See Refs. 47,48 for some earlier discussion of the Bell and Zeno problems.

Acknowledgements

The invitation to speak at the ICAAA 2005 and the warm hospitality shown me in Quy Nhon by Professors Nguyen Minh Chuong and Nguyen Minh Tri and the others there was greatly appreciated. The invitation to speak at the Institute of Mathematics in Hanoi afterward and the warm hospitality offered me there by Professor Le Tuan Hoa and his group was equally appreciated.

References

1. K. Gustafson, *Lectures on Computational Fluid Dynamics, Mathematical Physics, and Linear Algebra*, (World Scientific, Singapore, 1997).
2. K. Gustafson and D. Rao, *Numerical Range: The Field of Values of Linear Operators and Matrices*, (Springer, Berlin, 1997).
3. K. Gustafson, *Noncommutative Trigonometry*, Operator Theory: Advances and Applications: Wavelets, Multiscale Systems, and Hypercomplex Analysis (D. Alpay, ed.), Vol. 167 (2006), pp.127-155.
4. K. Gustafson, *Lin. Alg. & Applic.* **319**, 117 (2000).

5. K. Gustafson, *Num. Lin Alg. with Applic.* **34**, 333 (1997).
6. K. Gustafson and N. Sobh, *Computer Physics Communications*, **65**, 253 (1991).
7. K. Gustafson, *Linear and Multilinear Algebra* **37**, (1994) 139.
8. K. Gustafson, Antieigenvalues in Analysis, in *Fourth International Workshop in Analysis and its Applications*, Dubrovnik, Yugoslavia, June 1–10, 1990 (C. Stanojevic and O. Hadzic, eds.), Novi Sad, Yugoslavia (1991), pp. 57–69.
9. K. Gustafson, *Num. Lin Alg. with Applic.* **10**, 291 (2003).
10. K. Gustafson, *Lin. Alg. & Applic.* **354**, 151 (2002).
11. K. Gustafson, *Research Letters Inf. Math. Sci.* **8**, 105 (2005).
12. K. Gustafson, *International Statistical Review*, **74**, 187 (2006).
13. A. Einstein, B. Podolsky, and N. Rosen, *Physical Review* **47**, 777 (1935).
14. J. Von Neumann, *Die Mathematische Grundlagen der Quantenmechanik*, (Springer, Berlin, 1932).
15. J. Bell, *Physics* **1**, 195 (1964).
16. A. Aspect, J. Dalibard and G. Roger *Physical Review Letters* **49**, 1804 (1982).
17. E. Wigner, *American Journal of Physics* **38**, 1005 (1970).
18. J. Clauser, M. Horne, A. Shimony, and R. Holt, *Physical Review Letters* **23**, 880 (1969).
19. K. Gustafson, *The Geometry of Quantum Probabilities*, On Quanta, Mind, and Matter: Hans Primas in Context, (H. Atmanspacher, A. Amann, U. Mueller–Herold, eds.), Kluwer, Dordrecht (1999), pp. 151–164.
20. K. Gustafson, Bell's inequalities, in *The Physics of Communication, Proceedings of the XXII Solvay Conference on Physics*, (I. Antoniou, V. Sadvnichy, H. Walther, eds.), (World Scientific 2003), pp. 534–554.
21. B. Misra and G. Sudarshan, *Journal of Mathematical Physics* **18**, 756 (1977).
22. W. Guthrie, *A History of Greek Philosophy, V. 2: The PreSocratic Tradition from Parmenides to Democritus*, (Cambridge University Press, UK, 1965).
23. K. Gustafson, Reversibility and Regularity, to appear in *International Journal of Theoretical Physics.*
24. K. Sinha, *Helvetica Physica Acta* **45**, 621 (1972).
25. K. Gustafson, *On the "Counter Problem" of Quantum Mechanics*, unpublished (1974), pp. 14.
26. K. Gustafson, *Some Open Operator Theory Problems in Quantum Mechanics*, Rocky Mountain Mathematics Consortium Summer School on C^* Algebras, Bozeman, Montana, August 1975. Unpublished notes, (1975), pp. 7.
27. K. Gustafson, Irreversiblity questions in chemistry, quantum-counting, and time-delay, In *Energy Storage and Redistribution in Molecules*, (J. Hinze, ed.), (Plenum Press, 1983), pp. 516–526.
28. M. Namiki, S. Pascazio, and H. Nakazato, *Decoherence and Quantum Measurements*, (World Scientific, Singapore, 1997).
29. P. Facchi, V. Gorini, G. Marmo, S. Pascazio, and G. Sudarshan, *Physics Letters A* **274**, 12 (2000),.
30. P. Facchi, S. Pascazio, A. Scardicchio, and L. Schulman *Physical Review A* **65**, 012108 (2001).
31. S. Tasaki, A. Tokuse, P. Facchi, and S. Pascazio, *International Journal of*

Quantum Chemistry **98**, 160 (2004).
32. T. Kato, *Perturbation Theory for Linear Operators*, (Springer, Berlin, 1980).
33. Riesz, F. and Sz.-Nagy, B. *Functional Analysis*, (Ungar Publishing, New York, 1955).
34. K. Gustafson, *A Composition Adjoint Lemma*, Stochastic Processes, Physics, and Geometry: New Interplays. II (F. Gesztesy, H. Holden, J. Jost, S. Paycha, M. Röckner, S. Scarlotti, eds.), American Mathematical Society, Providence, 2000, pp. 253–258.
35. P. Exner, and T. Ichinose, Product formula related to quantum zeno dynamics, in *Proceedings of the XIV International Congress on Mathematical Physics*, (Lisbon, World Scientific, 2003). See also arXiv:math-ph/0302060.
36. C. Friedman, *Indiana University Mathematics J.* **21**, 1001 (1972).
37. P. Facchi, S. Pascazio, Quantum zeno subspaces and dynamical superselection rules, in *The Physics of Communication* (I. Antoniou, V. Sadovnichy, H. Walther, eds.), (World Scientific, 2003), pp. 251–286.
38. A. Schmidt, *Journal of Physics A* **35**, 7817 (2002).
39. K. Gustafson, *Partial Differential Equations and Hilbert Space Methods*, 3rd Edition, Revised, (Dover Publication, N. Y., 1999).
40. L. Viola, E. Knill, and S. Lloyd, *Physical Review Letters* **82**, 2417 (1999).
41. I. Prigogine, *Is Future Given?* (World Scientific, Singapore, 2003).
42. I. Antoniou, G. Lumer, (eds.) *Generalized Functions, Operator Theory, and Dynamical Systems*, Chapman and Hall, CRC Research Notes in Mathematics **399**, (London, 1998).
43. K. Gustafson, Rigged Spectral States: A Proclivity for Eigenvalues, in *Generalized Functions, Operator Theory and Dynamical Systems* (I. Antoniou, G. Lumer, eds.), CRC Research Notes in Mathematics **399**, London, 1998, pp. 48–55.
44. K. Gustafson, *Discrete Dynamics in Nature and Society* **8**, 155 (2004).
45. K. Gustafson, Semigroups and antieigenvalues, in *Irreversibility and Causality* (A. Bohm, H. Doebner, P. Kielanowski, eds.), Lecture Notes in Physics **504**, Springer, Berlin (1998), 379–384.
46. K. Gustafson, *J. Math. Anal. Appl.* **299**, 174 (2004).
47. K. Gustafson, The quantum zeno paradox and the counter problem, in *Foundations of Probability and Physics-2* (A. Khrennikov, ed.), (Växjo University Press, Sweden, 2003), pp. 225–236.
48. K. Gustafson, *International Journal of Theoretical Physics* **44**, 1931 (2005).